W0114976

F. Bodart and J. Vanderdonckt (eds.)

Design, Specification and Verification of Interactive Systems '96

Proceedings of the Eurographics Workshop
in Namur, Belgium,
June 5–7, 1996

Eurographics

SpringerWienNewYork

Prof. Francois Bodart
Jean Vanderdonckt, M. Sc.
Institut d'Informatique, Facultés Universitaires Notre-Dame de la Paix,
Namur, Belgium

Printed in Austria

Typesetting: Camera ready by authors
Printing: Druckerei Novographic, A-1238 Wien
Binding: Fa. Papyrus, A-1100 Wien

Graphic design: Ecke Bonk

Printed on acid-free and chlorine-free bleached paper

With 114 Figures

ISSN 0946-2767
ISBN 3-211-82900-8 Springer-Verlag Wien New York

Preface

This book is the final outcome of the 3rd International Eurographics Workshop on Design, Specification and Verification of Interactive Systems (DSV-IS'96), that was held in Namur (Belgium), from June 5 to 7, 1996. This workshop was the third of its kind, following two successful editions in Italy (Cararra, DSV-IS'94) and in France (Toulouse, DSV-IS'95).

Making systems easier to use implies an ever increasing complexity in managing communication between users and applications. Indeed an increasing part of the application code is devoted to the user interface portion. In order to manage this complexity, it is very important to have tools, notations, and methodologies which support the designer's work during the refinement process from specification to implementation.

The purpose of the workshop is to review the state of the art in this area, compare the different existing approaches to this field in order to identify the principal requirements and the most suitable notations, and indicate the meaningful results which can be obtained from them. Contributions are dedicated on topics such as :

- Foundations and Reference Models for Interactive Systems
- Verification of User Interfaces, application of theorem-provers
- Methodologies for abstract design, comparative studies of methods/description techniques
- Design methodologies for domain-oriented User Interfaces
- Specification languages and properties for Human-Computer Interaction, multi-modal interfaces, and multi-media user interfaces
- Formalisms for User Interface Design, Specification and Verification (e.g., State-Transition Diagrams, Transition Networks, Petri Nets)
- Temporal properties of interactive systems
- Cooperative properties of distributed interactive systems
- Formal description of user related properties
- Model-based approaches and tools for user interface design
- Task-based approaches and tools for user interface design
- Computer-Aided or Automated Generation of User Interfaces
- Links between Contextual Task Analysis and User Interface Design
- Links between Software Engineering and User Interface Design
- Design of Graphic Systems, Window Systems, User Interface Management Systems
- Constraint-Based Systems (e.g., Layout systems, Graphics)

Seventy-eight contributions have been submitted to this edition of the conference. Sixteen of them were accepted for inclusion in the present book, after a thorough review by an international programme committee. This event brought seventy-seven participants from fourteen different countries belonging to all major continents, thus simply doubling the number of participants of DSV-IS'95. This edition gathered people coming not only from North and South America but also from most European countries, Asia, Australia, and New Zealand. In order to foster discussions during working groups, only ten of the selected papers were presented orally during the workshop. The other papers have been included for discussion purposes.

Each day of the workshop was started by an invited speaker, presenting high-level insights on the conference topics. Each invited speaker brought his/her widely known experience to provide each of them a completely different viewpoint on the design, the specification and the verification of interactive systems.

The present book begins with these invited papers.

Pedro Szekely, who is the leader of the Humanoïd and the MasterMind projects, talked about some retrospective and challenges encountered for model-based interface development.

Véronique De Keyser highlighted how including human factors in user interface design, and especially in the domain of aeronautics, can be of high importance. In particular, bad user interface design is more likely to imply human errors.

Baudouin Le Charlier applied his thorough knowledge of abstract interpretation and application to user interface development : these kinds of techniques can be used not only for analyzing static programs, but also to verify properties of interactive systems. Using abstract interpretation for this purpose is probably an original contribution. The author shows how these techniques can be applied for several systems and techniques presented in the papers.

The presented papers were grouped into three broad topics, each one making up a chapter of this book : « Moving Towards Implementation », « Evaluating Formal Languages », « Analyzing Errors ». The remaining papers are grouped in the next chapter of this book called « Design, Specification and Verification » which is the mainstream of this book.

One day and half were devoted to the discussions in working groups. This year, two groups were formed and the reports of these working groups are available on-line on the Web server (http://www.info.fundp.ac.be/~jvd/dsvis/dsvis96.html) : « The Namur Principles: Criteria for the Evaluation of User Interface Notations » by Chris Johnson and « The Role of Formalisms » by

David Duke. The results from these groups were presented in a plenary session followed a by word of conclusion.

We would like to express our sincere gratitude to all people and organizations that helped us to make this event possible : Eurographics for their assistance in the organization process, the « Facultés Universitaires Notre-Dame de la Paix » (University of Namur) and its « Institut d'Informatique » (Institute of Computer Science) for their continuous support, the « Fonds National de la Recherche Scientifique » and the « Communauté Française ». We also would like to thank Ms. Van Kelecom from SUN Microsystems and Ms. Zyla from Siemens-Nixdorf : those two computers manufacturers have been our major sponsors. Finally, we thank the City of Namur and its promotion agency « Namur-Europe-Wallonie » for providing us international advertisement.

In particular, we would like to address special thanks to people from the « Institut d'Informatique » who considerably helped us to make this event an international success : Babette di Guardia-Vandenkerckhof, Vinciane André, Laura Oger, Anne-Marie Hennebert, and Jean-Marie Leheureux.

The titles and the abstracts of the papers accepted for DSV-IS'96 are available on-line at http://www.info.fundp.ac.be/~jvd/dsvis/dsvis96.html. The DSV-IS'96 Web site (http://www.info.fundp.ac.be/~dsvis96) has now moved to a more permanent location at http://www.info.fundp. ac.be/~jvd/dsvis. This site contains information relating to DSV-IS'94, DSV-IS'95, DSV-IS'96 and the call for participation for the next edition (http://www.info.fundp.ac.be/~jvd/dsvis/dsvis97.html).

We wish all the best to the forthcoming 4th International Eurographics Workshop on Design, Specification and Verification of Interactive Systems, to be held in Granada, Spain, 4-6 June 1997.

The DSV-IS'96 co-chairs

François Bodart and Jean Vanderdonckt

Contents

Analysing Errors

Design, Specification and Verification

Reports from Working Groups

Available on-line at http://www.info.fundp.ac.be/~jvd/dsvis/dsvis96.html

Programme Committee

G.	Abowd	Georgia Tech, Atlanta	(USA)
S.	Bagnara	University of Siena	(Italy)
R.	Bastide	LIS - University of Toulouse I	(France)
F.	Bodart	FUNDP Namur	(Belgium)
J.	Coutaz	CLIPS-IMAG Grenoble	(France)
V.	De Keyser	University of Liège	(Belgium)
A.	Dix	University of Huddersfield	(UK)
D.	Duce	Rutherford Appleton Laboratory	(UK)
E.	Fiume	University of Toronto	(Canada)
G.	Faconti	CNUCE - CNR	(Italy)
J.	Foley	Georgia Tech, Atlanta	(USA)
P.	Gray	University of Glasgow	(UK)
M.	Green	University of Alberta	(Canada)
R.	Jacob	Tufts University	(USA)
C.	Johnson	University of Glasgow	(UK)
M.	Harrison	University of York	(UK)
B.	Le Charlier	FUNDP Namur	(Belgium)
D.	Olsen	Brigham Young University	(USA)
P.	Palanque	LIS - University of Toulouse I	(France)
F.	Paternó	CNUCE - CNR	(Italy)
C.	Stephanidis	ICS-Forth	(Greece)
P.	Szekely	ISI, Univ. of Southern California	(USA)
J.	Vanderdonckt	FUNDP Namur	(Belgium)

Paper Reviewing Committee

D.	Carr	University of Luleå,	(Sweden)
S.	Chatty	CENA	(France)
M.F.	Costabile	University of Bari	(Italy)
D.	Duke	University of York	(UK)
I.	Herman	CWI	(The Netherlands)
J.	Landay	Carnegie Mellon University	(USA)
J.	Löwgren	University of Linkoping,	(Sweden)
T.	Moher	University of Illinois at Chicago	(USA)
L.	Nigay	CLIPS-IMAG Grenoble	(France)
M.	Noirhomme	FUNDP Namur	(Belgium)
G.	Reynolds	CWI	(The Netherlands)
C.	Rouff	NASA GSFC	(USA)
D.	Salber	CLIPS-IMAG Grenoble	(France)
J.-C.	Torres	University of Granada,	(Spain)
R.	Torres	IBM Software Solutions	(USA)

Retrospective and Challenges for Model-Based Interface Development

Pedro Szekely

Information Sciences Institute, University of Southern California
4676 Admiralty Way,
Marina del Rey, CA 90292, USA
Phone : +1-310-822-1511 (ext. 641) - Fax : +1-310-823-6714
E-mail : szekely@isi.edu

Abstract. Research on model-based user interface development tools is about 10 years old. Many approaches and prototype systems have been investigated in universities and research laboratories around the world. This paper proposes a generic architecture for these tools, reviews the different approaches in light of this architecture, and discusses their progress towards the goals of increasing the quality and reducing the cost of developing interfaces. The paper closes with a discussion of challenges for future model-based development tools.

Keywords. Model-based interface development, automatic user interface generation, user interface design.

1 Introduction

Model-based user interface development tools trace their roots to work on user interface management systems (UIMS) done in the early 1980's [27]. UIMSs seeked to provide an alternative paradigm for constructing interfaces. Rather than programming an interface using a toolkit library, developers would write a specification of the interface in a specialised, high-level specification language. This specification would be automatically translated into an executable program, or interpreted at run-time to generate the appropriate interface.

Many early UIMSs focused on dialogue specification [15]. They used state transition diagrams [18], grammars [30, 31] or event-based representations [41] to specify the interface responses to events coming from the input devices. The display aspects of the interface were typically specified outside the specification language, in call-back procedures that painted the screen as appropriate.

Some UIMSs used as their main specification the type and procedure declarations that defined the functional aspects of the application [3, 29]. Based on this information, they generated menus to invoke the procedures, and dialogue boxes to prompt users for the information needed to construct instances of the types.

Through the late 1980's and early 90's the specification languages became more sophisticated, supporting richer and more detailed representations that allowed the systems to generate more sophisticated interfaces. Today's systems use specifications of the tasks that users need to perform, data models that capture the structure and relationships of the information that applications manipulate, specifications of the presentation and dialogue, user models, etc.

The term *model-based interface development tools* refers to interface construction tools that use these rich representations to provide assistance in the interface development process. Tools range from automatic interface generation systems, generators of help systems for applications, interface evaluation tools, advisors, etc.

Even though model-based interface development tools are much more sophisticated than early UIMSs, they have not become popular in the commercial sector. Most software developers use interface builders, toolkits and a programming language to build the interfaces for interactive systems.

The main goal of this paper is to review the current progress in model-based tools, and discuss challenges for the next generation of user interface tools in general, and model-based tools in particular.

The paper is organised as follows. The next section will describe a general architecture of model-based tools that provides a way to classify model-based tools according to the components of the architecture that they emphasise. The sections after analyse the success of model-based work on automatic interface generation, high-level specification systems, help generation, and design advisors. The last part of the paper discusses new challenges for user interface software, including multi-platform support, intelligent support for the user, multi-modal interfaces and end-user tailoring. The paper closes with conclusions about the future of model-based tools.

2 Generic Model-Based Interface Development Architecture

Fig. 1 shows the typical components of a model-based interface development environments (MB-IDE). The rounded rectangles represent tools, the other shapes represent information produced or consumed by the various tools. The main components of the architecture are the *modelling tools*, the *model*, the *automated design tools*, and the *implementation tools*. Developers[1] use the modelling tools to build the model. The automated design tools are used to perform certain design activities that developers either choose or are forced to delegate to the system. The implementation tool transforms the model into an executable representation that is linked with application code, and delivered to the end-users. The following subsections discuss these components in more detail.

[1] This paper uses the term developer to refer to all the people involved in constructing an interactive application. When appropriate, the more specific terms such as task analyst, graphic designer, programmer, etc. will be used.

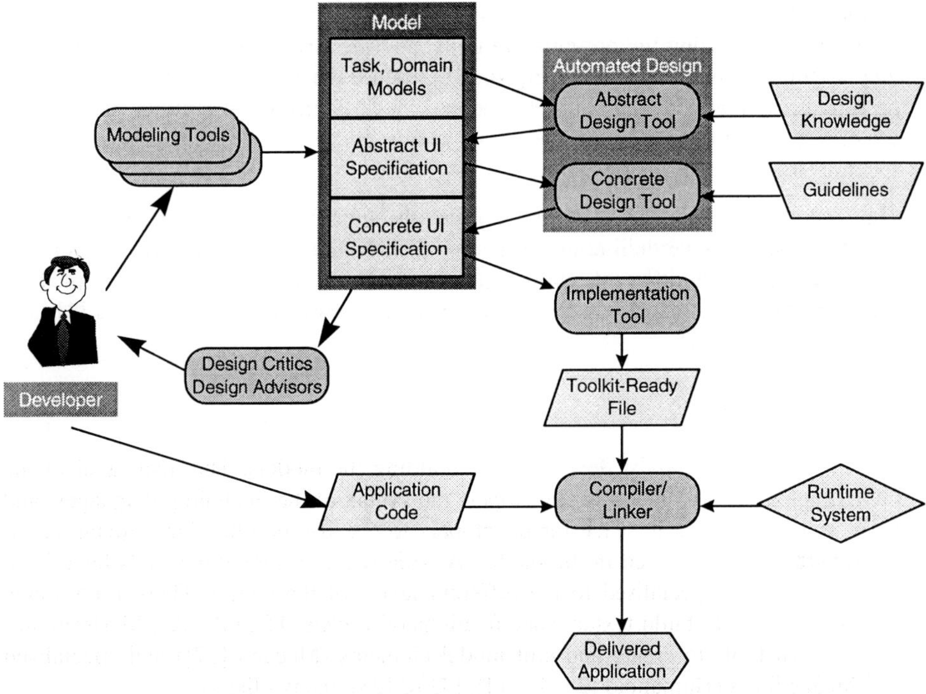

Fig. 1 Model-Based Interface Development Process

2.1 Model

The model is the main component of the system. The model typically organises information into three levels of abstraction. At the highest level are the task and domain model for the application. The task model represents the tasks that users need to perform with the application, and the domain model represents the data and operations that the application supports. Tasks models typically represent tasks by hierarchically decomposing each task into sub-tasks (steps), until the leaf tasks represent operations supplied by the application.

The second level of the model, called in this paper the *abstract user interface specification*, represents the structure and content of the interface in terms of two abstractions, a*bstract interaction objects (AIO), information elements* and *presentation units.* AIOs are low-level interface tasks such as selecting one element from a set, or showing a presentation unit. Information elements represent data to be shown, either a constant value such as a label, or a set of objects and attributes drawn from the domain model. Presentation units are an abstraction of windows. They specify a collection of AIOs and information elements that should be presented to users as a unit. In summary, the abstract user interface specification specifies in an abstract way the information that will be shown in each window, and the dialogue to interact with the information.

The third level of the model, called the *concrete user interface specification*, specifies the style for rendering the presentation units, and the AIOs and information elements they contain. The concrete specification represents the interface in terms of toolkit primitives such as windows, buttons, menus, check-boxes, radio-buttons, and graphical primitives such as lines, images, text, etc. In addition, the concrete specification specifies the layout of all the elements of a window.

The models of different MB-IDEs can differ substantially. Different MB-IDEs typically provide different modelling languages for specifying the contents of the model, and they also emphasise different levels of the model. For example, Mastermind [45] requires developers to explicitly specify all levels of the model, whereas Janus [1] only requires a data model.

2.2 Modelling Tools

The modelling tools assist developers in building the models. The main goal of the modelling tools is to hide from developers the syntax of the modelling languages, and provide a convenient interface for developers to specify the often large quantities of information that are stored in the model. A wide range of modelling tools have been developed, often specialised to the different levels of the model. These tools range from text editors to build textual specifications of models (ITS [48, 49], Mastermind), forms-based tools to create and edit model elements (Mecano [37]) and specialised graphical editors (Humanoid [25, 43, 44], FUSE [23], many others).

2.3 Design Critics and Advisors

Design critics are tools to evaluate designs. The model-based approach provides an excellent platform for constructing analytic design critics because models contain a rich representation of interface designs that these tools can analyse. Most design critics work with the concrete user interface specification layer of the model because in most cases they provide evaluations about detailed features of the interface (e.g., whether the interface provides a way to access all application functionality).

Design advisors are tools that suggest how to refine the abstract layers of the model into more concrete ones. Design advisors use a knowledge-base of design knowledge, typically represented as rules. The condition part of the rules identifies some aspect of a design (e.g., an AIO), and the action part of the rule specifies a way of refining/transforming the matched design element (e.g., the CIO to use for an AIO).

2.4 Automated Design Tools

Many MB-IDEs allow developers to only specify certain aspects of a model. These MB-IDEs feature automated design tools that compute the missing elements of the model from the information that developers do provide. For example, Janus [1] only requires developers to supply a domain model, and it features an automated design tool that automatically constructs both the abstract and concrete specifications of the interface. In contrast ITS and Mastermind [45] require developers to explicitly specify all levels of the model, so these systems do not offer automated design tools. What they do offer is the capability to re-use specifications. The following section discusses automated design tools in detail.

As shown in Fig. 1, automated design tools often use a repository of design knowledge or design guidelines that control the behaviour of the design tool. In most systems developers are not expected to modify the design knowledge, which is typically specified by user interface specialists and the architects of the MB-IDE[1].

2.5 Implementation Tools

The implementation tool translates the concrete specification of the interface into a representation that can be used directly by a toolkit or interface builder. There are essentially three kinds of implementation tools. Source-code generators (e.g., Mastermind) generate source code in a programming language, typically C++. UIMS generators (e.g., FUSE) generate a file that can be read by an existing UIMS or interface builder. Interpreters (e.g., ITS and Humanoid) do not generate an "implementation file", but rather interpret the model directly at runtime.

The last step in the interface generation process is to link the toolkit-ready-file with application specific code and a runtime library. This is typically done using the compiler and linker for the programming language used to implement the application. Interpreter-based systems such as ITS do not use the compiler and linker, but rather feature a runtime module that reads the models during runtime, and interprets the concrete specification of the interface.

Many MB-IDEs provide implementation tools that use the model to generate more than the user interface.

For example, Janus, FUSE, UIDE [11, 12] and Humanoid can generate significant parts of the help system for an application based on the information contained in the model. Janus not only generates the interface, but also generates the database schemas for an application, and much of the data management code.

Mastermind generates code for applications that allows other processes to connect to an application, and to request to be notified when certain tasks are completed, to be sent snapshots of the application state, and to remotely invoke application tasks. This facility supports the construction of agents that can assist users in various ways. This facility was used, for example, to build a history agent that keeps a history of all the tasks that the user has completed an allows users to re-invoke previously completed tasks.

As interfaces become more sophisticated, and users expect more services from their interfaces. The ability to provide such additional run-time services for free is one of the most attractive features of the model-based technology.

---

[1] ITS can be viewed as an automated design tool where developers have to explicitly build the design knowledge for each application or family of applications.

3 Retrospective

The following sections provide a retrospective of the main user interface design and construction problems that have been addressed using the model-based approach. These sections discuss the various approaches that have been used, and how well they solve the problems.

The retrospective section is organised into five main topics:

- *Automatic interface design.* This section discusses the main approaches for automating interface design and their limitations.
- *Specification-based MB-IDEs.* This section discusses MB-IDEs that do not try to automate interface design, but rather give developers convenient languages for expressing designs.
- *Help generation.* Many MB-IDEs feature components that automatically generate help. This section reviews the different approaches and comments on their success.
- *Modelling Tools.* This section discusses various approaches to modelling tools.
- *Design critics and advisors.* This section presents a categorisation of these tools and discusses their relative benefits.

Note. For each topic one or two tools are discussed in some detail. The chosen tools are not necessarily the best tools according to some metric, but rather illustrate a point well, and detailed papers have been published about them. The goal of this paper is to review the main approaches, not the individual tools.

3.1 Retrospective – Automatic Interface Design

The primary goal of many MB-IDEs is to automate as much as possible the design and implementation of a user interface. These MB-IDEs emphasise the domain and task models, and automatically generate the abstract and concrete user interface specifications from these models. Most MB-IDE in this category are oriented towards database applications and produce interfaces that allow the end-users to browse the database, to edit the contents of objects, to define new objects, and to delete objects.

This section argues that automating interface design is intrinsically difficult, so MB-IDEs should be very selective about the portions of the design that they choose to automate.

3.1.1 Structure of Automated Design Tools

Model Contents. MB-IDEs whose primary goal is to automatically design use mainly two kinds of models, a *domain model* that describes the structure and attributes of the information that the application provides, and a *task model* that describes the tasks that users need to perform. For example, tools like Janus, and early versions of Mecano, use only a domain model, whereas tools like Trident [46, 47], Adept [20], DON [22] and Modest [17] use primarily a task model, but also have a domain model.

The domain models of the automatic design tools are similar. They describe classes of objects, inheritance between classes, the attributes of each class together with their types and cardinality, and relationships between objects. In addition, the models typically allow the inclusion of user interface specific information. For example the model of object attributes often includes facets to indicate whether the attribute should be shown to the user, an ergonomic name, and other information to influence the choice of abstract interaction object to be used to specify the attribute.

The task models of these tools are also similar. Tasks are usually decomposed hierarchically, and information is included to specify the sequencing between the tasks (e.g., and, or, xor, parallel). Often, the task model includes references to the domain objects needed and produced in each task. The task model is used during automatic generation to determine the interface dialogue and to determine the information that should be shown in each window.

MB-IDEs in this category typically do not require developers to specify either the abstract or concrete specifications of the interface.

Design Process. Most automated design MB-IDEs use the following sequence of steps to automatically design an interface:

1. *Determine the presentation units.* This step essentially determines the windows that will be used, and what information will be shown in each window.
2. *Determine the navigation between presentation units.* This step computes a graph of presentation units that defines which units can be invoked from which other units.
3. *Determine the abstract interaction objects for each presentation unit.* The abstract interaction objects specify the behaviour of each element of a presentation unit in an abstract way (e.g., select one from set).
4. *Map abstract interaction objects into concrete interaction objects.* The concrete interaction objects represent the widgets available in the target toolkit.
5. *Determine the window layout.* This steps determines the size and position of each concrete interaction object.

The first three steps build the abstract user interface specification, and the last two build the concrete specification.

Post Editing. Once the concrete specification is built, and the implementation tool generates the "toolkit-ready" file, the developer has the opportunity to use and interface builder beautify the layout, change fonts, colours, add decorations, and perform other cosmetic enhancements.

3.1.2 Difficulties With Automated Design

Even though automatic design MB-IDEs can produce interfaces with little or no development effort, there is concern about the quality of the generated interfaces. There is substantial evidence to indicate that it is not feasible to produce good quality interfaces for even moderately complex applications from just a data and task models (together with simple annotations of the data model, such as flags that indicate whether object attributes are relevant to the user interface).

The chapters by Morten Harning [16] and by Stephanie Wilson and Peter Johnson [50] describe critical decisions that must be made in the design of an interface, which the automated design tools cannot currently make appropriately, and which do not seem feasible to automate.

Harning's paper contains an excellent example that illustrates the difficulty of automating steps 1 and 3. Harning's example is about a project management application where users want an interface to monitor progress in the various activities involved in a project. In this application there are four classes of objects represented in the data model: Employee, Project, Activity, Weekly Estimate, and Time Entry. Harning demonstrates using examples that of a good interface must satisfy the following properties:

- *Users need windows that show information drawn from multiple objects.* In the project monitoring example, the project display is based mostly on the Project object, but also shows attributes of the Employee and Activity objects. Furthermore, the example shows that the choice of attributes is task-dependent, and required developers to have a deep understanding of the user's tasks. This means that step 1 of the abstract design tool is hard, if not impossible to automate.

  This property is achieved in the interfaces generated using Trident. The Trident task model captures the information needed for each task, and the generation algorithm calculates how the information flows between tasks in order to determine what information to show in each presentation unit, and where to place it. Systems like Janus, which only use the data model do not satisfy this property.

- *Users do not want the raw information, but rather they need the information to be re-structured and summarised.* In the project monitoring example, users want a weekly report display that essentially combines the Activity and Weekly Estimate objects on a weekly calendar display that shows how much effort was spent on each activity during a specific week. Re-structuring and summarisation cannot be done without a deep understanding of the user's tasks, and again points to the difficulty of automating step 1.

  Another restructuring problem is that users want to see the names of people in the Project Leader field as "name (initials)". This means that rather than using two AIOs to present two different attributes, a single one should be used to present a combination of two attributes. This simple example suggests that the assignment of object attributes to AIOs (step 3) is also a hard problem.

- *Graphical displays are often more effective than tables and forms.* Harning's paper has an example of a graphical display that uses a plot with two curves to show how much time has cumulatively been spent on a project compared to the estimate of the time remaining to complete the project. This example shows that the set of AIOs need to be expanded to include more sophisticated elements such as plots. Of course, then the problem is how to select the appropriate one (step 3), how to set all its parameters, and then how to map it to concrete interaction objects (step 4).

There are two main approaches to automatic design, one based on task models, and the other based on the domain model. The task model approach performs better because task models have some of the information to satisfy the properties listed above. The domain model approach does not have access to such information, and can only produce simple interfaces, typically with one object per presentation unit.

The requirements listed above point to deep issues of interface design, and raise questions about the utility of completely automating the design process, especially steps 1 and 3. Even a small amount of developer involvement can have a huge difference. A simple calculation reveals the economics of the situation.

Most of the automatically designed interface force users to bring up several windows to view the information they need to perform a task, rather than a single window with all the information. Ignoring issues about time to assimilate improperly structured information and the error rates that can result, bringing up several windows and closing them can easily take 3 additional seconds. If users do this 20 times a day, in a year, one full day will be lost per worker. If an organisation has 40 users, 2 man months will be lost per year. Surely it is worth to have developers spend several weeks working on a design.

3.1.3 Discussion

The conclusion of this section is that none of the 5 steps should be completely automated. Rather, collaboration between developers and tools should be built in from the start. Tools should offer suggestions and alternatives. Developers make the decisions, accepting suggestions, choosing between alternatives or entering their own solutions.

This means that the abstract and concrete specification layers of the models should be available to the developers. The specification languages for these layers must allow developers to control all features of the interface that they want to control, no matter how low level. Emphasis should shift from automation to computer aided design.

A simple, and commonly used approach to computerised design aids is the post-editing approach. An automated generation tool generates a first draft of the design, and then the developer edits the draft to produce the final design. This approach has a serious shortcoming, namely that when developers change the model, they need to run the generator tool again, and the post-editing changes will be lost.

The post-editing approach has been used mainly to allow developers to beautify layouts. However, many MB-IDEs such as FUSE feature automatic generator of higher levels of abstraction, and run the risk of running into the same post-editor problems.

One solution to the post-editing problem is to record the changes performed during post-editing, and to reapply them to the output of the generation tools. This approach was used in early versions of Mecano, but it proved difficult to apply the changes reliably, especially when new elements were introduced to a design, or old elements were deleted.

A more robust solution requires a deep integration of the computerised advisor and the modelling tools. In this approach the advisor tools produce design alternatives and suggestions that developers can incorporate into an evolving design via the modelling tools. There is no batch generation process followed by a refinement phase, but rather an incremental evolution of the design, where the computerised advisors and the developers incrementally build the design.

Several MB-IDEs are moving away from automation in the direction of computerised advisors. For example, the Tadeus [38] system requires developers to specify steps 1 and 2 in a structure called a dialogue graph. Steps 3 and 4 are table driven. The system builds default tables with default entries, but developers can edit these tables and override any entry. Step 5 is done automatically, but Tadeus supports post-editing of the generated implementation file.

The FUSE system described in [23] also provides a specification language and tool (BOSS [39]) that lets developers specify the abstract interface specification, and many aspects of the concrete specification. In addition, FUSE provides a tool (FLUID [2]) that uses the task and domain model to produce specifications that can be fed to the BOSS tool to refine and produce an interface. It is unclear for the published papers whether and how FUSE avoids the post-editing problem.

Trident is perhaps the most sophisticated and robust system that combines automatic generation and computerised advice. Trident developed many different strategies and algorithms for performing each of the 5 steps listed above. For example, they developed six strategies for defining presentation units, and have tools that can automatically select and apply a strategy based on information contained in the task and domain model. Trident also offers developers the option of choosing a strategy, or performing the step by hand. However, it is unclear from the published literature on Trident whether it uses an integrated approach as described above.

3.2 Retrospective – Specification-Based MB-IDEs

MB-IDEs in this category seek to provide powerful interface specification languages. These languages provide effective layering or abstraction mechanisms that allow developers to express interface properties at a convenient level of abstraction to facilitate reuse and design modifiability. These languages also seek to give developers extensive control over all features of the interface, so that developers can express any design that they can think of. The goal is not to automate design, but rather to make it easy for developers to express designs, change designs, retarget designs to new platforms, new classes of users, new tasks, etc.

MB-IDEs in this category are oriented towards data management applications. Most business-oriented applications fall in this category, but many engineering and data visualisation applications do not, because they have interfaces whose graphical components are too complex to be expressed in their interface specification languages.

3.2.1 Structure of Specification-Based MB-IDEs

The structure of specification-based MB-IDEs is also compatible with the architecture shown in Fig. 1. They emphasise the model and the implementation tool, and typically do not have an automated design tool.

The modelling language of these MB-IDEs have facilities for developers to express models at the three different levels of abstraction shown in Fig. 1. The models of these MB-IDEs typically feature a data model, but not always a task model. The data model is used mostly in the implementation tool to generate the binding between the interface objects and the application data, so that the interface objects can access the application objects to retrieve the pieces of information that will be displayed (e.g., access the name field of a person object).

The modelling languages to specify the abstract and concrete user interface specifications are designed to maximise reuse. Even though the goals of the different MB-IDEs in this category are the same, the features of the modelling languages are different. For this reason, this section will not attempt to describe these languages in general terms, but rather uses the well known ITS system as an example. Other MB-IDEs in this category include BOSS, Humanoid and Mastermind.

3.2.2 ITS

The ITS system was developed by IBM research, and was used to construct several large applications such as the information kiosks for the Seville world fair, a purchasing system for a large corporation, an insurance industry application, and many others.

ITS has modelling components corresponding to the three levels of modelling shown in Fig. 1. The domain model is called a *data pool,* there is no task model, the abstract specification is called *content specification*, and the concrete specification is called a *style specification.*

The data pool definition language (domain model) supports the specification of structured objects and sequences of objects, like the domain model in many other MB-IDEs. The following is an example of the data pool specification for an airline reservation system.

```
list listname = flights, numrecords = 10
      field destination, rangename = cities, size = 20
      field departure_time, size = 10
      field departure_date, size = 20
      field airline, rangename = airlines, size = 20
      field number_stops, size = 5
```

The content specification (abstract user interface specification) of an interface consists of a collection of frames. Frames can contain lists, forms, choices, information blocks, and nested frames. These elements specify the information that will be presented to the user. Top-level frames correspond to presentation units. Lists and forms specify which elements of the data pool are to be shown in a frame. Information blocks specify static pieces of information to be shown in a frame. Choices indicate sets of alternatives that can be chosen by the user, and correspond to AIOs. Each element specification can be elaborated using an extensive set of attributes that specify the interface content in detail. The following is a fragment of the content specification for the airline reservation example. This frame specifies that five flights are to be displayed, and specifies which fields of the flights object to display.

```
frame id = check_today, action  = getlist, listname = flights, value = flights.data
      list listname = flights, number = 5
          list-item field = destination, message = "To"
          list-item field = departure_time, message = "Departure"
          list-item field = departure_date, size = 20
          list-item field = airline, message = "Carrier"
      frame message = "To search for selected flights"
...
```

The style specification (concrete user interface specification) specifies the mapping from AIOs to CIOs.

To quote from Wiecha's paper, "a style is a co-ordinated set of decisions on the appearance and behaviour of the interaction techniques used in a family of applications". Styles are specified using rules. The condition part of the rule can test any of the attributes of a frame or its children. The action part of the rule selects the CIO to use, and specifies values for the attributes.

Typically, the rule set for an application consists of general rules that apply to families of frames (e.g., there could be a rule for displaying choices as radio buttons), and specific rules that match specific frames defined in the content (e.g., a rule for the check_today frame defined above). General rules are reused in multiple applications and within a single application. Specific rules are used to specify the features of a particular interface that make it different from the generic case.

The following is an example of a style rule. It specifies that if the content is a choice, then construct a vertical group of a title, and something else, depending on which of the nested conditions match. If only one element can be chosen, then the second component is a vertical group, or a collection of horizontal groups, one for every choice. The horizontal group consists of a dingbat to indicate radio buttons, and a message. Note that this rule does not completely specify the display of choices. Other rules may be used to determine the attributes of the unit types used within this rule (VertGroup, HorzGroup, Dingbat and Message).

```
:conditions source = choice
      unit type = VertGroup
          unit type = Title
          :eunit

          :conditions kind = 1_and_only_1
                   unit type = VertGroup
                            unit type = HorzGroup, replicate = all
                                     unit type = Dingbat
                                     :eunit
```

```
                                    :unit type = Message
                                    :eunit
                              :eunit
                        :eunit
                  :econditions
                  ...
                  ...
                :eunit
          :econditions
```

The implementation tool of ITS consists of the rule interpreter and the run-time support system that fires the rules appropriate rules when actions are invoked and the contents of the data pool change.

3.2.3 Discussion

The main difference between specification-based systems such as ITS, and automated design tools such as Janus is one of philosophy. In specification-based MB-IDEs the modelling language is open, whereas in automated design tools it is closed. In automated design tools, developers can only control the design using a few attributes that the tool developers chose to export for that purpose, limiting the developers' ability to control the design of interfaces, and ultimately limiting the quality of the interfaces that can be generated.

Even though ITS is a specification-based MB-IDE, developers do not specify all the features of every individual window. The main point of ITS is that developers should not have to do that. Developers using ITS must specify the abstract user interface specification completely, that is, they have to specify the abstract interface for every different *kind* of window. As argued in the previous section, this is good because the abstract interface is precisely the hardest aspect to generate automatically. However, developers using ITS do not have to specify the concrete user interface specification completely. There is no automated designer to do it, but developers can reuse rule sets from libraries that contain the abstract to concrete mapping for significant portions of the interface specification. This reuse capability enables specification-based MB-IDEs to incorporate many of the cost savings capabilities of automated designers, while overcoming the most serious problems.

Other specification-based MB-IDEs such as Humanoid and Mastermind share the design philosophy of ITS, but differ in the nature of the modelling language. In a large logistics application developed using Humanoid, the developers were able to identify about 13 different families of windows to account for the more than 100 different windows that the system provided. Developers modelled those 13 windows so they did not have to specify each window separately, as would appear to be necessary with a pure specification-based system. However, the design of the 13 windows was according to user requirements, and it would not have been possible to design those windows automatically.

The BOSS system, briefly described in Loczewski's and Schreiber's chapter, is another example of a specification-based MB-IDE. BOSS is also a module of the FUSE system, which is a mixture between automated designer, as implemented in its FLUID module, and a specification system.

3.3 Retrospective – Help Generation

Many MB-IDEs [23, 26, 32, 33, 42] have the ability to automatically, or semi-automatically generate a help system for an application based on model information used to construct the user interface in the first place.

Cartoonist [42] was the first system to provide a compelling demonstration of help generation. Cartoonist allowed the user to ask "how do I do X?" questions, where X could be any of the actions of an application. In response, it would show an animation showing the exact actions that the user needed to perform with the mouse and keyboard to invoke the action. A typical example would show the mouse selecting an object (if one was not selected), then pulling down the appropriate menu, filling out a dialogue box, and finally clicking the OK button.

Cartoonist used the UIDE interface models. The abstract interface specification of UIDE describes the actions that users can perform. The action specification contains pre-conditions that specify the contexts in which the action can be performed, and post-conditions that specify how actions modify the context. The concrete specification models the mapping between actions and concrete interaction objects. Using this information, Cartoonist was able to construct a plan with the sequence of interaction techniques that needed to be invoked in order to perform an action. Cartoonist could even determine what other actions need to be invoked before in order to modify the context to satisfy the preconditions of the action being explained. This allowed the user to ask for help at any time, even when the context was not appropriate to perform the action.

Humanoid also generated a help system for an application based on the model [26]. The help system provided hypertext help to explain the information displayed in a region selected by the user (e.g., paper.txt represents a file), and explain all the commands that the user could issue (e.g., paper.txt can be selected by clicking with the left button, and then the commands delete, and grep can be applied to it). An important contribution of the Humanoid help system is that it used an example-based technique to assist developers in specifying the text of the help windows. Humanoid first generated text automatically, but developers could select text fragments to edit the wording, and then Humanoid would interact with the developer to find an appropriate place in the model to store the edited text fragment. Placement in the model determined the contexts in which the text fragment would appear.

The chapter by Contreras and Saiz in [8] illustrates how the knowledge in the models can be used to automatically generate software tutors, and how the tutors can be customised to different classes of users with different tutoring needs and preferences.

The chapter on the FUSE system, also describes how the information in a model can be used to construct a help system. FUSE, like Cartoonist, produces context sensitive help using the model information. It uses a different style of modelling and also delivers the help in HTML pages rather than using animation.

3.4.1 Discussion

The ability to generate help systems using the information contained in the model is one of the main benefits of the model-based technology. All of today's applications feature a help system, and significant development effort must be devoted towards implementing it. Context-sensitive help is especially difficult to implement because it must reference internal data structures of the interface in order to query the current context of the interface.

The next sections argue that it is precisely the ability to generate runtime services such as help, that give the model-based technology an edge over conventional technologies for implementing interfaces. Using conventional technologies, each runtime service must be separately designed and implemented. Using the model-based technology the services are generated for free, or for a small incremental cost. The reason is that the services use the same information that is used to build the interface in the first place. In addition, as an interface design evolves, the services automatically evolve with it to remain consistent with the design.

3.5 Retrospective – Modelling Tools

Interestingly, ITS, the most widely used MB-IDE does not have a graphical modelling tool. Developers must learn the syntax of the modelling language, and enter the models using a text editor. The creators of ITS found that developers learn the syntax of the language quickly, and that the lack of a modelling tool is not an obstacle to using the tool. They also report (personal communication) that a syntax directed editor was built, but developers refused to use it.

The lesson to be learnt from this experience is that it is false that some tool is better than no tool. A text editor is a powerful tool that is always available. Its most attractive features are users know how to navigate with it, that it is very fast, that it provides cut and paste, effective search mechanisms, global replace, the ability to easily comment out pieces of a design, etc.

However, experience with widely used CASE tools, and expert system shells such as Nexpert Object [28] and Kappa [36] suggest that well engineered graphical tools for building models are useful for the development of large applications. They can be better than text editors, but they must be well engineered, and designed to support large applications.

Most MB-IDEs feature simple forms-based interfaces for creating and editing model entities. Some MB-IDEs such as FUSE and Adept provide visual modelling tools. These tools have not been extensively used, so it is early to comment about their usability for developing large applications.

An interesting approach to modelling tools is embodied in a tool called Grizzly Bear [13]. This tool tries to hide from developers the intricacies of the models by providing an interface that looks like a traditional interface builder or a drawing editor. The interface provides a palette of building blocks and a drawing area where developers can draw pictures of the interface. Grizzly Bear builds models by demonstration. It extracts model entities from the example interfaces that developers draw. It can generalise different pictures into different classes, and most importantly, it can infer dialogue fragments from before and after snapshots of an interface. Grizzly Bear was used to completely build the model for a simple drawing editor based on demonstrations of how the editor should work. An interesting feature of this tool is that it shows developers a textual view of the model as it is being constructed. This view helps novice developers learn the modelling language, and allows experienced developers to edit the textual representation directly. Grizzly Bear represents the first step towards this kind of tool, and further progress needs to be made before such a tool is ready for serious application development.

3.6 Retrospective – Design Critics and Advisors

Much work on design critics and advisors has been done in the context of model-based tools [4, 10]. The reason is that in order to evaluate a design, and automated critic has first to analyse the design to determine what it does. The models provide rich information for critics and advisors to do their work.

The following kinds of evaluation tools have been investigated.

Property verification. The tool verifies that a design satisfies certain properties (e.g., all application functionality is reachable). Some tools [11, 32] can only verify a set of pre-defined properties encoded in a knowledge-base. More powerful tools [35] allow developers to specify the properties to be verified.

End-user simulation. These tools [21] simulate a user interacting with an application, and make predictions about times to perform tasks, learning times and likely errors.

Summative evaluation. These tools produce numbers that can be used to rank designs. An example of such a tool is AIDE [40], a tool to compute metrics based on a theory of layout quality. Work on such tools is still very preliminary. The chapter by Comber and Maltby [7] describes experiments designed to validate the results of some of these tools.

Many property verification tools [24] are designed to detect violations of standard user interface guidelines (e.g., File menu should have the mnemonic F). These tools play a similar role to spelling checkers in word processors: they detect surface problems that show a lack of professionalism. They do not detect problems related to the semantics of the interface, but nevertheless, they are very useful.

Most style-guide verification tools are not model-based, but rather take as input the toolkit ready file used in well known toolkits (e.g., resource files for Windows, UIL files for Motif). The limitations of these tools are discussed in the paper by Farenc et. al. [9]. The problem is that the toolkit-ready file does not contain enough information about a design to verify many of the style rules. In the context of ERGOVAL, 44% of the rules can be automatically verified using the toolkit-ready file, and up to 78% could be automated if the evaluation tool had access to appropriate information. The model-based approach to interface development should allow tools to get closer to the 78% limit. For example, Farenc et. al. illustrate the limitations of toolkit-ready files with the rule that states that "for any input, if there are any acceptable values, such values must be displayed.". Such a rule can be automated in the context of most MB-IDEs because their models contain information about the acceptable values for inputs, and information about how the inputs are displayed.

There are a few notable examples of design critics aimed at more fundamental design issues, addressing issues similar to grammar and document content in word processing. These critics require very detailed models, more detailed than the models currently being used in most MB-IDEs. One example of such a tool is NGOMSEL [21], which belongs to the end-user simulation category of design critics. NGOMSEL takes as input a detailed task model where the leaf tasks represent interaction techniques (CIO). It can simulate a user interacting with the application, and predict how long it will take an expert user to complete a high level task. NGOMSEL can also make predictions about features of an interface that users will find difficult to learn.

Another example of a sophisticated design critic is embodied in the work of Fabio Paterno [35]. His critic is a property verification critic that uses detailed models of an application specified using the LOTOS [34] notation. His system allows developers to specify complex properties using a notation based on temporal logic. One of Paterno's papers [35] discusses an interesting example about an air traffic controller application that uses a message area to display messages to the user. The last message to arrive is shown in the message area, and the previous ones are queued until the operator gets around to view them. This design could lead to subtle timing problems where operators delete the wrong message, skip viewing a message, etc. His paper shows how required properties of this interface can be verified (e.g., the user can read a message several times), or how undesirable effects can occur (e.g., user unwittingly deletes the wrong message). The expense of building the complex models required by this critic can be justified in safety critical applications such as air-traffic control.

Much work remains to be done before these advanced critics become a useful tool for developers. These critics require detailed models that are time-consuming to build, and expressed in specialised notations that most developers do not know. However, work is in progress to integrate these tools with MB-IDEs (NGOMSEL with Mastermind [5], Paternó is working on an implementation tool for his notation). Once this work is complete these design critics will have a more substantial impact on the design and development of interfaces.

An interesting question is the extent to which MB-IDEs can render style-guide verification tools unnecessary because the kinds of errors that they detect cannot be committed when using an MB-IDE.

Automatic generation MB-IDEs provide one answer to this question. The design algorithms of these tools are based on style-guides, so they will automatically be obeyed. Most violations will be due to exceptions specifically coded in the design algorithms.

Design advisors provide a different answer to this question, in the context of specification-based MB-IDEs. Design advisors can be viewed as pro-active critics. Rather than telling designers what they did wrong, they try to steer designers away from poor design choices. The most attractive feature of automated design advisors is that they complement specification-based MB-IDEs so that developers do not have to construct specifications on their own, but are assisted by advisors whose knowledge-bases codify expert knowledge and wisdom about interface design.

The work on design advisors has not yet reached a level of maturity that allows a critical discussion of their approach and effectiveness. Two well known systems are Trident and Expose [14].

4 Challenges and Opportunities

The main opportunities for model-based interface technology lie ahead because it is better suited than traditional technology to meet the new interface challenges that technology is creating.

Faster machines and networks enable more an more sophisticated applications, providing users with more capabilities and more information, but at the same time overwhelming them with more commands and options. Interfaces will need to become more intelligent to assist users in performing their tasks, to help them come up to speed in a new application, to allow users to customise them to make them effective for the particular tasks that users perform most often.

Laptops are commonplace. Smaller portable devices such as PDAs and pagers are getting linked to the networks and provide the ability to access the same information that is available via workstations and laptops. The need will arise for applications that scale across a wide range of devices to provide users with the same or a scaled down version of the workstation functionality. Scaled up versions will also be needed to take advantage of wall-sized displays.

New modalities such as speech, natural language, hand-writing recognition are maturing. Applications will need to reconfigure their interfaces to take advantage of whatever modalities are available on the user's platform.

The following sections discuss why the model-based technology is well positioned to meet these challenges, and give some suggestions on how MB-IDEs need to evolve.

4.1 Challenge 1 – Task-Centred Interfaces

The main difficulty that users face when interacting with an application is to figure out how to use the capabilities of the application to perform desired tasks. Applications often offer many dozens of commands and options, so it is difficult for users to learn and remember the sequence of commands needed to perform a task. Many of the most popular and complex applications such as Microsoft Office and its competitors attempt to cope with this problem by offering *task assistants*. For example, Microsoft Excel has assistants to construct charts, to pivot tables, to create templates, etc. Microsoft Word has assistants to format tables, to format documents, to correct spelling, to do mail merge, etc. The typical behaviour of an assistant is to analyse the current context (e.g., the array of selected cells in a spreadsheet), and then ask users a sequence of questions about how they want the task performed, and finally perform the task for the user. Assistants make certain tasks easy to perform, even if the limit the set of options that users have.

Related to task assistants are *guidance systems*. Guidance systems have two main components, an indexing component that helps users find the topic they need guidance on, and a component that component that guides the user in performing the task. For example, Microsoft's answer wizard, a kind of guidance system, allows users to index in several ways: they can use keywords to find topics, of they can browse the hierarchy of topics. Guidance is given to users using the task assistant technology, traditional hypertext help windows, enhanced hypertext windows with buttons to invoke relevant application functionality.

Today's task assistants and guidance systems are implemented separately from the interface, most surely at a significant development cost. Developers of these systems must, at least informally, build a model of the tasks that users are expected to perform. For the task assistants they must encode in detail all the steps for performing the task, taking into account all the contingencies that arise from the different contexts in which the assistant is invoked. For the guidance systems, developers must encode a comprehensive model of the tasks, including the words that can be used to index them, the steps for each task, pointers to application commands that perform particular steps, etc.

One of the challenges and opportunities for model-based technology is to partially automate the generation of task assistants and guidance systems. Many MB-IDEs use a task model to assist with the design of the interface, and also to control the dialogue at runtime. Such task models already contain much of the information needed for task assistants and guidance systems. They already contain a representation of all the tasks, the steps to perform each task, sequencing constraints, information needed for each task, etc. The abstract and concrete interface representation contain the information that links tasks to the interaction techniques that invoke the various steps of a task. It seems quite sensible to enhance the task model representation to include any additional information needed for the task assistants and guidance systems, and to generate these services from the model.

As mentioned in a previous section, significant progress has been made in this direction. What is needed is to make the transition from an interesting feasibility demonstration, to a robust, high quality implementation. Current demonstrations of help generation work for some of the tasks, not all, generate poor quality text full of the internal names of objects (e.g., start1stConnection), and for the most part, have never been user tested or formally evaluated.

The comparison between automated design tools and specification-based tools is relevant here. Automatic interface generation systems offer interesting demonstrations, but only systems like ITS become successful, because they provide developers with appropriate control over the design. Likewise, model-generated task assistants and help generation systems will achieve high enough quality only if developers of these systems can exert *complete* control over the text that is produced, and significant control over the format.

4.2 Challenge 2 – Multi-Platform Support

Most of the user interface tools developed during the late 1980's and early 90's were designed for a canonical platform featuring a mouse, a keyboard, and a 13 inch colour monitor. Today's platforms are stretching the limits of the canonical platform, often yielding hard to use interfaces. Large, high resolution monitors cause the displays of some applications to become unusable because the icons and text become hard to see, and hard to point at with the mouse. Smaller displays, such as laptop 9 inch displays result in some applications using almost all the screen space for menus, toolbars, dialogue boxes, leaving users a tiny window to perform their work. As argued before, the situation will get much worse once radically smaller (PDA, pager) and larger devices become popular (wall displays).

Interfaces developed using traditional interface builders and toolkits are hard to adapt to different platforms because developers must redesign each window for each new platform. As the set of platforms proliferates, this becomes expensive.

Model-based technology offers a much better approach. For qualitatively similar devices (e.g., workstation and laptop), changes in the AIO to CIO mapping, and the CIO parameters are typically enough to appropriately scale the interface. More radical changes can be done by redesigning the abstract interface specifications. The important point to bear in mind is that in a system like ITS the amount of work is proportional to the number of style rules, which is typically much smaller than the number of windows.

ITS has demonstrated the usefulness of this approach by refining style rules to port interfaces to use a touch screen rather than a mouse. The change involves making the target areas larger, and increasing the spacing between adjacent target areas.

One of the interesting challenges in this area is to develop techniques to scale interfaces to radically different platforms, such as PDAs and pagers. Fig. 2 shows an example of a first step in this direction. The figure shows simple adaptations explicitly represented in Mastermind's model that cause an interface to adapt to changes in screen size by progressively removing less important information as the available space becomes smaller. The fist adaptation causes the first column of scrolling areas to be replaced by buttons that bring up pop-up windows with the same information. The second adaptation causes some headings to disappear and remaining heading fonts to become smaller.

4.3 Challenge 3 – Interface Tailoring

Interface tailoring refers to the ability to customise and optimise an interface according to the context in which it is used. Interfaces can be tailored to tasks that different segments of the user population need to perform most often, to the level of use and experience of users, to the physical abilities of users, to platform characteristics, etc.

There is a whole spectrum of tailoring possibilities. Interface tailoring can happen at the factory, that is, developers produce several versions of an application tailored according to different criteria. Tailoring can also be done at the user's side, for instance, by system administrators or experienced users. In the extreme, individual users

Workstation

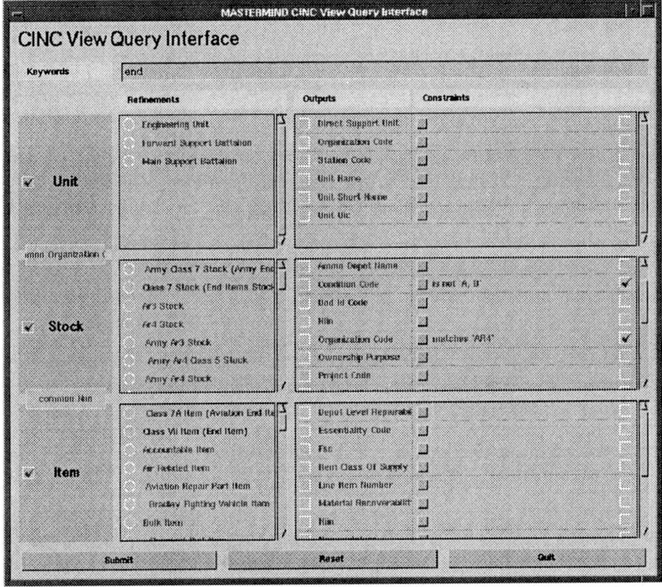

Laptop

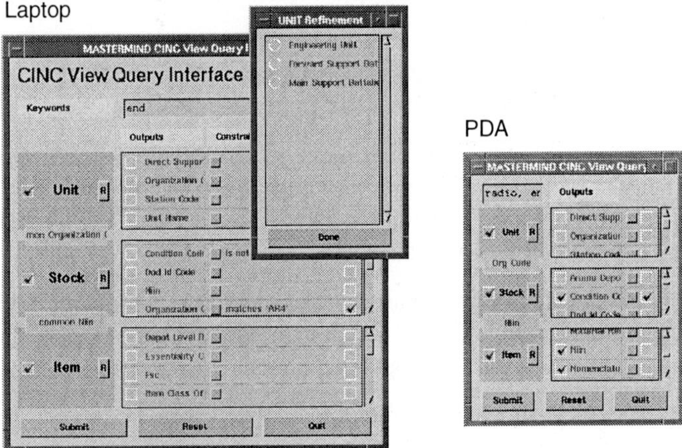

Fig. 2 Scaling an Interface to Multiple Platforms in MASTERMIND

might tailor the interfaces themselves, or the interface could adapt on its own by analysing the user's patterns of use.

No matter when tailoring happens, and what interface features are tailored, tailoring involves modifying the interface design. The simplest level of tailoring happens at the concrete level of an interface specification where features such as the layout, colours and fonts of an interface are changed. More sophisticated tailoring can happen at the abstract interface specification where the dialogue gets modified, for example to shortcut certain steps, to rearrange the order for performing steps, etc. At the highest level, new tasks might be defined by composing existing tasks.

Many model-based interface tools address some aspects of interface tailoring. For example, the FUSE system presents examples of how an interface can be tailored according to the user's level of experience. However, most of the examples are about factory tailoring, where developers construct the rules that define how the interfaces should adapt depending on certain contextual information such as a simple user model.

However, it should be possible to use the automatic interface generation capabilities of many MB-IDEs to support end-user, or administrator-level tailoring of interfaces. Such a facility would be a compelling example of the benefits of the model-based technology.

4.4 Challenge 4 – Multi-Modal Interfaces

New input modalities such as speech, natural language and pen gestures have matured to the point where they can be effectively used in practical applications. Currently, applications that take advantage of these modalities are custom built without much tool support.

Building interfaces that combine these modalities with traditional graphical elements is hard for several reasons:

- To incorporate speech and natural language developers must define the lexicon and perhaps the grammar for parsing and interpreting natural language sentences.
- Speech, natural language and pen input are intrinsically ambiguous. No matter how good the recognisers get, they will always produce a set of alternative interpretations with levels of confidence, rather than a single certain interpretation. Interfaces must be designed to cope with this uncertainty.
- Users can speak and point at the same time, and the interpretation of the inputs depends on their relative timing. In addition, the inputs from the various modalities can refer to each other (e.g., put this file <click> there <click>).

New output modalities such as 3D graphics are also becoming cheaper and commonplace.

Currently, not many model-based interface systems are addressing the construction of interfaces that use these modalities. The model-based interface community runs the risk that the architectures and tools that are being developed will not work with these modalities.

One notable exception is the work by Phil Cohen [6]. His system provides an open architecture for the development of multi-modal user interfaces. The system uses a blackboard architecture that allows an open-ended set of agents to collaborate. Agents collaborate to perform user tasks, to disambiguate natural language requests, etc.

5 Conclusion

Much progress has been made towards demonstrating that the model-based approach provides a viable and effective new technology for developing user interfaces. Model-based systems have evolved from simple proof of concept prototypes that were used on toy applications, to powerful systems that address the construction of interfaces for realistic applications (ITS, Trident, Janus, Mastermind, etc.) The models of many MB-IDEs have been integrated with mainstream software engineering modelling techniques such as OOA (Janus), ER models (Trident, GENIUS [19]), making it easier to use these tools together with other well established software engineering methodologies.

As a community, we need to make progress in two fronts. The first is to build compelling demonstrations of the benefits of model-based tools. The interesting demonstrations of help generation, platform scalability, design critics need to be proven in more realistic settings with realistic applications.

The second front is to address the challenges being posed by new technology developments. As discussed in the last section, these challenges are in fact opportunities for the model-based technology. The challenges point towards solutions where models play an important role, so the model-based technology is well positioned to address them.

References

1. Balzert, H., Hofmann, F., Kruschinski, V., Niemann, C.: *The JANUS Application Development Environment-Generating More than the User Interface.* In: Vanderdonckt J. (ed.): Proceedings of CADUI'96. Namur: Presses Universitaires de Namur 1996 (pp. 183-207).

2. Bauer, B., *Generating User Interfaces from Formal Specifications of the Application*, In: Vanderdonckt J. (ed.): Proceedings of CADUI'96. Namur: Presses Universitaires de Namur 1996 (pp. 141-157).

3. Beshers, C.M., Feiner, S.K.: *Scope: Automated Generation of Graphical Interfaces.* In Proceedings of UIST'89. New York: ACM Press 1989 (pp. 76-85).

4. Bodart, F., Hennebert, A.-M., Leheureux, J.-M., Vanderdonckt, J.: *Computer-Aided Window Identification in TRIDENT.* In Nordbyn K., Helmersen P.H., Gilmore D.J., Arnesen S.A. (eds.): Proceedings of INTERACT'95, London: Chapman & Hall 1995 (pp. 331-336).

5. Byrne, M.D., Wood, S.D, Sukaviriya, P., Foley, J.D, Kieras, D.E.: *Automating Interface Evaluation.* In Adelson, B., Dumais S., Olson J. (eds.): Proceedings of CHI'94. New York: ACM Press 1994 (pp. 232-237).

6. Cohen, P.R., Cheyer, A., Wang, M., Baeg, S.C.: *An Open Agent Architecture.* In AAAI Spring Symposium (pp. 1-8).

7. Comber, T., Maltby, J.: *Investigating Layout Complexity.* In: Vanderdonckt J. (ed.): Proceedings of CADUI'96. Namur: Presses Universitaires de Namur 1996 (pp. 211-229).

8. Contreras, J., Saiz, F.: *A Framework for the Automatic Generation of Software Tutoring.* In: Vanderdonckt J. (ed.): Proceedings of CADUI'96. Namur: Presses Universitaires de Namur 1996 (pp. 171-182).

9. Farenc, Ch., Liberati, V., Barthet, M.-F.: *Automatic Ergonomic Evaluation: What are the Limits?* In: Vanderdonckt J. (ed.): Proceedings of CADUI'96. Namur: Presses Universitaires de Namur 1996 (pp. 159-170).

10. Fischer, G., Nakakoji, K., Ostwald, J., Stahl, G., Sumner, T.: *Embedding Computer-Based Critics in the Context of Design.* In Ashlund S., Mullet K., Henderson A., Hollnagel E., White T. (eds.): Proceedings of INTERCHI'93. New York: ACM Press 1993 (pp. 157-164).

11. Foley, J.D.: *History, Results and Bibliography of the User Interface Design Environment (UIDE), an Early Model-based System for User Interface Design and Implementation.* In Paternó F. (ed.): Proceedings of DSV-IS'94. Berlin: Springer-Verlag 1995 (Focus on Computer Graphics Series, pp. 3-14).

12. Foley, J.D., Kim, W.C., Kovacevic, S., Murray, K.: *UIDE - An Intelligent User Interface Design Environment.* In Sullivan J.W., Tyler S.W. (eds.): Intelligent User Interfaces. New York: ACM Press 1991 (pp. 339-384).

13. Frank, M.: *Grizzly Bear: A Demonstrational Learning Tool For A User Interface Specification Language.* In van der Veer G.C., Bagnara S., Kempen G.A.M. (eds.), Proceedings of UIST'95. New York: ACM Press 1995 (pp. 75-76).

14. Gorny, P.: *EXPOSE - An HCI-Counseling for User Interface Design.* In Nordbyn K., Helmersen P.H., Gilmore D.J., Arnesen S.A. (eds.): Proceedings of INTERACT'95, London: Chapman & Hall 1995 (pp. 297-304).

15. Green, M.: *A Survey of Three Dialogue Models.* ACM Transactions on Graphics, Vol 5, No. 3, 244-275 (July 1986).

16. Harning, M.: *An Approach to Structured Display Design - Coping with Complexity.* In: Vanderdonckt J. (ed.): Proceedings of CADUI'96. Namur: Presses Universitaires de Namur 1996 (pp. 121-138).

17. Hinrichs, T., Bareiss, R., Birnbaum, L., Collins, G.: *An Interface Design Tool based on Explicit Task Models.* In Tauber M.J., Bellotti V., Jeffries R., Mackinlay J.D., Nielsen J. (eds.): Companion Proceedings of CHI'96. New York: ACM Press 1996 (pp. 269-270).

18. Jacob, R.J.K.: *A Specification Language for Direct-Manipulation User Interfaces.* ACM Transactions on Graphics, Vol. 5, No. 4, 283-317 (October 1986) .

19. Janssen, C., Weisbecker, A., Ziegler, J.: *Generating User Interfaces from Data Models and Dialogue Net Specifications.* In Ashlund S., Mullet K., Henderson A.,

Hollnagel E., White T. (eds.): Proceedings of INTERCHI'93. New York: ACM Press 1993 (pp. 418-423).

20. Johnson, P., Johnson, H., Wilson, S.: *Rapid Prototyping of User Interfaces Driven by Task Models.* In J. Carroll (ed.): Scenario-Based Design: Envisioning Work and Technology in System Development. London, John Wiley & Sons 1995 (pp. 209-246).

21. Kieras, D.E.: *A Guide to GOMS Model Usability Evaluation Using NGOMSL.* In Helander M., Landauer T. (eds.): The handbook of human-computer interaction. Amsterdam: North-Holland 1996.

22. Kim, W.C., Foley, J.D.: *Providing High-level Control and Expert Assistance in the User Interface Presentation Design.* In Ashlund S., Mullet K., Henderson A., Hollnagel E., White T. (eds.): Proceedings of INTERCHI'93. New York: ACM Press 1993 (pp. 430-437).

23. Lonczewski, F., Schreiber, S.: *The FUSE-System: an Integrated User Interface Design Environment.* In: Vanderdonckt J. (ed.): Proceedings of CADUI'96. Namur: Presses Universitaires de Namur 1996 (pp. 37-56).

24. Löwgren, J., Nordqvist, T.: *Knowledge-Based Evaluation as Design Support for Graphical User Interfaces.* In Bauersfeld P., Bennett J., Lynch G. (eds.): Proceedings of CHI'92. New York: ACM Press 1992 (pp. 181-188).

25. Luo, P., Szekely, P., Neches, R.: *Management of Interface Design in Humanoid.* In Ashlund S., Mullet K., Henderson A., Hollnagel E., White T. (eds.): Proceedings of INTERCHI'93. New York: ACM Press 1993 (pp. 107-114).

26. Moriyón, R., Szekely, P., Neches, R.: *Automatic Generation of Help from Interface Design Models.* In C. Plaisant (ed.): Proceedings of CHI'94. New York: ACM Press 1994 (pp. 225-231).

27. Myers, B.A.: *User Interface Software Tools.* ACM Transactions on Computer-human Interaction, Vol. 2, No. 1, 64-103 (March 1995).

28. *Neuron Dataelements Environment.* http://www.neurondata.com/

29. Olsen, D.R.: *A programming language basis for user interface managment.* In Bice K., Lewis C. (eds.):.Proceedings of CHI'89. New York: ACM Press 1989 (pp. 171-176).

30. Olsen, D.R.: *SYNGRAPH : a Graphical User Interface Generator.* Computer Graphics, Vol. 23, No. 3, 43-50 (July 1983).

31. Olsen, D.R.: *MIKE: The Menu Interaction Kontrol Environment.* ACM Transactions on Information Systems, Vol. 5, No. 4, 318-344 (1986).

32. Palanque, P., Bastide, R.: *Contextual Help for Free with Formal Dialogue Design.* In Alty J.L., Diaper D., Guest S. (eds.): Proceedings of HCI'93. Cambridge: Cambridge University Press 1993.

33. Pangoli, S., Paternó, F.: *Automatic Generation of Task-oriented Help*. In van der Veer G.C., Bagnara S., Kempen G.A.M. (eds.), Proceedings of UIST'95. New York: ACM Press 1995 (pp. 181-187).

34. Paternó, F., Faconti, G.: *On the Use of LOTOS to Describe Graphical Interaction*. In Monk A., Diaper D., Harrison M.D. (eds.): Proceedings of HCI'92. Cambridge: Cambridge University Press 1992 (pp. 155-174).

35. Paternó, F., Mezzanotte, M.: *Formal Verification of Undesired Behavious in the CERD Case Study*. In Bass L., Unger C. (eds.): Engineering for Human-Computer Interaction, Proceedings of EHCI'95. London: Chapman & Hall 1995 (pp. 213-226).

36. *PowerModel® The Object Power Tool*. http://www.intellicorp.com/power-model.html

37. Puerta, A.: *The Mecano Project: Comprehensive and Integrated Support for Model-Based Interface Development*. In: Vanderdonckt J. (ed.): Proceedings of CADUI'96. Namur: Presses Universitaires de Namur 1996 (pp. 19-36).

38. Schlungbaum, E., Elwert, T.: *Automatic User Interface Generation from Declarative Models*. In: Vanderdonckt J. (ed.): Proceedings of CADUI'96. Namur: Presses Universitaires de Namur 1996 (pp. 3-18).

39. Schreiber, S.: *Specification and Generation of User Interfaces with the BOSS-System*. In Blumenthal B., Gornostaev J., Unger C. (eds.): Proceedings of EWHCI'94. Berlin: Springer-Verlag 1994 (Lecture Notes in Computer Sciences, vol. 876, pp. 107-120).

40. Sears, A.: *AIDE: A Step Toward Metric-Based Interface Development Tools*. In van der Veer G.C., Bagnara S., Kempen G.A.M. (eds.), Proceedings of UIST'95. New York: ACM Press 1995 (pp. 101-110).

41. Singh, G., Green, M.: *Automating the Lexical and Syntactic Design of Graphical User Interfaces: The UofA* UIMS*. ACM Transactions on Graphics, Vol. 10, No. 3, 213-254 (July 1991).

42. Sukaviriya, P., Foley, J.D.: *Coupling a UI Framework with Automatic Generation of Context-Sensitive Animated Help*. In Proceedings of UIST'90. New York: ACM Press 1990 (pp. 152-166).

43. Szekely, P., Luo, P., Neches, R: *Facilitating the Exploration of Interface Design Alternatives: The Humanoid Model of Interface Design*. In Bauersfeld P., Bennett J., Lynch G. (eds.): Proceedings of CHI'92. New York: ACM Press 1992 (pp. 507-514).

44. Szekely, P., Luo, P., Neches, R.: *Beyond Interface Builders: Model-Based Interface Tools*. In Ashlund S., Mullet K., Henderson A., Hollnagel E., White T. (eds.): Proceedings of INTERCHI'93. New York: ACM Press 1993 (pp. 383-390).

45. Szekely, P., Sukaviriya, P., Castells, P., Muthukumarasamy, J., Salcher, E.: *Declarative interface models for user interface construction tools: the Mastermind ap-*

proach. In Bass L., Unger C. (eds.): Engineering for Human-Computer Interaction, Proceedings of EHCI'95. London: Chapman & Hall 1995 (pp. 120-150).

46. Vanderdonckt, J.: *Automatic Generation of a User Interface for Highly Interactive Business-Oriented Applications.* In Plaisant C. (ed.): Companion Proceedings of CHI'94. New York: ACM Press 1994 (pp. 41 & 123-124).

47. Vanderdonckt, J.: *Knowledge-Based Systems for Automated User Interface Generation: the TRIDENT Expierence.* Technical Report RP-95-010. Namur: Facultés Universitaires Notre-Dame de la Paix, Institut d'Informatique 1995. Available at http://www.info.fundp.ac.be/cgi-bin/pub-spec-paper?RP-95-010.

48. Wiecha, C., Bennett, W., Boies, S., Gould, J., Green, S.: *ITS: A Tool for Rapidly Developing Interactive Applications.* ACM Transactions on Information Systems, Vol. 8, No. 3, 204-236 (July 1990).

49. Wiecha, C., Bennett, W.: *Generating Highly Interactive User Interfaces.* In Bice K., Lewis C. (eds.):.Proceedings of CHI'89. New York: ACM Press 1989 (pp. 277-282).

50. Wilson, S., Johnson, P.: *Bridging the Generation Gap: From Work Tasks to User Interface Designs.* In: Vanderdonckt J. (ed.): Proceedings of CADUI'96. Namur: Presses Universitaires de Namur 1996 (pp. 77-94).

Human Factors in Aeronautics

V. De Keyser and D. Javaux

Work Psychology Department - Faculty of Psychology and Educational Sciences
University of Liège - bld du Rectorat, 5
B-4000 Liège 1 (Belgium)
E-mail : vdekeyser@ulg.ac.be, javaux@vm1.ulg.ac.be

Abstract. In this paper, we will discuss the human factors that must be taken into account in aeronautics, concentrating on the technological evolution this sector is undergoing and the cognitive demands which are therefore put on pilots. This will oblige us to clearly define the status of human error and to understand its processes, as well as the phenomena of judgment and causal attribution by which it is marked. We will also analyze the question of how error is corrected within the framework of established technical systems. Finally, the predictability and, therefore, the certification of such technical systems will be questioned.

Keywords

Aeronautics, human error, error recovery, cooperative distributed control, modes, situation awareness, complexity, predictability

1 Introduction

In this paper, we will discuss the human factors that must be taken into account in aeronautics, concentrating on the technological evolution this sector is undergoing and the cognitive demands which are therefore put on pilots. This will oblige us to clearly define the status of human error and to understand its processes, as well as the phenomena of judgment and causal attribution by which it is marked. We will also analyze the question of how error is corrected within the framework of established technical systems. Finally, the certification of such technical systems will be questioned. But in addition of these points, one detects just beneath the surface a question which concerns all specialists in the human factor: what do we know about the way in which human beings function, reason, and adapt when working with these technical systems? Faced with such complexity, are we able to contribute anything more to aeronautics than simplistic models, which are based on social or cognitive sciences, but which fail to take into account a situation, its variables, and their specificity ? It is ultimately this question, with all the indecision and obscurity it entails, which will be central in the following report.

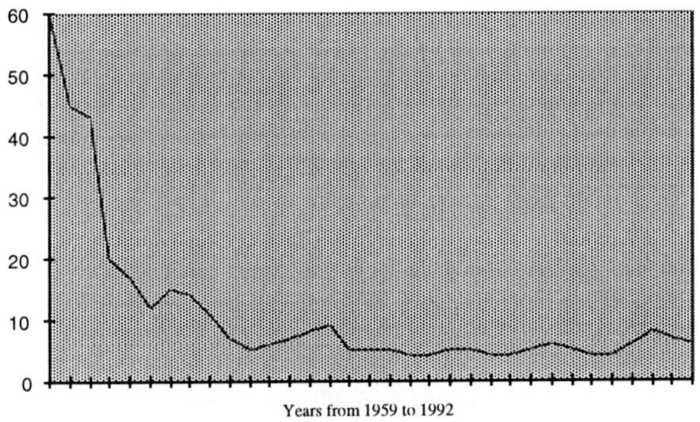

Fig. 1 *Annual rate of commercial air accidents, represented by millions of manoeuvres at airports (take-offs, landings; information from Boeing, Statistical Summary of Commercial Jet Aircraft Accidents, Worldwide Operations, 1959-1991). The current average rate is around one accident per one million manoeuvres. It varies by a factor of 10, according to geographical zones: from five accidents per million, at the highest rate (Eastern countries, Africa, certain Asian countries) to .5 accidents per million at the lowest rate (United States). This rate does not account for accidents due to sabotage or military actions. Note that this rate was reached in the 1970s and has not declined, despite the considerable improvements in technology and training which have since been made[Amalberti, 1996]*

The success of specialists in the human factor is based, at least partially, on a misunderstanding, that is to say, on the conviction of certain leaders in the field that human error is clearly identifiable and categorisable. These specialists claim that, by means of strict procedures and security devices which limit the choices made by people, as well as by means of automatic devices which free them from some of their tasks, one can reduce human error quite considerably. Error would thus be, in some sense, inherent to human nature, hidden within a person. This person must, therefore, be helped, guided, or provided with foolproof safeguards. All this is not entirely unreasonable. The research that has been conducted on mental overload, vigilance, the chronobiological effects of jet-lag, and leadership within teams and the enormous effort that has been expended on selection and training are not without interest. Nevertheless, if one accepts Boeing's statistics [Amalberti, 1996] - see fig.1, for the past twenty years, in spite of these studies and more and more sophisticated technical systems, the rate of accidents per take-off has remained unchanged, and routine errors, such as those related to communication, have been becoming more frequent.

If a significant breakthrough in security is not made, even with a static rate of accidents per take-off, the increase in air traffic in coming years will cause more frequent air catastrophes, which by their number will become more visible to the public. Airlines will thus have more difficulty competing with other organizations offering rapid forms of transportation for short and average distances (the railways, for example). This is why, in spite of airline executives' continuing skepticism regarding the contributions of ergonomists, and of the human sciences in general, to the issue - see fig.2 [Baud, 1996], these latter are more and more in demand. An example of the interest in the subject is the Human Factor Network, created by the General Management of Civil Aviation (DGAC) in France, which has founded a series of human science research centers centered around the Airbus Industry. The network is less concerned with the question of human error than with that of the future "Joint Cognitive System" which is to be created out of more and more advanced technical systems. The network is also concerned with a social and technical environment which extends far beyond the on-board equipments because it includes, apart from air control, other aircraft.

> *"A proper understanding of pilots' needs and of the operational factors influencing their work is, in our view, of greater value for cockpit design than the applications of rules determined on the basis of purely academic studies, which, however, we always read with great interest. Professional ergonomists have formalized certain rules, but these are usually drawn up a posteriori. Therefore they tend to be too general for such useful application as providing practical guidelines in design work. In some cases they can even prove to be counterproductive" [Baud, 1996].*

Fig. 2 *Excerpt from Baud (1996).*

2 Human error and its correction

The profusion of methods which may be used to categorize and then to collect and analyze data on human error indicates the complexity of the phenomenon and the arbitrariness which sometimes governs the label imposed on an act or judgment that has led to undesirable consequences. In the already outdated categorization made by Gutmann and Swain [1983], or even in the more recent one made by Hollnagel [1993], the emphasis is placed on the visibility of the act; whether it be a commission or an omission error, a delay, an inversion or a drift, it is always evaluated in relation to a planned procedure, from which a person has deviated despite his or her intention to follow it. The matter is clear--the standard procedure must not be violated. If someone deviates from it without intending to, it is an error; if he or she does so intentionally, that is a violation.

The problem is more complicated when one introduces two aspects into this judgment of error: the situation, its complexity and variability, and the process which, beginning with an initial stimulus which motivates a behavior or judgment, ends in undesirable effects, an incident or catastrophe such as we have discussed. In fact, the idea that a procedure can serve as a norm rests on an extremely rigid interpretation of the world; in this world, events occur regularly enough that the people involved have learned to automatically deal with them and preliminary studies have been done to establish the procedures which should be followed. When a situation becomes increasingly variable, this system of planning ahead can possibly be maintained, but only by becoming more complicated.

Far from being linear, plans which have been pre-established by research departments ramify, taking different directions, with conditional branches and interruptions - see fig. 3 [Javaux, 1995, 1996]. We have seen, in recent railway catastrophes, the difficulties faced by mechanics, lost in the maze of a series of somewhat contradictory maintenance regulations, which were laid out in different manuals, and placed under time pressure because of the network's disorganization following a breakdown [De Keyser, 1995]. Can we still, then, talk of human error, when what is involved in following up on a procedure surpasses normal abilities to process information?

Furthermore, the idea of respecting a plan also presupposes that all the steps of the situation have been foreseen, assuming a sort of complete and hermetic world, incompatible with the complexity of interactions that can arise between the different participants and the different technical parameters of the system. And when someone is confronted with an unforeseen condition, the situational planning that he or she performs – that is to say, his or her instantaneous re-evaluation of the situation and the decision that he or she makes – often cannot be judged as an error or a success until afterwards, when the consequences of the action are known. What we call an error is very often only a social judgment [Hollnagel, 1993], marked by a causal attribution, once we reconsider an accident in retrospect. In fact, we focus on the series of factors, human error being only one of these, leading to the accident, without taking into account the role of probability and all the possible situations which, even with the same errors, could have produced different outcomes.

These considerations lead Woods et al. [1994] to declare: "Studies in a variety of fields show that the label "human error" is prejudicial and unspecific. It differs rather than advances our understanding of how complex systems fail and the role of human practitioners in both successful and unsuccessful system operations. The investigation of the cognition and behavior of individuals and groups of people, not the attribution of the error itself, points to useful changes for reducing the potential for disaster in large complex systems. Labeling actions and assessments as errors identifies a symptom, not a cause; the symptom should call forth a more in-depth investigation of how a system comprising people, organizations and technologies both functions and malfunctions" (p.3) [Rasmussen et al., 1987; Reason, 1990; Hollnagel 191b; 1993].

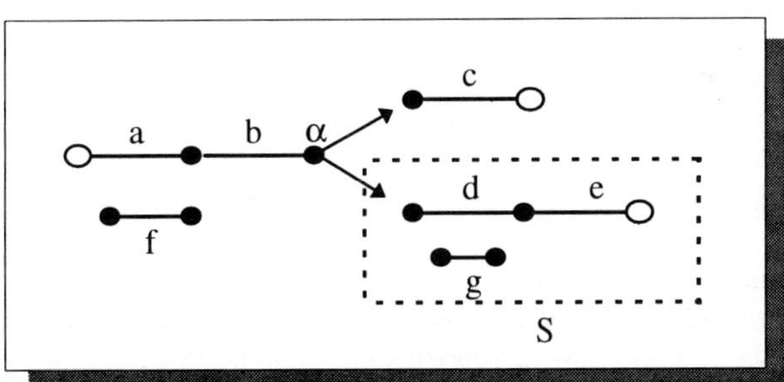

Fig. 3 *Complex plans as they can be found in real-world situations [Javaux, 1995; 1996]*

This brings us, therefore, to the "Joint Cognitive System"--with the conditions under which it functions and the adaptable regulations which make it possible, fortunately, to correct so many human errors. In this respect, two concepts are important: the first being that of "latent failures" and the second that of "recovery". According to research projects conducted in mines and the steel industry under the direction of the ECSC (European Coal and Steel Community), as early as 1967 [Faverge, 1967], accidents, like incidents, errors, or breakdowns, were considered to be the symptom of a more general malfunctioning of the human-mechanical interaction, and considered to be a probabilistic phenomenon.

Faverge had demonstrated what he called the problems of reliability--that is to say, factors that, when present in a given situation, increased the risk of an accident without inevitably causing one. Among others, there were organizational factors, systems design, connections between different services, etc. Twenty years later, in his book about human error, Reason [1990] comes back to this notion and, in a vivid way - see fig. 4, illustrates both the probabilistic aspect of a catastrophe or an accident and the multiple failures which must occur concurrently if a commonplace error is to have disastrous consequences. In fact, ever since the principle of prevention first began to have as great an influence on the organization of technical systems as on their design, it is usually possible to correct the potentially dangerous human errors that do remain. In various ways, several factors influence the correction process.

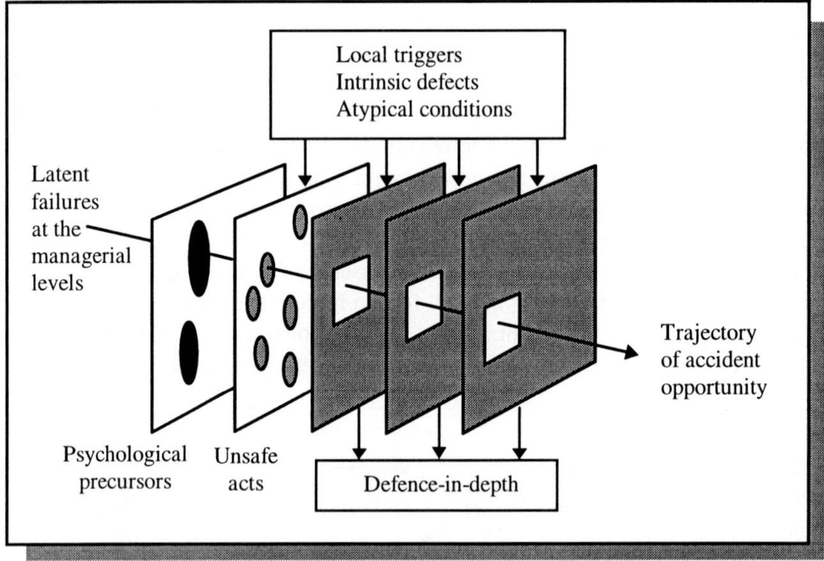

Fig. 4 *The dynamic of accident causation. The diagram shows a trajectory of accident opportunity penetrating several defensive systems. This results from a complex interaction between latent failures and a variety of local triggering events. It is clear from this figure, however, that the chances of such a trajectory of opportunity finding loopholes in all the defenses at any one time is very small indeed [adapted from Reason, 1990].*

2.1 The nature of the error

In his typology of common errors, Reason [1990] makes a distinction between "slips" and "mistakes". Slips are related to lapses of attention or memory; the person in question knows what he or she is supposed to do or say, but because of unexpected circumstances – distractions, changes in the environment, etc. – he or she does not follow his or her original intention.

Mistakes on the other hand are related to the limits of one's knowledge, reasoning, and/or judgment when faced with a problem; the action taken fails to resolve this problem. Error recovery seems much easier in the case of a slip than in that of a mistake; usually, it is the person who has committed the error who then corrects it--notices that he or she has left the keys in the apartment, that he or she has entered the wrong code into the computer, etc. But in any case, because the intention is clear, the deviation from this intention is obvious. If a pilot enters a command input which does not correspond to the desired and stated goal, it is easy for the copilot to correct it, by a routine "cross-check". Mistakes, however, pose many more problems. According to Woods [1982], Rizzo et al. [1986], De Keyser & Woods [1990], mistakes are rarely corrected in time by the person who made them. Generally, a "fresh observer" (that is to say, a person who has not taken part in the decision) is needed to counteract a judgment or decision based on false assumptions or erroneous inferences.

For this reason, in certain high-risk industries such as nuclear energy, emergency plans rely on various diagnoses to be made, based on the same information, but in three different locations and by different groups of people: in the control room, by the work team; in an emergency committee which meets within the power station, but includes people who are not implicated in the incident; and outside of the station, by external experts.

2.2 The rate of situation deterioration

Woods et al. [1994] point out that certain incidents occur very quickly; suddenly things go wrong, and it is easy for the person who is supposed to be in control to realize that there is a problem. But in other cases, deviations from normal procedure are more insidious; certain parameters disintegrate slowly, and the operator unknowingly slips into a deteriorating situation, without having realized its gravity. Such errors in judgment are difficult for the operator alone to correct in time. Once again, a fresh observer, or technical methods which we shall discuss later, may be needed in order to formulate a correct evaluation of the situation.

2.3 The absence or delay of feedback after an action

In order to detect an error, the connection between a cognitive process and an action must be expressly linked to a visible result [De Keyser, 1995; Hoc, 1996, Norman, 1989]; in short, there must be quick and clearly identifiable feedback. This is the whole point of making a technical system transparent, a matter which is raised here and will be addressed later, with regards to modes in automated aircraft. In fact, as soon as:

- there is a delay between an action and its consequences;
- the person who commits the error is not the one who suffers its consequences;
- actions are distributed within a complex system,

the identification and, therefore, the correction of an error become extremely difficult.

3 Controlling actions within a complex system of distributed roles

Numerous authors – Norman [1985], Sheridan & Hennessy [1984], Rasmussen [1982], Bainbridge [1987, 1988] and the whole French-speaking ergonomy school – have been interested in the control of dynamic processes, in an attempt to clearly define them and to highlight the cognitive processes which they bring into play: supervision, detection, diagnosis, planning, error recovery, etc. The control of dynamic processes depends upon the existence of a kind of loop which applies to a set of variables that form part of the process. This "loop" would bring into play, on the one hand, the collection of information (either directly or by means of sensors which detail the condition of the variables) and, on the other hand, commands direct or indirect effectors which allow for the modification of the condition of the variables. Access to the process, whether it be for the purpose of collecting information, analyzing it, or performing a command, may, according to the case in question, be either direct (by manual control) or mediated by several levels of automatons and computers.

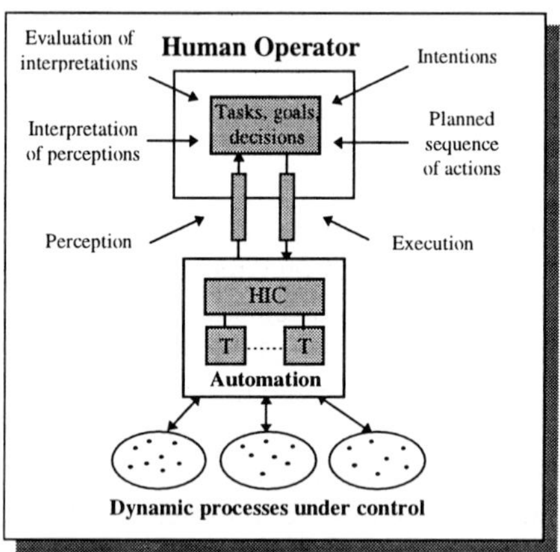

Fig. 5 *Distributed control in supervisory control situations (Sheridan & Hennessy, 1984). A set of dynamic processes is controled by a distributed cooperative system made of a human operator and several computerized agents : the different tasks to be executed upon the dynamic processes are allocated or distributed between the human and the automation. The human operator is in charge of the allocation of tasks. He or she is thus involved in two simultaneous control loops : one on the dynamic processes and one on the automation itself. The cognitive processes underlying this dual control task are described (drawn from Norman, 1988). The automation is made of a HIC (human interactive computers) that mediates the interaction with the operator and of several TIC (task interactive computers) that close control loops on the dynamic processes. (Distributed) information processing (either of human or automated nature) is depicted in grey.*

In the latter case, the human being is no more than the regulator of the system; this is the "supervisory control" described by Sheridan & Hennessy [1984], which allows for numerous variants, depending on whether the person remains (either entirely or partly) in the control loop or whether he or she is completely external to it - see fig. 5. Every industrial sector is subject to this kind of technological evolution. In aeronautics, however, the sophistication of information technology has been pushed very far-- beyond the popular stereotype illustrated by the question "Is there a pilot in the airplane?" We must ask ourselves about the kind of distributed control that is found in this sector today, about the complexity that it creates for flight crews, and about its impact on the predictability and anticipation of the flight's course. As a rule, analysis of the control distribution should not be limited to the crew and the technical devices, because air control and the flight environment are also factors to be considered; in order to simplify this report, however, we will nevertheless limit our discussion to the former elements. We will focus principally on the problem of the auto-pilot flight modes and, in order to illustrate this problem, we will use as an example an air catastrophe which seems to implicate an error of this type, that of the A320 at Mont St. Odile near Strasbourg in 1992. This example will allow us to come back to the problems of "latent failures" and "recovery" which we have already mentioned.

3.1 The Airbus A320 accident at Mont St. Odile, January 20th, 1992 [Monnier, 1992]

On January 20th, 1992, an Airbus A320 flown by the airline Air Inter crashed at night on Mont St. Odile near Strasbourg, leaving 87 victims. The administrative inquiry following the accident did not highlight the technical failure of the aircraft, certainly no more so than in the air disasters which have implicated Airbus aircraft in the past few years[1]. In fact, the most likely hypothesis accepted not so much by the administrative as by the judicial inquiry, centers on an error of the type committed by a pilot, or a problem of "situation awareness". At the moment of approach, believing to be in auto-pilot descent mode FPA (or FLIGHT PATH ANGLE, a vertical descent mode coupled with horizontal navigation mode TRK or TRACK in the A320), the pilot appears to have entered a descent input of 3.3. This should, in that case - FPA mode active -, have led to a rather normal descent angle of 3.3°.

But, the auto-pilot *was not* in FPA descent mode : in order to comply with radar guidance at the moment of approach, the pilots had changed the horizontal navigation mode to HDG (or HEADING), and this mode happens to be coupled automatically with vertical descent mode V/S (or VERTICAL SPEED) : the vertical descent mode had changed from FPA to V/S, this being unnoticed by either of the pilots.

The 3.3 input entered on approach (that is, after the selection of horizontal mode) no longer meant a descent angle of 3.3° (corresponding roughly to a descent rate of 700 to 800 feet per minute) but a much faster rate of 3,300 feet per minute. Altitude was lost five times faster than expected and this was not detected by the pilots. As a consequence, the airplane hit Mont St. Odile at 18 h 20 local time, killing 87 people. The VCR black box containing the communication of the pilot and copilot shows that up to the moment of impact and, in spite of certain displays on the FCU and the PFD (primary flight display) screen, the crew were unaware of the imminence of the catastrophe. Because they were unaware of the situation, we can say that the initial auto-

---

1 Amalberti (1996) lists ten accidents which have occurred in the Airbus "family" over the past six years, with no known cause of breakdown (p. 40) - Op. Cit.

pilot mode error had been replaced by one of fixation, the two members of the flight crew being insensitive to certain information appearing on their screens. In many respects, Mont St. Odile is a textbook case and, unfortunately, we do not have enough space here to discuss the ergonomics of the cockpit, the composition of the flight crew, the role of radar guidance, emergency procedure, and so on.

We will confine ourselves to the problem of auto-pilot modes and the paradoxical role they play in the distribution of flight control; whilst assisting the crew at times of normality, they may increase the complexity of control in an abnormal situation.

3.2 Description of the process

As we shall see, modes are computer aids which enable the flight crew to delegate some of the tasks involved in a given procedure. How, though, can we describe this at once in terms of variables and of relationships between these variables? While most aeronautical manuals dealing with the human factor describe in detail the control and command mechanisms of the cockpit (and, notably, glass cockpits), they fail to mention the model(s) for the operation of the aircraft itself, which lie(s) behind these mechanisms.

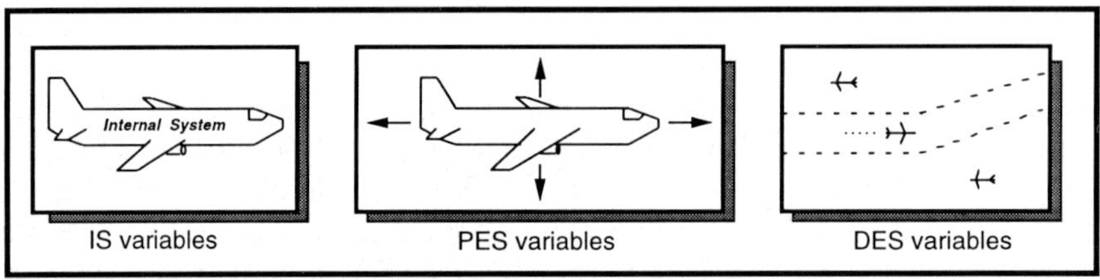

Fig. 6 *Analysis of the situation: the three "processes" under control [Javaux et Figarol, 1995]*

Javaux and Figarol[1995] make a heuristic breakdown of this process into three classes of variables under the control of pilots - see fig. 6.

- The variables which correspond to the technical and concrete systems of the plane (IS, Internal Systems). These systems are, as it were, internal to the plane (e.g., the structure of the airplane, different hydraulic and pneumatic circuits, and so on). The thrust of the engines, fuel consumption, and the configuration of the flaps or landing gear are examples of typical internal variables, the malfunction of which is cause for alarm and is followed by recovery procedures.

- Variables describing the airplane as a dynamic system, driven in a movement relative to its immediate spatial and aerodynamic environment (PES). These variables concern the aerodynamics of the airplane (lift, drag, angle of attack) and its orientation in space (pitch, roll, yaw). Speed (indicated and true airspeed) fits equally into this category of variables. It is principally these variables which are the object of restrictive and active automated protections on the Airbus A320, with the goal of maintaining the reliability of the system.

- Variables which characterize the position of the plane in air-space and its integration into the air-traffic (DES). These variables are the main target of the auto-pilot modes: i.e., intelligent flight-aids or agents assisting the flight crew in the most economical and safest management of the flight path, as long as it rests within the range of foreseen variabilities.

The main idea is that each class of variables constitutes a separate dynamic process (in the sense of controlled processes), with its own set of associated tasks, constraints and resources dedicated to it. There is every reason to think that the complexity of the situation and of the tasks is distributed in today's glass cockpits differently according to these three categories. Systems management (IS) and navigational management (DES) during a normal flight have been so greatly simplified - by means of automated resources - that the presence of an on-board flight engineer and navigator is no longer deemed indispensable. However, problems remain and impact piloting mainly through the auto-pilot and navigational errors (particularly in the vertical dimension)..

3.3 The complexity of the auto-pilot modes

The flight crew, comprising pilot and copilot, distributes flight control. On the one hand, the crew members are dealing with a physical and social environment which may be a source of variability and of alternative commands: air traffic control, atmospheric conditions, other airplanes, and so on.

On the other hand, the crew also delegates certain tasks to the auto-pilot (e.g. fly the aircraft on a 180° heading at flight level 330 - i.e. 33,000 ft) or cooperates with the different computerized agents available on the flight deck. These agents are the auto-pilot modes, which may in theory be deactivated if the plane encounters an abnormal situation. In reality, though, the complexity of these on-board flight aids has increased to the extent of forming, for some pilots, an impenetrable layer of computer technology which prevents prediction of the aircraft's behavior and knowledge of the actual situation. This is the problem of situation awareness encountered at St. Odile.

In fact, there are two ways of considering these modes: according to a paradigm of utilization or according to a paradigm of cooperation. According to the first paradigm, an auto-pilot mode is basically a function which regulates or makes a transition into a state that is mostly regarded as static: the state of the variable under control (e.g. altitude) only changes under the effects of actions issued by the pilots (either manually or through the automation, e.g., the altitude of the plane only changes when pilots decide it). An auto-pilot mode is thus *used* to perform a task of transition (e.g., to climb from 10,000 to 15,000 ft). The relation between man and automation resorts from a kind of master-slave relationship, the automation being used and controlled by the pilots. In this case, the criteria for mode interfacing come under the same category as those proposed by the classical field of human-computer interaction (HCI), under the generic label of "usability". A second paradigm consists in seeing modes as independent agents, and interaction on the part of the flight crew is no longer a question of control and utilization, but of cooperation: modes and human beings *cooperate*, performing a task together, which is more complete and more complex than their respective sub-tasks.

Consequently, the criteria, and in general the methods of evaluation and design which determine the interfacing of the auto-pilot modes according to the paradigm of utilization, are no longer applicable. One of the essential differences is to be found at the level of supervision: an auto-pilot mode is active over a long period of time. Supervision of a mode cannot be reduced - as in the case of the interfacing of transition functions - to a simple verification that the expected transition has actually taken place (primary feedback). It also involves to monitor *how* the transition takes place and the consequences that it entails (secondary feedback on the side-effects). A continuous evaluation of the development of the situation is necessary, and this regular reevaluation[1] is made all the more difficult by the perceptual, motor, and cognitive promptings which can mobilize the crew over the course of time.

The complexity of modes is not only a question of the types of regulations and transitions they effect, the classes of variables on which they act, their large number, or their interrelationships (coupling, exclusion, sequencing)[2], but equally of the different interactions that they presuppose with the flight crew and the cognitive constraints associated with this. At any given moment, the auto-pilot modes can be inactive, armed (that is, selected but not active), or active. Moreover, we can distinguish between two main axes of function - see fig. 7. The first emphasizes the opacity of the target i.e. the objective of the task assigned to the mode. Thus, when modes are *selected*; the goal to be attained or maintained by the plane is assigned by the pilots.

When the modes are *managed*, the goal is determined by the FMS (Flight Management System). It is, of course, more difficult to understand where the plane is, and which agents it has put into play, in the second case than in the first. A second axis concerns the the execution of the task when the auto-pilot is used: either it is the pilot who, via the Flight Director, achieves the execution of the task computed by the automation, or it is done automatically, by the auto-pilot itself. This is the organization found in the double-entry table in figure 7. If one adds that these squares of figures can be distinct for horizontal and vertical navigation modes, that gives us a good idea of the complexity of human-computer cooperation progressively introduced into aeronautics. The ergonomics of the cockpit has tended to make perceptible certain characteristics of the technological component: e.g., the selected auto-pilot mode appears on the FCU (Flight Control Unit), the use of the "bird" or the bars differentiates between the HDG or TRK modes in horizontal navigation, etc.

But, generally speaking, vertical flight modes have received little attention and sensory prompting is mainly central vision. Moreover, the internal dynamics of the modes, their interrelationships, and the swift transition from one to another are not transparent or salient. We are certainly not dealing here with a logic of ecological interfacing, bridging the gulf between the processes to be controlled and the human

---

1 Problems with which are known to have caused catastrophic fixation errors--cf. De Keyser V. & Woods D.D. 1990. Fixation errors: Failures to revise situation assessment in dynamic an risky systems. In A.G. Colombo and A. Saiz de Bustamente (Eds). *System reliability assessment.* Dordrechts, The Netherlands : Kluwer.

2 Thus, in the Airbus A320, a lateral navigational mode is coupled with a vertical navigational mode; as soon as one is selected, the other automatically follows.

being. This leads us to what is at once a problem of situation and of error recovery [Sarter & Woods, 1991, 1995], if an error should occur.

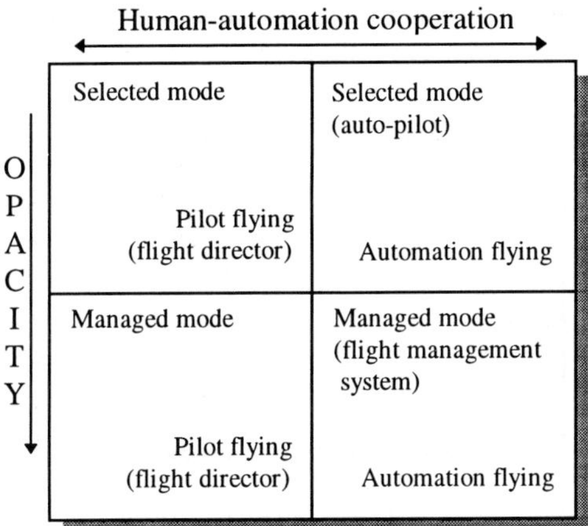

Fig. 7 *Description of auto-pilot modes, based on the opacity of the target and the delegation of control and execution.*

3.4 Cognitive complexity and situation awareness

The importance of auto-pilot modes in today's practice leads us to consider them as potential sources of cognitive complexity in mordern glass-cockpits. Some of their characteristics, namely autonomy and authority (Sarter & Woods, 1995), make them rather independant agents which constitute with the pilots a "joint" (Woods & al, 1994) or "distributed" (Hutchins & Klausen, 1991) cognitive system achieving the flying task. The allocation or distribution of tasks between the differents agents (including themselves) now resorts to the pilots. Being tasks managers rather than "skilled" pilots is a new role for them and this seems to be a potential source of complexity. In order to assess this complexity, we have started a series of researches at the University of Liège.

We consider – following the approach led by Kieras & Polson (1985) – cognitive complexity as a computational complexity of the cognitive processes underlying the achievement of a task in a specific situation. In particular, we attempt to evaluate the cognitive complexity of the task allocation problem by means of a decomposition into simpler but integrated sub-tasks.

Achieving the allocation of tasks between the agents of the distributed cooperative system involves for the pilots to (see fig. 8):

- evaluate the current and future global tasks, that is the tasks allocated to the aircraft seen as an integrated cooperative system, e.g. tasks assigned by air traffic control (1);
- evaluate the state of the aircraft and its environment in the present and the future (2);

- evaluate the distribution of the tasks between auto-pilot modes and pilots (3);
- make decisions concerning possible modifications to be made in tasks distribution (4);
- implement possible modifications to tasks distribution by means of communication and interactions with the interfaces (5).

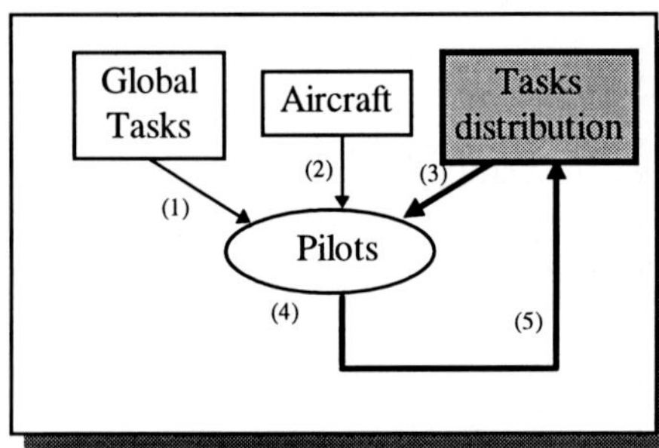

Fig. 8 *Decomposition of the task allocation problem into 5 sub-tasks*

Without going into much details, let say that we try to derive computational (user) models of the managerial task from an analysis of the sub-tasks, the peculiarities of auto-pilot modes and their interfaces, and knowledge of the way pilots interact with them. The next step of this research is to characterize the computational complexity of each of the elementary – cognitive – sub-tasks, in order to explain known problems, incidents or accidents related to modes, as well as to pinpoint weak points in modes behavior, architecture or interfacing.

On the other hand, without making reference here to all the literature on situation awareness, one may, according to the same sort of task division logic, say that in order to build and maintain awareness of the situation on-board the plane and in order to satisfy the conditions for error recovery described by Woods et al. [1994], it would be necessary to be able to assure the following related tasks:

- maintaining a correct representation of the condition of the aircraft, as long as it does not change;
- correctly modifying the representation of the situation of the airplane, as soon as this is changed through the action of the aircraft's own dynamics (e.g. relate vertical speed with altitude changes);
- corrrectly modifying the representation of the situation of the airplane, as soon as this is changed through the action of the pilots;
- correctly modifying the representation of the situation of the airplane, as soon as this is changed through the action of the automation (auto-pilot modes).

But what does "correctly modifying" mean? The value of these related tasks is that they each mark out the informational content which must be available to the navigational crew in order to maintain situation awareness, but they do not tell us how these representational modifications should be introduced, according to which modalities and in which forms. In fact, they skirt around the debate over the human cognitive model. The analysis of the Mont St. Odile catastrophe and that of numerous other fixation errors show that signals warning the people involved of the disastrous turn of events were not interpreted as precursors of an accident, but rather ignored. They were not in fact included in a complete and systematic reevaluation of the range of variables and possibilities which come into play in this situation.

4 Predictive or descriptive models?

Once we have established the various tasks to be carried out, whether directly, through delegation or in cooperation with the automation, as well as some of their characteristics such as opacity, temporal pressure, affected sensory modality, and degree of experience in the field of the crew members, can we define the cognitive complexity of the distributed control and, as a consequence, can we predict the performance of pilots faced with the kind of system? The answer to this question is essential for the certification of the new generation planes, for their designing cycle is so long – almost a decade – and their cost so high that last-minute adjustments must be kept to a minimum. Up to now modeling attempts start from a structural description and decomposition of the tasks at hand and consider their cognitive complexity in terms of computational complexity derived from the cognitive processes (e.g., production rules) to be applied in order to achieve them.

This type of research is similar to the one used in GOMS [Card, Moran & Newell 1983], Kieras [1988] and more recently to the one by Kieras and Meyer [1995], with their EPIC architecture. This last mentioned project attempts to establish quantitative measurements that could be used to predict the subject's performance. It lays more stress on sensory and motor modalities than on attentiveness or memory as weak points in the interaction between humans and machines in a multiple task environment. Nowadays however, experimentation bears only on quite simple tasks which are a far cry from the complexity we have only tentatively suggested above.

In the field of car industry, where certification is also required, research projects have attempted to describe cognitive complexity, starting from the works of Kieras and Polson [1985]: DIADEM is a software which makes it possible to assess in a predictive way the performance of a driver using a number of supports derived from the new technologies and adapted from the field of aeronautics while driving, an instance of these is the HEAD UP display. These projects are interesting but, as we have seen, their applications are yet very limited.

A more pragmatic approach [Amalberti 1996; Sarter and Woods 1992, 1994] consists in analyzing the behavior of pilots who use the technical contrivances to be tested. Through considering the heuristic strategies they use, the pitfalls they meet and the mistakes they make, the research team can deduce the ergonomic specifications which the new devices must meet if they are to be more user-friendly. The model thus arrived at is not an a priori theoretical model but a *descriptive model deduced from the subjects' behavior*; the stress is laid on the heuristic strategies and on the cognitive compromises that have to be resorted to in order to carry out the task.

An essential aspect of these compromises is the management of time which operates both on short and on long term delays, both on reactive planning and on anticipation. Equally important of course are the competence and the meta-knowledge of subjects, which enable them to act with some degree of confidence and certainty within their safety limits; within these limits they will take calculated risks, but they are reluctant to go beyond. If we come back to the question of modes it is clear that many pilots will feel that their non mastering of the complexity of the devices takes them beyond the limits – which vary from one subject to another – within which they feel safe because they have a clear situation awareness and can easily predict their plane's behavior. As a consequence, the pragmatic approach emphasizes the need for early experimentation of prototypes using scenarii that do not necessarily involve catastrophes but that do include some of the difficulties that will actually have to be met. Criteria for performance, attitude, social acceptance of the devices to be tested, mental strain, etc. are developed without any theoretical construct except a framing model for this kind of competence which describes the possible heuristic strategies and cognitive limitations. So far, this approach has not devoted much interest to multimodal perception or to the motor constraints entailed by multiple tasks.

As a matter of fact, the two approaches have so far been developed almost independently. In theory though they could very well be combined. This is what has been attempted by the GEM research project (Generic Evaluation Methodology for Integrated Driver Support Applications). This is what we are attempting to do here at the University of Liège.

5 Conclusion

Problems are raised by the growing complexity and opacity which results from the increased computer sophistication in the latest generation of planes. One can even suspect that without their being the only cause they may well have played a part in recent air crashes and several incidents. In this respect latent failures such as described by Faverge [1967] and Reason [1990] make the process of retrieving errors and the regular reassessment of a situation more difficult; this has been clearly shown by Gras et al.'s inquiries into pilots' perception of automation [1995]. However, increasing the transparency of the computer system to make it easier for the crew to master it cannot be carried too far. Indeed, total transparency would inevitably take away from the degree of attentiveness which is necessary to carry out certain tasks. We thus need to combine two approaches which we think cannot be dissociated.

The first consists in modeling the cognitive complexity of the tasks to be carried out and the perceptual and motor constraints associated with them; this is in fact the approach followed by Kieras and Polson [1985].

The second approach defines a minimal amount of competence relating to automation and specifies the degree of transparency that is needed and the form it must have depending on the subjects, on the cooperation between them, on the tasks themselves, and on an environment which is known to be dynamic.[1] This approach is de-

---

1 These conditions are necessary for pilots to know, for instance, when and how to disengage automation when their planes comes out of a nominal flight, an operation which many still fear today.

veloped by the models that describe the pilots' behavior. Recent research on situation awareness and problems related to modes shows that this transparency must not be conceived of in visual terms only, that it should combine various sensory modalities, including touch. People should be made to draw on all their resources in a joint cognitive system which exceeds the merely cognitive dimension.

Acknowledgments

This research could be carried out thanks to a contract agreement with Direction Générale de l'Aviation Civile en France, within the frame of a study group on human factors.

References

Amalberti, R. (1996). *La conduite de systèmes à risques.* Le Travail Humain. Paris: Presses Universitaires de France.

Bainbridge, L. (1987). *Ironies of automation.* In Rasmussen, J., Duncan, K. & Leplat, J. (Eds), New Technology and human error. Chichester, England: Wiley.

Bainbridge, L. (1988). *Types of representation.* In Goodstein, L.P., Anderson, H.B. & Olsen, S.E. (Eds), Tasks, errors and mental models. London : Taylor & Francis.

Baud, P. (1996). *Introduction to human factors at Airbus Industrie.* Second Human Factors Symposium. Regional Conference America's. Miami, February 26th-29th 1996

Card, S.K., Moran, T.P. & Newell, A. (1983). *The Psychology of Human-Computer Interaction.* Hillsdale, N.J.: Lawrence Erlbaum Associates.

De Keyser, V. (1995). *Time in ergonomics research.* Ergonomics, 38, 8, 1639-1660.

De Keyser, V. & Woods, D.D. (1990). *Fixation errors : Failures to revise situation assessment in dynamic and risky systems.* In Colombo, A.G. & Saiz de Bustamente, A. (Eds). System reliability assessment. Doordrecht, The Netherlands: Kluwer.

Faverge, J.M. (1967). *Psychosociologie des accidents de travail.* Paris: Presses Universitaires de France.

GEM (1995). *Generic Evaluation Methodology for Integrated Driver Support Application. telematics for Transport Services* - V2065 GEM, CEC - DG XIII.

Gras, A., Moricot, C., Poirot-Delpech, S. & Scardigli, V. (1994). *Face à l'automate.* Paris: Publications de la Sorbonne.

Hoc, J.M. (1996). *Supervision et contrôle de processus.* La cognition en situation dynamique.

Hollnagel, E. (1993). *Human reliability analysis : context and control.* London: Academic Press.

Hollnagel, E. (1993). *The phenotype of erroneous actions.* Int. J. Man-Machine Studies, 39, 1-32.

Hollnagel, E. (1991a). *Cognitive ergonomics and the reliability of cognition.* Le Travail Humain, 54, 4.

Hutchins E., Klausen T.. (1991b). *Distributed Cognition in an Airline Cockpit,* Internal Report. Department of Cognitive Science, University of California, San Diego, La Jolla.

Javaux, D. (1995). *Modelling situations.* In M. Richelle, V. De Keyser, G. d'Ydewalle, A. Vandierendonck (Eds.), 'Temporal Reasoning and Behavioral Variability'. Report 1990-1994 of the IPA Convention n° 31. Work Psychology Dept: University of Liège.

Javaux, D. & Figarol, S. (1995). *The Macro-Pilot and the Distributed Nature of the Flying Task.* In T.B. Sheridan (Ed.), Proceedings of the International Federation of Automatic Control - Man-Machine Systems (IFAC-MMS) Conference, 27-29 June, MIT, Boston: IFAC.

Javaux, D. (1996). *Introduction à la formalisation des tâches. La dimension temporelle.* In J.M. Cellier, V. De Keyser, C. Valot (Eds.), 'Gestion du temps dans les environnements dynamiques', Paris: Presses Universitaires de France. To appear.

Kieras, D.E. (1988). *Towards a practical GOMS model methodology for user interface design.* In Helander, M. (Ed.), Handbook of Human-Computer Interaction. Amsterdam: North Holland Elsevier.

Kieras, D.E. & Polson, P.G. (1985). *An approach to formal analysis of user complexity.* Int. J. of Man-Machine Studies, 22, 365-394.

Kieras, D.E. & Meyer, D.E. (1995). *An overview of the EPIC Architecture for Cognition and Performance with Application to Human Computer Interaction.* EPIC Report n° 5 (TR 95/ONR - EPIC-5). December 5. University of Michigan.

Monnier, A. (1992). *Rapport préliminaire de la Commission d'enquête administrative sur l'accident du Mont Saint-Odile du 20 janvier 1992.* Paris, France: Ministère de l'Equipement, du Logement, des Transports et de l'Espace.

Norman, D.A. (1985). *New views of human information processing : implications for intelligent decision support systems.* In Hollnagel, E., Mancini, G. & Woods, D.D.

(Eds), Intelligent Decision Aids in Process Environments. San Miniato: NATO Advanced Study Institute.

Norman, D.A. (1988). *The Design of Everyday Things.* New York: Double Day Currency.

Norman, D.A. (1989). *The "problem" of Automation. Inappropriate Feedback and Interaction. Not "Overautomation".* ICS Report 8964.

Rasmussen, J. (1981). *Models of mental strategies in process plant diagnosis.* In Rasmussen, J. & Rouse, W. (Eds.), Human Detection and Diagnosis of system Failures. New York: Plenum.

Rasmussen, J., Duncan, K. & Leplat, J. (1987). *New Technology and human error.* Chichester, England: Wiley.

Reason, J. (1990). *Human error.* Cambridge, UK: Cambridge University Press.

Rizzo, A., Bagnara, S. & Visciola, M. (1986). *Human error detection processes.* In Mancini, G., Woods, D.D. & Hollnagel, E. (Eds). 'Cognitive Engineering in Dynamic Worlds'. Ispra, It. : CEC Joint Research Center.

Sarter, N.B. & Woods, D.D. (1991). *Situation awareness: a Critical but Ill-Defined Phenomenon.* Int. Journal of Aviation Psychology, 1, 43-55.

Sarter, N. & Woods, D.D. (1992). *Pilot interaction with cockpit automation. I. Operational experiences with the Flight Management System.* Int. Journal of Aviation Psychology, 2, 303-321.

Sarter, N. & Woods, D.D. (1994). *Pilot interaction with cockpit automation. II. An experimental study of pilot's mental model and awareness of the Flight Management System (FMS).* Int. Journal of Aviation Psychology, 4 (1), 1-28.

Sarter, N.B. & Woods, D.D. (1995). *How in the world did we get into that mode ? Mode error and awareness in supervisory control.* Human Factors (Special Issue on Situation Awareness).

Woods, D.D., Johannesen, L., Cook, R.I. & Sarter, N. (1994). *Behind Human Error: Cognitive Systems, Computers and insight.* Whright-Patterson AFB, OH: CSERIAC.

Swain, A.D. & Guttman, H.E. (1983). *Handbook of Human Relianility Analysis with Emphasis on Nuclear Power Plant Applications.* NUREG/CR 1278. Albuquerque, NM: Sandia National Laboratories.

Woods, D.D. (1982). *Operator Decision Behavior during the Steam Generator Tube Rupture at the Ginna Nuclear Power Station.* Research Report 82-IC57-CONRM-R2. Pittsburgh, Pen. : Westinghouse R & D Center.

Abstract Interpretation and Application to Interactive System Verification

Baudouin Le Charlier

Namur University (FUNDP)
Institut d'Informatique
Rue Grandgagnage, 21
5000 Namur (Belgique)
E-mail: ble@info.fundp.ac.be

Abstract. Abstract interpretation is widely known as a methodology for building static analyses of programs and is mainly used for program optimization; it can also be used for verification and is thus of interest for interactive system verification. The aim of this paper is twofold: Firstly, a survey of the methodology of abstract interpretation is presented; secondly, some favourite formalisms and notations from the interactive system verification area are reviewed at light of abstract interpretation. We argue that systematic use of abstract interpretation in interactive system verification could help improving and unifying current techniques.

Keywords

Abstract Interpretation, Verification of Interactive Systems

1 Introduction

Abstract interpretation [1, 11, 12] is mainly known as a methodology to build static analyses of programs. In the original framework of P. and R. Cousot, imperative (flow-chart) programs were first considered. Subsequent researches have tackled all other programming paradigms including functional programming [1, 38], logic programming [3, 28, 30, 34], concurrent programming [20, 35], and constraint logic programming [31, 32]. *Interactive systems* can be viewed as concurrent processes involving both humans and computers. As a consequence, most researches in interactive system verification (e.g., [2, 42, 45]) are based on formalisms, languages, and techniques originally introduced for concurrent system specification and verification. None of these researches seems to be based on abstract interpretation however, although such methodology is widely applicable and could be used both to unify and to improve on several techniques currently used in the interactive system verification area. The aim of this paper is twofold:

Firstly, we want to summarize the salient technical aspects of abstract interpretation in order to provide a basic introduction to the field; secondly, we want to show how the theoretical and practical achievements of abstract interpretation could be applied to interactive system verification.

Abstract interpretation was originally developed as a *semantics-based* technique for static analysis of programs. Such technique allows us to derive properties that are shared by possibly infinitely many program executions, in a fully automated way. This can be contrasted with other analysis techniques such as syntactic analysis (parsing), dynamic analysis, and mechanized program proving:[1] Syntactic analyses do not address the meaning of the program but only its form; dynamic analyses are performed at run time and are thus unable to derive general information about the program; finally, program proving techniques, because they are based on formalized mathematics, i.e., on complex and demanding formal systems, usually need to be human driven. Because analyses obtained by abstract interpretation are fully automated, they are simpler—at least more focused—than what can be obtained by means of general theorem provers. As a consequence, abstract interpretation has been mainly used in conjunction with program optimization and transformation (for partial evaluation, for instance). Nevertheless, as the theory of abstract interpretation evolves and as, at the same time, computers become more and more powerful, more complex analyses become attainable by the means of abstract interpretation (see, e.g., [9, 10, 29]). Hence, abstract interpretation appears to be (or to be becoming) a serious alternative to general theorem proving for computer system verification.

In abstract interpretation, full automation is possible because *approximate computation* is systematically used. As a matter of fact, many problems that are computationaly intractable or even undecidable become efficiently solvable when approximated: By accepting that the analysis returns *"I don't know"* in some cases, we gain both acceptable response time and precise answers in many practical situations.

Although abstract interpretation was originally designed for and applied to programming languages, its applicability is by no way restricted to such (i.e., executable) languages. The only requirement that a formal language has to fulfil in order to be analyzable is to be equipped with a well-defined *semantics*. Current works on interactive system verification involve a variety of languages and formalisms based, to name a few, on first-order logic, typed higher-order logics (e.g., HOL), set theory (e.g., Z, VDM), modal and temporal logics (e.g., CTL), Petri nets, concurrency theory (e.g., LOTOS, UAN), and synchronous languages (e.g., LUSTRE). Abstract interpretation can be applied to most such formalisms since most have clear semantic foundations. Using abstract interpretation can

[1] In practice, the distinctions are not so clear cut but, since we want to emphasize the salient ideas, we somewhat simplify the presentation.

bring the following benefits.

1. Some existing verification techniques—for instance, those based on graph theory, Petri nets, and model-checking—can be first recasted and then generalized in the framework of abstract interpretation, leading to more powerful tools.
2. General theorem provers—based on resolution, type theory, or special computational logics such as Boyer and Moore's—can be replaced in several interesting cases by abstract interpretation based static analyses.
3. As argued in [16], there is a need in interactive system verification for integrating currently used formal notations and verification techniques. Abstract interpretation provides a way to achieve such integration on a semantic basis: It should be possible to view the various semantics of different specification languages (e.g., temporal logic, user action notation, and petri-nets) as different abstractions of a single "concrete" semantic domain for interactive systems; based on this, a common system integrating several formalisms could be build.

The rest of this paper is organized as follows: Section 2 is an introduction to abstract interpretation; Section 3 reviews some formalisms, languages, and techniques proposed in the proceedings of DSV-IS'94, DSV-IS'95, and DSV-IS'96: We explain how abstract interpretation can be used in such contexts and how it could improve on some previous verification methods.

2 Abstract Interpretation

2.1 Outline of the Main Intuitions

In order to understand the intuition underlying abstract interpretation, let us imagine a program P written in some imperative programming language (e.g., Pascal). An execution of P consists of a possibly infinite sequence of basic execution steps, which all compute some value by applying some operation to some previously computed values. We call such execution a *concrete execution*; similarly, we call *concrete values* and *concrete operations* the values and operations involved in a concrete execution. Let us now suppose that we are interested in analyzing the program P so as to understand whether or not the program behaves as intended by the programmer or by the user. Obviously, this cannot be done by observing a single execution of the program since in general (possibly infinitely) many inputs are acceptable and lead to different concrete executions. Moreover, even in a single concrete execution, the same basic operation can be executed a huge number of times, preventing an exhaustive analysis. The key intuition of abstract interpretation is that the problem can be overcome by replacing all relevant concrete executions by a single (so-called) *abstract execution* of the program P. Abstract execution is similar to ("mimics") concrete executions but works on *properties (descriptions)* of concrete values instead of working on the actual values. Thus the slogan of abstract interpretation is that we can automatically derive global properties of programs by *calculating with properties instead of calculating with values.* The idea is the same as in the program correctness literature (see, for instance, the works of Floyd, Dijkstra, and Gries) where predicate transformers—the so-called weakest precondition *wp* and strongest postcondition *sp*—allow us, in theory, to automate program verification. The main difference between abstract interpretation and predicate transformers lies in the way properties are represented: In abstract interpretation, properties are represented by means of (so-called) *abstract values*, which can be arbitrary objects—in general, not predicates—equipped with an ad hoc semantic interpretation: While predicates are logical and mathematical formulae, which naturally exhibit their meaning, abstract values are chosen and designed so as to be efficiently manipulatable; their meaning is thus largely arbitrary—albeit precisely defined. As a consequence of the efficiency requirement, a domain of abstract values (a so-called *abstract domain*) is in general not able to express all properties of interest: Such domain is (often much) less expressive than an interpreted predicate calculus. However, we require that every property of interest can be *approximated* in some way; in the standard version of the theory, we even assume that there is a *best* approximation to every property. The last mentioned assumption allows us to perform abstract computations systematically because it is then (at least theoretically) possible, for every concrete operation, to design a corresponding abstract operation, which, given abstract descriptions of the inputs, computes the best possible abstract description of the output of the operation. Finally, since we have complete freedom to choose the objects of the abstract domain, we can define a vast range of analyses corresponding to different compromises between efficiency (i.e., time and space needed to perform the analysis) and accuracy (i.e., precision of the information resulting from the analysis).

The paradigm of abstract interpretation does not apply to programming languages only. What is basically required is that the meaning of every possible construct of the language is defined by a precise and finite set of rules. Then we can always choose a notion of "approximate meaning" for the constructs of the language and we can subsequently adapt, modify, and rephrase the "concrete" rules so that the approximate meaning of any construct can be computed automatically. The technical details can be quite involved however; thus we do not further elaborate on the basic intuitions; several examples are provided later on.

2.2 Mathematical Framework and Simple Abstract Domains

We first consider some simple abstract domains and we use them to illustrate the favourite mathematical structures in abstract interpretation: complete lattices and Galois insertions.

Abstraction and Concretization Let D be some basic data domain. The elements of D (concrete values) can be for instance Booleans, numbers, or strings. (In the examples below, we assume that $D = \mathbf{Z}$, where $\mathbf{Z}$ denotes the set of all integers.) In order to abstract a computation over D, we need to choose an abstract domain A, whose elements represent some "interesting" properties of the concrete values. Formally, the meaning of every abstract value is given by a *concretization function* $Cc : A \to \wp(D)$ that maps each abstract value to the set of all concrete values verifying the property represented by the abstract value. It is also convenient to require that each concrete value has a best approximation, which is given by means of the *abstraction function* $Abs : D \to A$. Finally, in order to be able to "mimic" a concrete computation over the abstract domain, we need replacing every concrete operation by an abstract counterpart. Let us consider, for instance, a binary operation $o_D : D \times D \to D$. Its abstract counterpart $o_A : A \times A \to A$ should be such that, for all $d_1, d_2 \in D$,

$$o_A\langle Abs(d_1), Abs(d_2)\rangle \;=\; Abs(o_D\langle d_1, d_2\rangle). \tag{1}$$

Example 1 Sign Analysis. Consider the following domains and functions:

$$\begin{aligned}
D &= \mathbf{Z};\\
A &= \{\ominus, \odot, \oplus\};\\
Cc(\ominus) &= \{d \mid d \in \mathbf{Z} \wedge d < 0\},\\
Cc(\odot) &= \{0\},\\
Cc(\oplus) &= \{d \mid d \in \mathbf{Z} \wedge d > 0\};\\
Abs(d) &= \ominus \text{ if } d < 0,\\
&= \odot \text{ if } d = 0,\\
&= \oplus \text{ if } d > 0.
\end{aligned}$$

Let us assume that the operation o_D is just the operation $\times$, i.e., integer multiplication. Let us denote its abstract counterpart by $\times_A$. We have

$$\begin{aligned}
&\ominus \times_A \ominus = \oplus \times_A \oplus = \oplus,\\
&\ominus \times_A \oplus = \oplus \times_A \ominus = \ominus,\\
&a \times_A \odot \;= \odot \times_A a \;= \odot \quad (\forall a \in A).
\end{aligned}$$

The definition of the operation $\times_A$ is nothing but the *"rule of signs"* teached in elementary school. Thus the calculation

$$(-237 \times 732{,}547) \times (-35 \times (-891))$$

can be abstracted to

$$(\ominus \times_A \oplus) \times_A (\ominus \times_A \ominus) \;=\; \ominus \times_A \oplus \;=\; \ominus.$$

The abstract result allows us to conclude that the concrete result is negative without necessitating to perform the concrete multiplications. Moreover, the conclusion is valid not only for the particular concrete calculation but also for infinitely many other computations, i.e., for all computations of the form

$$(-v \times w) \times (-x \times (-y)),$$

where v, w, x, y denotes strictly positive integers.

Approximation Ordering Not all abstract operations can be defined as elegantly as the operation $\times_A$ above. Assume, for instance, that we want to define the abstract counterpart $+_A$ of the concrete operation $+$ over integers. We should have, at the same time,

$$\oplus +_A \ominus = Abs(237 + (-25)) = Abs(212) = \oplus,$$

and also

$$\oplus +_A \ominus = Abs(25 + (-237)) = Abs(-212) = \ominus,$$

which is contradictory. Thus we need to augment the abstract domain with additional abstract values and to replace Relation (1) by a more applicable relation. This can be achieved if we require not only that every concrete value has a best approximation but also that every *set* of concrete values has a best approximation. Thus we modify the signature of the abstraction function, which becomes $Abs : \wp(D) \to A$, and we require the two following properties.

1. Every set c of concrete values is approximated by $Abs(c)$, i.e.,
   $c \subseteq Cc(Abs(c))$.
2. Moreover, $Abs(c)$ is the best approximation of c, i.e.,
   $\forall a \in A :\; c \subseteq Cc(a) \;\Rightarrow\; Cc(Abs(c)) \subseteq Cc(a)$.

It is also desirable to avoid redundancy in A (i.e., to eliminate multiple elements denoting the same property); to this aim, we can require that the function Abs is onto or, equivalently, that the function Cc is one-to-one. Then, we can define an ordering on A by the following identity:

$$\forall a_1, a_2 \in A :\; a_1 \leq a_2 \;\Leftrightarrow\; Cc(a_1) \subseteq Cc(a_2).$$

Galois Insertions The structure $\langle \wp(D), \subseteq, A, \leq, Abs, Cc \rangle$ defined in the previous section enjoys the property of being a *Galois insertion.* In general, a Galois insertion is a structure $\langle C, \sqsubseteq_C, A, \sqsubseteq_A, Abs, Cc \rangle$ such that the following conditions hold.

1. The pairs $\langle C, \sqsubseteq_C \rangle$ and $\langle A, \sqsubseteq_A \rangle$ are partially ordered sets.
2. The functions $Abs : C \to A$ and $Cc : A \to C$ are monotone; moreover, Abs is onto and Cc is one to one.
3. For every $c \in C$, $c \sqsubseteq_C Cc(Abs(c))$.
4. For every $a \in A$, $a = Abs(Cc(a))$.

In a Galois insertion, C is called the *concrete* domain and A is called the *abstract* domain. Remember that, in the case of our simple domain, $C = \wp(D)$. Thus, C is not the actual domain of computation but rather the domain of all possible *properties* of the values used in the concrete computation (a property can be identified to the set of all values that verify the property). For more complicated semantic domains, the concrete domain may not be a power set but its elements still conceptually denote properties of the concrete objects of computation. More complex domains are considered later on. It must also be noticed that the four above conditions are in fact redundant. For minimal characterizations and equivalence results, see [12].

Example 2 Sign Analysis (Continued). The smallest abstract domain for sign analysis that contains $\{\ominus, \odot, \oplus\}$ and is a Galois insertion is defined by

$$
\begin{aligned}
C &= \wp(\mathbf{Z}); \\
A &= \{\bot, \ominus, \odot, \oplus, \top\}; \\
Cc(\bot) &= \emptyset; \\
Cc(\ominus) &= \{d \mid d \in \mathbf{Z} \wedge d < 0\}, \\
Cc(\odot) &= \{0\}, \\
Cc(\oplus) &= \{d \mid d \in \mathbf{Z} \wedge d > 0\}; \\
Cc(\top) &= \mathbf{Z}; \\
Abs(c) &= \bot \text{ if } c = \emptyset, \\
&= \ominus \text{ if } c \neq \emptyset \ \wedge \ (\forall d \in c : d < 0), \\
&= \odot \text{ if } c = \{0\}, \\
&= \oplus \text{ if } c \neq \emptyset \ \wedge \ (\forall d \in c : d > 0), \\
&= \top \text{ otherwise.}
\end{aligned}
$$

Complete Lattices An interesting property of Galois insertions is that the abstract domain $\langle A, \sqsubseteq_A \rangle$ is a complete lattice whenever the concrete domain $\langle C, \sqsubseteq_C \rangle$ is. The interest of complete lattices stems from the fact that any set of *abstract* values always has a best approximation, called the least upper bound of the set. This property is convenient to define abstract operations in a natural—unique—way. Since concrete domains are often naturally defined as complete lattices—for instance, power sets ordered by inclusion are complete lattices—, it is natural to require that abstract domains are also complete lattices, explaining why this mathematical structure is a favourite one in abstract interpretation.

Designing Abstract Operations In the context of Galois insertions, abstract operations can be designed systematically in two steps, as follows. Let $o_D : D \times D \to D$ be an operation defined on the actual computational domain (we assume that $C = \wp(D)$). Firstly, we define the so-called *collecting* version $o_C : C \times C \to C$ of the operation by the equality

$$o_C\langle c_1, c_2\rangle = \{o_D\langle d_1, d_2\rangle \mid d_1 \in c_1 \wedge d_2 \in c_2\} \ (\forall c_1, c_2 \in C).$$

Secondly, we "abstract" the collecting operation o_C into a function $o_A : A \times A \to A$ by the following other equality

$$o_A\langle a_1, a_2\rangle = Abs(o_C\langle Cc(a_1), Cc(a_2)\rangle) \ (\forall a_1, a_2 \in A).$$

Notice that the last definition uses both the abstraction function and the concretization function, explaining their fundamental role: Thanks to the concretization function, all possible concrete values corresponding to a_1 and a_2 are first determined, giving the sets $Cc(a_1)$ and $Cc(a_2)$; then the set $o_C\langle Cc(a_1), Cc(a_2)\rangle$ of all possible results is obtained thanks to the collecting operation o_C; finally, the set of all possible results is abstracted in the best possible way through the abstraction function. There is nevertheless a subtle point here: The previous definition of the operation o_A does not provide a practical way to *compute* the value $o_A\langle a_1, a_2\rangle$ since in general none of the functions Abs, Cc, o_C are computable (because they manipulate infinite sets). Thus, the "definition" of the abstract operation o_A is only a starting point from which a more practical definition of the operation remains to be derived. The framework of abstract interpretation at least ensures that such operation exists but exhibiting a practically computable definition still requires some creativity. Not surprisingly, many researches in the field of abstract interpretation are devoted to systematizing the derivation of practical definitions from the "theoretical" one (see, e.g., [9, 12]).

Example 3 Sign Analysis (Continued). Let us partially show how the practical definition of the operation $+_A$ can be derived for the abstract domain of Example 2. The collecting operation $+_C$ is given by the equality

$$c_1 +_C c_2 = \{d_1 + d_2 \mid d_1 \in c_1 \wedge d_2 \in c_2\} \ (\forall c_1, c_2 \subseteq \mathbf{Z}).$$

Then we can derive, for instance,

$$
\begin{aligned}
\bot +_A \ominus &= Abs(Cc(\bot) +_C Cc(\ominus)),\\
&= Abs(\emptyset +_C \{d \mid d \in \mathbf{Z} \wedge d < 0\}),\\
&= Abs(\emptyset),\\
&= \bot;\\
\ominus +_A \odot &= Abs(Cc(\ominus) +_C Cc(\odot)),\\
&= Abs(\{d \mid d \in \mathbf{Z} \wedge d < 0\} +_C \{0\}),\\
&= Abs(\{d_1 + d_2 \mid d_1 \in \mathbf{Z} \wedge d_1 < 0 \wedge d_2 = 0\}),\\
&= Abs(\{d \mid d \in \mathbf{Z} \wedge d < 0\}),\\
&= \ominus; \text{ etc.}
\end{aligned}
$$

Safe Abstract Operations The mathematical framework of Galois insertions ensures that optimal abstract operations exist. For complex abstract domains, it can nevertheless be difficult to derive a computable definition of the optimal operation. Moreover, such definition can be fairly involved and can contain complex case analyses, which are needed for optimality but actually provide low additional accuracy. In practice, optimal abstract operations are often better replaced by less accurate versions, which satisfy the following weaker requirement:

$$Abs(o_C\langle Cc(a_1), Cc(a_2)\rangle) \leq o_A\langle a_1, a_2\rangle \quad (\forall a_1, a_2 \in A).$$

This requirement can be equivalently expressed, using the concretization function only, as

$$o_C\langle Cc(a_1), Cc(a_2)\rangle \subseteq Cc(o_A\langle a_1, a_2\rangle) \quad (\forall a_1, a_2 \in A),$$

or, using the abstraction function only, as

$$Abs(o_C\langle c_1, c_2\rangle) \leq o_A\langle Abs(c_1), Abs(c_2)\rangle \quad (\forall c_1, c_2 \in C).$$

Such operations o_A are said to be *safe* abstractions of the corresponding operations o_C and o_D. The two last mentioned characterizations of safe abstract operations suggest that alternative (weaker) frameworks—based on a concretization function only or on an abstraction function only—are possible (see Reference [13]).

2.3 Abstract Domains

Other Sample Abstract Domains The variety of possible abstract domains is potentially unlimited. In this section, we briefly outline some abstract domains previously proposed in the literature (see [12] for a list of references) as well as some well-known verification techniques that can be viewed as particular cases of abstract interpretation.

Interval Analysis Sign analysis can be generalized to *interval analysis*, which amounts to abstracting a set c of integers by the least interval containing c, i.e., by $[\min(c)..\max(c)]$. This abstract domain is infinite, which makes difficult to ensure the convergence of the analysis (see Section 2.6)

Dimension Analysis Dimension analysis is used by physicists to quickly check that a physics formula is plausible: every physical quantity (e.g., *force, resistance,* ...) is measured according to a specific unit (e.g., *Newton, Ohm,* ...). Every specific unit is reducible to a canonical expression involving the four fundamental units: *meter, kilogram, second, amp.* For instance, we have

$$Newton = \frac{kilogram \times meter}{second^2}.$$

Thus, given a formula such as $f = m \times a$ (force equals mass times acceleration), we can first replace physical quantities by their corresponding units and then we can compute the values of both sides in terms of fundamental units. If the values are different, the formula is certainly erroneous; otherwise, it is plausible.

Type Analysis Type analysis in typed programming languages, is similar to dimension analysis; it basically consists of computing the values of the program's expressions assuming that the values of simple operands are their declared type and that every operator is a function from types to types.

Abstracting Cartesian Products and Finite Domain Functions: Abstract Stores and Abstract Environments Designing abstract domains for basic data domains is not enough to achieve full analysis of a formal language. The constructs of such languages may contain variables (more generally, identifiers), which are bound to some value at run time (for a programming language) or represent some arbitrary value ranging in a given domain (for a specification language). The concrete evaluation of such constructs requires to consider a *store*,[2] which is a function from a finite set $\mathbf{I}$ (of variables, identifiers, memory locations, ...) to the set D of semantic values. There are several ways to abstract stores: Some abstractions simply abstract the values of individual variables separately; other abstractions keep tracks of the dependencies between the values of the variables. Let A be an abstract domain corresponding to the standard domain D. Then, an abstract store can be a function $\sigma_A : \mathbf{I} \to A$. Assume for instance that $\mathbf{I} = \{x, y\}$, an abstract store is then a finite function such as $\{x \leadsto \ominus, y \leadsto \odot\}$, which specifies that x is negative and that y is equal to 0. Such abstract domain is unable to keep track of dependencies between x and y. Dependencies are expressed by means of constraint domains. For instance, we can define an abstract domain SV whose elements are sets of unordered pairs

[2] or an *environment*, or both; we only consider the case of a flat store here.

$\{x, y\}$ expressing that the values of x and y are always equal. The domain SV can be used in conjunction with almost any standard domain D: We only need an equality operation defined over D. For specific domains D, more powerful dependencies can be expressed. Let us for example assume that D is a set of numbers (i.e., we have $D \subseteq \mathbf{R}$). Then, we can express dependencies between the variables of $\mathbf{I}$ by means of subspaces of $\mathbf{R}$. Sample abstract domains found in the literature consider affines subspaces—where dependencies are expressed by (canonical) sets of linear equations [46]—and convex hulls [18].

Combining Abstract Domains Combination of abstract domains is an important research area in abstract interpretation because

1. it may be necessary to perform several simultaneous analysis to ensure the correctness of a specific one; for example, the aliasing problem in programming languages requires to enhance individual value analysis with equality constraints;
2. besides correctness issues, interacting analyses may improve the accuracy of each other;
3. combining abstract domains may increase the complexity of abstract operations up to unmanageableness and may also induce efficiency problems.

Some techniques such as the *reduced product* [12] and the *open product* [9] have been proposed to systematize abstract domain combination; due to lack of space it is not possible to explain those techniques here.

Abstract Domains for Structures and Sequences Compound data structures (e.g., lists, terms, records,...) and sequences can be abstracted by means of formal grammars (e.g., regular expressions, context-free grammars, syntactic diagrams). The resulting abstract domains can be fairly complex (see, e.g., [10, 22]). Such domains are however worth considering in the context of interactive system verification, where temporal properties are especially important to verify. Such properties express requirements about the sequence of states reached by the system along time. In order to verify temporal properties, we can use abstract interpretation to abstract the set of all sequences corresponding to all possible behaviours of the system. In a further step, the abstraction can be used to prove (or disprove) that the system enjoy the temporal properties (see Section 3.2).

2.4 Abstract Semantics

A key idea of abstract interpretation is that any analysis of a formal text (e.g., program, formula, specification, graphical object, ...) can be viewed as the *calculation* of its *abstract semantics*. The abstract semantics is a mathematically well-defined object that "reifies" the (approximate) meaning of the formal text under consideration. It is convenient to separate the definition of the abstract semantics from its actual calculation. We first consider the former issue.

Although other approaches are possible (and can be more practical in some situations), the classical method to the design of an abstract semantics proceeds in three steps that we describe in turn.

Identifying the Standard Semantics The method of abstract interpretation only applies to languages that are equipped with a precisely defined semantics. Most of widely used formal languages enjoy a well-accepted semantics, which can be operational, denotational, model theoretic, axiomatic, ... In many cases, several related semantics are available. Our first task is thus to identify the semantics of the language that is best suited to our needs, i.e., to the kind of properties we want to verify. This semantics is called the *standard semantics* of the language. Notice that in some cases, our search for a standard semantics may fail. Then, we can start our work by defining a proposal standard semantics. Of course, the validity of our subsequent work will depend on the adequacy of our proposal to the understanding of other users.

Building the Collecting Semantics The second step in the methodology is similar to and generalizes the definition of collecting operations (see Section 2.2). The standard semantics of a formal language is in general expressed in terms of individual values, not in terms of properties. However, given a class of properties we are interested to derive, it is always possible to reexpress ("to lift") the semantics of the formal constructs in terms of the properties. Such semantics is called a *collecting semantics.*

Let us illustrate this on an example borrowed from [11]. Consider a simple imperative programming language. Its standard (operational) semantics defines a transition relation between states, where a state s is of the form $\langle p, \sigma\rangle$, the letter p denotes a program point,[3] and σ denotes a program store. It is natural to "lift" this operational semantics to a collecting semantics that maps each program point p to the set S_p of all program stores possibly computed at this point. The collecting semantics is no longer an operational one: Since the "collecting stores" S_p are sets and since they can possibly be infinite, a fixpoint semantics needs be used: The collecting semantics is defined as the least solution—with respect to inclusion—of a finite set of equations:

[3] We assume for convenience that program points are denoted by the numbers 1, ..., n, where n is the number of program points in the program.

$$S_1 = f_1\langle S_{q_1^1}, \ldots, S_{q_{m_1}^1}\rangle,$$
$$S_2 = f_2\langle S_{q_1^2}, \ldots, S_{q_{m_2}^2}\rangle,$$
$$\ldots \qquad \ldots$$
$$S_n = f_n\langle S_{q_1^n}, \ldots, S_{q_{m_n}^n}\rangle,$$

where every function f_p is a monotone and continuous function that defines the set of program stores at the program point p in terms of the sets of program stores at the program points $q_1^p, \ldots, q_{m_p}^p$, leading to p. Importantly, each function f_p can—and should—be expressed as a combination of a few basic collecting operations.

Deriving the Abstract Semantics The abstract semantics is derived straightforwardly from the collecting semantics: It is sufficient to replace the collecting domain (e.g., in the previous section, sets) by some abstract domains and to replace the collecting operations by the corresponding abstract operations—which should be monotonic. The abstract semantics is just the least solution of the set of abstract equations derived from the equations of the collecting semantics.

Notice that an abstract semantics is normally *generic* since it is parameterized on the abstract domain. Many different analyses of the same formal language can be based on the same collecting semantics: We only have to change the abstract domain and to provide the corresponding abstract operations; no additional reasonings about the standard and concrete semantics are needed.

Genericity of abstract semantics nevertheless has limitations: The variety of possible analyses is restricted to the properties entailed by the collecting semantics. It is therefore sometimes necessary to define several collecting semantics, each of them corresponding to a whole class of possible analyses, for the same language. Let us consider the case of a simple imperative language, once again. The collecting semantics previously defined forgets the dependencies between variable values at *different* program points. Expressing and deriving such dependencies requires a different collecting semantics based on *sets of sequences of states.*

2.5 Computing the Abstract Semantics

Given a particular—implemented—abstract domain, computing the abstract semantics of a given formal text amounts to solving a *fixpoint problem.* Efficient algorithms devoted to this class of problems can be found in the literature (e.g., [25, 27, 29, 40]). We only summarize the main issues of this research area here.

Two families of fixpoint algorithms are worth distinguishing. *Bottom-up algorithms* [8, 40] compute the abstract semantics entirely. Specific properties of the formal text can then be verified by querying the abstract semantics. *Top-down algorithms* [25, 26, 27, 28] only compute (roughly speaking) the part of the abstract semantics that is relevant to derive a specific property: The abstract semantics is computed on demand. Both approaches have their own pros and cons; they can be prefered depending on the particular generic abstract semantics, the chosen abstract domain, and the context of use. Bottom-up and top-down fixpoint algorithms can be respectively related to bottom-up computation of logic programs in deductive data bases and to top-down execution of logic programs as implemented by Prolog (see, e.g., [44]). However fixpoint algorithms for abstract interpretation involve additional optimization techniques to avoid redundant computations.

2.6 Finite Versus Infinite Abstract Domains: Widening Techniques

Termination of abstract interpretation algorithms is considered highly desirable since we wish to achieve a completely automatic analysis. Generic fixpoint algorithms are guaranteed to terminate when the abstract domain (i.e., all potential abstract semantics of a formal text) is finite or has the "ascending chain property." But, for arbitrary abstract domains, the algorithms may loop. In order to clarify the different possibilities, let us consider the analysis of a simple imperative program by means of three different abstract domains A_1, A_2, A_3, defined as follows.

- An element of A_1 maps every program variable to an element of the sign domain $\{\bot, \ominus, \odot, \oplus, \top\}$ (see Example 2). This domain is finite since the number of variables and the number of program points are finite.
- An element of A_2 is a (canonical) system of linear equations on the variables of the program; such system defines an affine subspace containing all possible program stores corresponding the analyzed executions. This domain is infinite but it enjoys the ascending chain property, i.e., due to the monotony of abstract operations, the dimension of the affine subspace increases at each fixpoint iteration until the fixpoint is reached; hovever the whole underlying space has finite dimension; thus the number of iterations is bounded.
- An element of A_3 is (some representation of) a closed convex polyhedron (see [18]) containing all possible stores. This domain does not enjoy the ascending chain property since there exist infinite sequences of (bigger and bigger) embedded closed convex polyhedra.

Due to the above mentioned convergence problems, arbitrary infinite abstract domains are considered useless by some authors (see, e.g., [24]). But since general domains are more expressive than restricted ones, other researchers prefer to work with the former ones at the price of introducing additional approximation techniques—called *widening operators* [11, 14]—to ensure convergence. The design of good widening operators is a fairly creative activity, which amounts to the discovery of clever heuristics; examples from the literature (e.g., [10, 18, 33]) show that this approach can be rewarding.

2.7 Concluding Remarks About Abstract Interpretation

Abstract interpretation is usually advocated as a general methodology for designing and implementing static analyses. But, due to the numerous variants[4] to and the wide generality of the approach, we should better characterize abstract interpretation as a "paradigm", i.e., as a number of powerful albeit not entirely formalizable ideas, which give rise to complete, useful formalizations in many application areas.

Alternatively, we can also think of abstract interpretation as a large bunch of techniques that we can reuse to solve new problems. (Section 3 explores its use for interactive system verification.) From this viewpoint, the mathematical framework of abstract interpretation as well as the technicalities of some previously developed analyses can sometimes be found difficult by newcomers in the field. This drawback should not be underestimated: Although no serious research could be undertaken by people lacking a good scientific education, we believe that there is a need for professional level generic tools based on abstract interpretation. The theoretical material to build these tools exist and some experimental systems exist as well (e.g., the generic system *GAIA*, see [28]), but real professional systems are still to come. Such systems would play—with respect to abstract interpretation—a role similar to LEX and YACC—with respect to parsing theory.

Finally, let us briefly compare abstract interpretation with two other verification techniques: model checking and general theorem proving. *Model checking* deals with *decidable* theories. Many techniques used in abstract interpretation—for instance, fixpoint computation—are also used in model-checking, but in a more restricted way since no approximation technique is used: A decision problem is answered by "yes" or "no"; the answer "I don't know" is not considered acceptable. However, the previous distinction is misleading since, in model checking, decidable theories are often used to solve problems that approximate more difficult—often undecidable—problems. Thus, the main advantage of abstract interpretation is that we are allowed to tackle the original problem more directly and to introduce approximations that are better tailored to this specific problem. At the opposite of model checking, *general theorem proving* is in theory more powerful than abstract interpretation since any mathematical reasoning can be automated "in principle." From a theorem proving standpoint, abstract interpretation can be seen as a uniform proof technique based on reduction to

[4] See Reference [13]

normal form and on a fixpoint induction principle. (Abstract objects can be seen as canonical formulae since they express properties of concrete objects.) Such uniform proof technique is less powerful than techniques based on formal logic but can be fully automated. On the contrary, general theorem provers are less efficient and require user assistance, which can be a drawback in some real life (e.g., industrial) contexts.

3 Applying Abstract Interpretation to Interactive System Verification

In the next part of this paper, we suggest possible uses and benefits of abstract interpretation for interactive system verification. Various formal languages and notations are currently used in this research area (see, for instance, [2, 42, 45]). Our primary goal here is thus to show how abstract interpretation can support analyses of interactive systems based on these notations. We also suggest that abstract interpretation could be used as a unifying paradigm for verification techniques based on multiple formalisms. This part of the paper is organized as follows: Section 3.1 summarizes our view of the interactive system verification problem while Section 3.2 reviews some specification languages and some verification techniques. In the latter section, we aim at showing how the languages can be analyzed by means of abstract interpretation and how the verification techniques compare with abstract interpretation; the choice of the reviewed topics is mostly a matter of taste, with no attempt to exhaustivity.

3.1 Verification of Interactive Systems

Basic Concepts An *interactive system* consists of concurrent processes of which some are human behaviours while the others are executed by computers. Moreover, in an interactive system, humans and computers typically interact through a graphical interface. *Automated verification* of an interactive system requires and is applied to a formal description of the system or of some part of it. A complete description should represent both the human and the computer part; in many cases however, only the computer part of the system is formally described. When only the computer part is formalized, general properties of the system, such as reachability of some key states, can be verified without paying attention to how the user must collaborate to reach the states. When the human behaviour is also described, we are basically faced with a concurrent system verification problem.

Formal Verification and Semi Formal Descriptions Various formalisms and languages are currently used to model interactive systems. In a recent paper

[16], P. Gray and C. Johnson compare three different notations: computational tree logic (CTL), (extended) user action notation (X)UAN, and Petri nets. Petri nets and CTL are completely formal notations since they consist of both a formal syntax and a precise mathematical semantics. On the contrary, UAN is a *semi-formal* notation: Its syntax is formal but its semantics, which aims at reflecting the *asynchronous interaction style* in user interface design, is only intuitively and partially defined. The obvious "problem" with semi-formal notations is that formal verification techniques can be applied only to the syntactical aspects of descriptions written in the notation; thus verifications are not very deep. Truly formal notations allows us in principle to verify important semantic properties of the system such as safety properties (e.g., absence of deadlocks) and reachability properties (e.g., key states can always be attained).

Importantly, it can be observed from the literature of interactive system verification [6, 16] that (truly) formal notations are in fact often used in the same way as semi-formal notations, i.e., as a tool to help the specifier structuring and disambiguating his/her informal descriptions. Such descriptions are *not* completely formal—albeit the notation is— because the formal text is understood at the intuitive level as a reformulation of a natural language sentence, regardless of the actual semantics of the notation. From this standpoint, formal and semi-formal notations used in interactive system verification have the following pros and cons. Semi-formal notations provide the specifier with more syntactical constructs and more domain specific concepts; thus specifications are more concise and are easier to understand—although at an intuitive level. It can also be argued that such specifications provide better support for discussion to people involved in an interactive system design. In contrast, the benefit of specifications written in a completely formal notation stems from the fact that they can be analyzed by means of verification tools. But, since the specifications are written at the intuitive understanding level, severe discrepancies between the specifier or user's understanding and the actual formal meaning of the formal text may exist. Hence formal analyses have to be interpreted with care: They are useful to detect problems and misunderstandings but they cannot be trusted when no formal error is found.

It should be clear that high level confidence in a computer system—interactive or not—can be obtained only by a combination of rigorous informal reasonings and automated formal analyses. Informal reasonings aim at (1) providing a high level, intuitively understandable, and natural model of the system and at (2) translating desirable properties of the natural model into formal descriptions suited for automated analysis. The above mentioned translation process must rely on precise representation conventions, which are also needed to interpret the results of formal analyses. The respective roles of formal and informal reasonings in computer system verification is nicely illustrated by the minimal colouring problem in graph theory: The fact that four colours are enough to colour adjacent countries with different colours on a map has been proven by checking a large number of actual configurations on a computer. This is exactly what automated verification is about. Importantly, a lot of preliminary reasonings have been done in order to reduce the original problem to this finite set of configurations. Similar preliminary reasonings are needed in *any* rigorous approach to formal verification.

Finally, abstract interpretation provide us, if not with a systematic method, at least with a powerful paradigm to the design of formal analyses: Whatever formalism is used to model the system to be analyzed, the semantics of the formal description can be viewed as an abstraction of a detailed natural model of the system. Moreover, abstraction and concretization functions are useful mathematical tools to relate the formal description to the natural model.

3.2 A Review of Some Notations Used for Interactive System Verification

Petri Nets Petri Nets provide a graphical notation to describe concurrent systems; they are widely used in interactive system verification (see, e.g., [16, 23, 41]). Roughly speaking, a Petri net is a graph with two kinds of nodes called *conditions* and *actions*, respectively. The concrete semantics of a Petri net can be defined as a set of possible executions where conditions are *enabled* by the presence of *tokens* and where actions are *fired* as soon as their preconditions, i.e., the conditions leading to the action, are enabled. Several variants to the basic notion have been proposed in order to handle more complex and/or specific situations. It is intuitively obvious that Petri nets are able to model concurrent systems thanks to the token paradigm. Moreover, since Petri nets are basically graphs, many algorithms to analyze the modelled systems are available.

Analyzing Petri Nets by Abstract Interpretation. Reachability analysis is a main application of Petri nets. The aim is to determine whether some distinguished state is always reachable from some configuration or to prove, on the contrary, that some "bad" state cannot occur. The corresponding algorithms can be viewed as particular fixpoint computations and hence as particular abstract interpretations. However, explicit use of abstract interpretation may improve the applicability of such algorithms when, for instance, tokens may range over infinite domains and when action firing may depend on elaborated conditions on the values [36]. Such generalized Petri nets can potentially reach infinitely many different states. Thus they cannot be analyzed by traditional algorithms. Such Petri nets can be analyzed by abstract interpretation, provided that concrete domains of tokens are replaced by adequate abstract domains. Existing domains based on intervals, linear equations, and convex hulls can be reused in such contexts (see [12, 18]). Another improvement made possible thanks to abstract interpretation is the possibility of performing (possibly infinitely) many analyses at the same time: Instead of analyzing the states reachable from (or leading to) a given unique configuration, we can analyze the states reachable from an abstractly defined set of initial configurations. For instance, such abstract description could be the number of initial tokens in every condition, an interval containing this number, a linear equation over the number of tokens, etc.

Beyond reachability analysis, other analyses based on different abstractions of Petri nets can be designed. For instance, we can design abstract domains that abstract either the action or the condition components of a Petri net, describing either the set of all possible action sequences or the set of all sets of simultaneously enabled conditions. Such abstraction could be used to compare different views of the same system (as modelled, e.g., in [36]) and to point out differences of understanding, in a precise and clear way, by delivering a description of the differences.

Petri Nets as Concrete and Abstract Domains Not only can Petri nets be used as a notational device, but they can also serve at the semantic level, i.e., as semantic values. For instance, a variant of Petri nets called *contextual nets* is used by H. Montanari and F. Rossi to describe the concrete semantics of the concurrent constraint language CC (see, e.g., [37]). The considered contextual nets are *acyclic* and they are intended faithfully to describe all the concurrency potentially available in a CC program. The same semantic domain—or some improvement of it—could be used to describe the semantics of other languages used for interactive system modelling (e.g., UAN). Such domain could be used as a common semantic basis for a range of description languages such as Petri nets, UAN, LOTOS, ... , allowing us to combine different specification styles. Notice that, as observed in [37], abstract interpretation can be used to derive other semantics that forget some details about the concurrent semantics but are easier to reason about (e.g., an interleaving semantics, a proof theoretic semantics, etc.). Thus, in such context, we can still use the specific tools available with the individual languages but we have additionally gained a common semantic basis to interpret the multiple and/or complementary descriptions as well as the results of the corresponding formal analyses.

Petri nets or variants of Petri nets are also worth considering at the abstract level. For instance, finite (non acyclic) Petri nets can be used to finitely abstract acyclic—possibly infinite—Petri nets. Such abstraction could once more be used to combine, normalize, and/or compare formal descriptions possibly expressed in different formalisms.

User Action Notation User action notation (UAN) [16, 21] is a domain specific concurrent language especially crafted to describe human-computer interaction through a graphical interface. This description language is based on the so-called asynchronous interaction style and is only semi-formal: Only the syntax of the language is precisely defined, yet it is open since the specifier may possibly introduce new basic operations.

The fact that no complete mathematical semantics has been provided yet for UAN is easily explained by the richness of the language containing many basic concepts as well as many structuring mechanisms. Thus it can be argued that nobody would actually use such semantics if it should eventually be defined. As a matter of fact, complex programming languages can hardly be given a complete formal semantics. Moreover, very few programmers refer to such a semantics to reason about their programs. Nevertheless, a precise semantic definition is required at least to derive a correct implementation of the language. Similarly, it could be worth providing UAN with a true concurrency semantics, based perhaps on contextual nets (see, e.g., [37]). The true concurrency semantics could then be abstracted through abstract interpretation into simpler semantics underlying available verification tools (e.g., interleaving semantics, computational tree semantics). Moreover, such true concurrency semantics of UAN could be abstracted to possibly several operational semantics allowing us to produce provably correct (maybe prototype version of) user interfaces.

Temporal Logic Temporal logic (e.g., CTL*) is relevant for interactive system verification, since it allows us to state desirable properties of concurrent systems such as fairness, absence of deadlocks, etc. Abstract interpretation can improve applicability and usefulness of temporal logic notations at two levels.

1. Temporal logic is adequate to state properties that are either axioms imposed to the system or are desirable consequences of the specifications, which we may wish to verify automatically. Nevertheless, temporal logic may not be convenient to describe an interactive system entirely [16]. The problem of combining temporal logic with other notations such as Petri nets or UAN lies in the lack of a common semantic domain. A solution to the problem is to consider the branching time semantic model of temporal logic as an abstraction of a common concrete semantics. Thus, temporal formulae can be—indirectly—interpreted over the common semantics while being verified by means of available verification tools based for instance on model checking.
2. Model checking techniques (see, e.g., [47]) have proven useful to verify validity and satisfiability of some classes of temporal formulae. Alternative techniques based on abstract interpretation can complement, replace and/or generalize model checking. Consider, for example, the propositional temporal formula

   $$p \wedge \bigcirc(\Box(q \Rightarrow p) \wedge \Diamond q).$$

A model of such formula is an infinite sequence of truth value assignments to p and q. For example, the following sequence is a model of the previous formula:

$$< \{p \leadsto true, q \leadsto false\}, \{p \leadsto true, q \leadsto false\}, \{p \leadsto true, q \leadsto true\}, \ldots >$$

(The "..." stands for infinitely many occurrences of $\{p \leadsto true, q \leadsto true\}, \ldots$.) We can define the collecting semantics of such temporal formula as the set of all its models. The collecting semantics can be abstracted in various ways: For example, we can use regular expressions over the Boolean algebra of propositional formulae to approximate sets of sequences of truth assignments (see [7]). Hence, we can derive, for the above formula, the following regular expression:[5]

$$\{p\} . \{q \Rightarrow p\}^* . \{q \wedge p\} . \{q \Rightarrow p\}^\omega .$$

Such domains could be used to approximate the meaning of complex temporal specifications and to check the validity/satisfiability of other temporal formulae with respect to the approximate meaning.

Theorem Proving General theorem proving is another natural approach to the verification of interactive systems. For instance, in [4, 5], P. Bumbulis et al. advocate the use of the HOL logic to validate graphical interfaces.

General theorem proving and abstract interpretation can be seen as two alternative approaches to verification: They achieve a different compromise between efficiency and expressivity. In a theorem proving approach, much more properties—expected from the analyzed system—are expressible since properties are described by means of arbitrary formulae typically embodying full arithmetic. In contrast, abstract domains express a restricted set of—relevant—properties, which are represented in an ad hoc way, suited to efficient computation. As a consequence, property approximation is systematically applied in abstract interpretation. No such approximations are needed in theorem proving because most properties are formalizable. Another consequence of this different balance between expressivity and efficiency is that analyses are straightforwardly performed in abstract interpretation—through abstract operation executions and fixpoint computations—while nondeterministic search and lemma guessing are required in a theorem proving approach. The two last mentioned problems may cause theorem proving to be too inefficient when used in a fully automatic way. Thus, theorem provers are human driven in general.

As an example, let us compare the work of Bumbulis et al. [4, 5] for interactive system verification with an abstract interpretation approach. These authors propose to use a variant of Dijkstra's guarded commands to model graphical

[5] The subexpression $\{q \Rightarrow p\}^\omega$ denotes an infinite sequence of formulae $\{q \Rightarrow p\}$.

interface behaviours. Since they are interested in proving temporal properties of the interface, they model the semantics of a command as a set of possibly infinite sequences of alternate states and actions. Notice that such modelling amounts to defining a collecting semantics of the command language. Technically, sets of sequences are represented by predicates over sequences; within the abstract interpretation approach, such predicates would be replaced by elements of an abstract domain. The semantics of basic commands is given by primitive predicates in [4, 5], while the semantics of compound commands is defined by predicate transformers. In an abstract interpretation context, basic constructs would be specific abstract values, while the abstract semantics of compound commands would be defined by means of a few abstract operations (fixpoint definitions are needed for repetitive commands).

In [4, 5], an invariant property of a simple graphical interface is formally proven. Such properties could be automatically computed with an abstract domain expressing invariant properties of variables by means of linear equations and inequations. More powerful domains combining numerical constraints and restricted forms of temporal logic are also possible.

We conclude this section with two remarks.

1. Theorem proving and abstract interpretation techniques can be combined into a single system: A specific class of formulae can be identified and manipulated in a specific way. Using abstract interpretation in such context generalizes the use of decision procedures—for subtheories—as additional inference rules, in theorem provers. This approach is also related to constraint logic programming (CLP) where (possibly approximated) constraint solving is combined with Prolog inference engine (see, e.g., [43]).
2. In a verification context such as [4, 5], there is an alternative to the usual fixpoint approach of abstract interpretation: It is possible merely to verify some key properties—for instance, invariants—asserted by the specifier (see [19]). Since this approach is computationally less expensive, it makes possible to use more expressive abstract domains. More work is required from the specifier however and, moreover, fixpoint computation can be eliminated only if a complete set of assertions is provided. Nevertheless, a combination of verification and fixpoint computation is possible.

Synchronous Programming *Synchronous programming*, as examplified by the language LUSTRE, constitutes another approach to *reactive system* design and verification (see, e.g., [17]). We focus on the language LUSTRE here since it is used by B. d'Ausbourg et al. for interactive system verification, in a recent paper [15].

In the language LUSTRE, variables denote infinite sequences of values, not single values. As a consequence, the language can be used both to describe algorithms and to specify assertions about algorithms. This is not possible in a usual imperative language because the sequence of values assigned to a variable, by a program along time, cannot be explicitely manipulated by another program. The concepts of LUSTRE can be summarized in a few sentences. "Programs" are structured into *nodes*, which specify sequence transformers, i.e., a node transforms a finite number of input sequences (input variables) into a finite number of output sequences (output variables). Most primitive operations perform componentwise: For instance, $x + y$ denotes the sequence $< x_1 + y_1, x_2 + y_2, \ldots, x_i + y_i, \ldots >$ where x denotes $< x_1, x_2, \ldots, x_i, \ldots >$ and y denotes $< y_1, y_2, \ldots, y_i, \ldots >$. The two additional operations `pre` and `->` are available to combine sequences in a more elaborated way; we have $\texttt{pre}(x) =< \texttt{nil}, x_1, x_2, \ldots, x_i, \ldots >$ and $x- >y =< x_1, y_2, y_3, \ldots, y_i, \ldots >$, where `nil` denotes an "undefined" value. When designing and verifying a reactive system in LUSTRE, some nodes are used to model the system itself while other nodes, called *observers*, model the desirable properties of the system. The nodes modelling the system specify how the sequences of inputs to the system are transformed into the sequence of the corresponding reactions of the system. Observers typically receive both the input and output sequences of the system and check that an invariant is respected by the sequences, i.e., they return a sequence that should be identically true.

Abstract interpretation can be (actually is) used both for optimizing and verifying LUSTRE programs [17]. At the implementation level, abstract interpretation is used to transform the LUSTRE program into an efficient automaton (see [17]): A first inefficient automaton consists of a single "big" loop, which computes the next component of all sequences from the preceding ones. This loop involves many tests, which can be simplified by splitting the entry point of the single loop into several different points corresponding to specific conditions on component values. These conditions can be inferred by abstract interpretation. Finally, the original automaton is partially evaluated according to the conditions holding at the splitting points. At the verification level, it can be checked that output sequences corresponding to observers contain the Boolean value *true* only. Direct analyses of the optimized automaton—not based on observers—are also possible (see [17]).

In [15], B. d'Ausbourg et al. propose to use LUSTRE to verify interface specifications. They first translate interface specifications from the description language UIL to the language LUSTRE, according to the *interactor* paradigm, i.e., the UIL description is translated into nodes modelling interactors. This phase can be viewed as a form of abstract interpretation called "abstract compilation" (see, e.g., [8]). The LUSTRE description of the system is then verified by means of observers. The verification is automated through a model checking tool. Once more such verification tool can be viewed as performing a particular form of abstract interpretation. More powerful verification techniques explicitely based on abstract interpretation are possible.

In conclusion, it could be argued that the LUSTRE approach to interactive system verification is remarkably simple, elegant, and powerful; notice however that programming in LUSTRE is an especially cumbersome task because sequences must be manipulated globally (i.e., we are not allowed to manipulate the components explicitely). Thus, the verification is made easier because more work is required from the programmer/specifier.

References

1. S. Abramsky and C. Hankin, editors. *Abstract Interpretation of Declarative Languages.* Ellis Horwood Limited, West Sussex, England, 1987.
2. R. Bastide and P. Palanque (Ed.). *Proceedings of the Second Eurographics Workshop on Design, Specification, Verification of Interactive Systems.* Bonas, France, June 1995. (Informal proceedings).
3. M. Bruynooghe. A practical framework for the abstract interpretation of logic programs. *Journal of Logic Programming,* 10(2):91–124, February 1991.
4. P. Bumbulis, P.S.C. Alencar, D.D. Cowan, and C.J. Lucena. Combining Formal Techniques and Prototyping in User Interface Construction and Verification. In *[2]*, 1995.
5. P. Bumbulis, P.S.C. Alencar, D.D. Cowan, and C.J. Lucena. Validating Properties of Component-based Graphical User Interfaces. In *[45]*, 1996.
6. D.A. Carr. Toward More Understandable User Interface Specifications. In *[45]*, 1996.
7. R. Cleaveland, Purush Iyer, and D. Yankelevich. Optimality in Abstraction of Model Checking. In *[39]*, 1995.
8. M.-M. Corsini, K. Musumbu, A. Rauzy, and B. Le Charlier. Efficient Bottom-up Abstract Interpretation of Logic Programs by means of Constraint Solving over Symbolic Finite Domains. In Penjam J. and M. Bruynooghe, editors, *Proceedings of the Fifth International Workshop on Programming Language Implementation and Logic Programming (PLILP'93),* volume 714 of *Lecture Notes in Computer Science,* Tallin, August 1993. Springer-Verlag.
9. A. Cortesi, B. Le Charlier, and P. Van Hentenryck. Combination of abstract domains for logic programming. In *Proceedings of the 21th ACM SIGPLAN-SIGACT Symposium on Principles of Programming Languages (POPL'94),* Portland, Oregon, January 1994.
10. A. Cortesi, B. Le Charlier, and P. Van Hentenryck. Type analysis of prolog using type graphs. *Journal of Logic Programming,* 23(3):237–278, June 1995.
11. P. Cousot and R. Cousot. Abstract interpretation: A unified lattice model for static analysis of programs by construction or approximation of fixpoints. In *Conference Record of Fourth ACM Symposium on Programming Languages (POPL'77),* pages 238–252, Los Angeles, California, January 1977.

12. P. Cousot and R. Cousot. Abstract interpretation and application to logic programs. *Journal of Logic Programming*, 13(2–3), 1992.
13. P. Cousot and R. Cousot. Abstract interpretation frameworks. *Journal of Logic and Computation*, 2(4):511–547, 1992.
14. P. Cousot and R. Cousot. Comparison of the Galois Connection and Widening/Narrowing Approaches to Abstract Interpretation (Invited Paper). In M. Bruynooghe and M. Wirsing, editors, *Proceedings of the Fourth International Workshop on Programming Language Implementation and Logic Programming (PLILP'92)*, Lecture Notes in Computer Science, Leuven, August 1992. Springer-Verlag.
15. B. d'Ausbourg, G. Durrieu, and P. Roche. Deriving a formal model of an interactive system from its UIL description in order to verify and to test its behaviour. In *[45]*, 1996.
16. P. Gray and C. Johnson. Requirements For The Next Generation Of User Interface Specification Languages. In *[2]*, June 1995.
17. N. Halbwachs. About Synchronous Programming and Abstract Interpretation. In [29], pages 179–192, 1994.
18. N. Halbwachs. Verification of Linear Hybrid Systems by Means of Convex Approximations. In [29], pages 223–237, 1994.
19. J. Henrard and B. Le Charlier. FOLON: An environment for Declarative Construction of Logic Programs (Extended Abstract). In M. Bruynooghe and M. Wirsing, editors, *Proceedings of the Fourth International Workshop on Programming Language Implementation and Logic Programming (PLILP'92)*, Lecture Notes in Computer Science, Leuven, August 1992. Springer-Verlag.
20. M. Hermenegildo and F. Rossi. Strict and Non-Strict Independent And-Parallelism in Logic Programs: Correctness, Efficiency, and Compile-Time Conditions. *Journal of Logic Programming*, 1991. (also published as Technical Report Computer Science Dept, Universidad Politecnica de Madrid, Spain, Sept 1991).
21. D. Hix and H.R. Hartson. *Developing User Interfaces.* John Wiley and Sons, London, 1993.
22. G. Janssens and M. Bruynooghe. Deriving descriptions of possible values of program variables by means of abstract interpretation. *Journal of Logic Programming*, 13(4), 1992.
23. C. Johnson. The Evaluation Of User Interface Notations. In *[45]*, 1996.
24. R.B. Kieburtz and M. Napierala. Abstract semantics. In S. Abramsky and C. Hankin, editors, *Abstract Interpretation of Declarative Languages*, chapter 7, pages 143–180. Ellis Horwood Limited, 1987.
25. B. Le Charlier, O. Degimbe, L. Michel, and P. Van Hentenryck. Optimization Techniques for General Purpose Fixpoint Algorithms: Practical Efficiency for the Abstract Interpretation of Prolog. In Cousot P. and all, editors, *Proc of the Third International Workshop on Static Analysis (WSA'93)*, number 724 in Lecture Notes in Computer Science, Padova, September 1993. Springer-Verlag.
26. B. Le Charlier, K. Musumbu, and P. Van Hentenryck. A generic abstract interpretation algorithm and its complexity analysis. In K. Furukawa, editor, *Proceedings of the Eighth International Conference on Logic Programming (ICLP'91)*, Paris,

France, June 1991. MIT Press.

27. B. Le Charlier and P. Van Hentenryck. A general top-down fixpoint algorithm (revised version). Technical Report 93-22, Institute of Computer Science, University of Namur, Belgium, (also Brown University), June 1993.
28. B. Le Charlier and P. Van Hentenryck. Experimental Evaluation of a Generic Abstract Interpretation Algorithm for Prolog. *ACM Transactions on Programming Languages and Systems (TOPLAS)*, January 1994.
29. B. Le Charlier (Ed.). *Static Analysis: Proceedings of the First International Static Analysis Symposium.* Number 864 in Lecture Notes in Computer Science. Springer-Verlag, September 1994.
30. K. Marriott and H. Søndergaard. Bottom-up abstract interpretation of logic programs. In R.A. Kowalski and K.A. Bowen, editors, *Proceeding of Fifth International Conference on Logic Programming (ICLP'88)*, pages 733–748, Seattle, Washington, August 1988. MIT Press.
31. K. Marriott and H. Søndergaard. Analysis of Constraint Logic Programs. In *Proceedings of the North American Conference on Logic Programming (NACLP-90)*, Austin, TX, October 1990.
32. K. Marriott and P. Stuckey. The 3 R's of optimizing Constraint Logic Programs: Refinement, Removal, and Reordering. In *Proceedings of the 20th ACM Symposium on Principles of Programming Languages (POPL'93)*, Charleston, South Carolina, January 1993. ACM Press.
33. L. Mauborgne. Abstract Interpretation using TDGs. In [29], pages 363–379, 1994.
34. C.S. Mellish. Abstract Interpretation of Prolog Programs. In S. Abramsky and C. Hankin, editors, *Abstract Interpretation of Declarative Languages*, chapter 8, pages 181–198. Ellis Horwood Limited, 1987.
35. N. Mercouroff. *Analyse Sémantique des Communications entre Processus de Programmes Parallèles.* PhD thesis, Ecole polytechnique, Paris, France, September 1990. In French.
36. T. Moher, V. Dirda, R. Bastide, and P. Palanque. Monolingual, Articulated Modeling of Users, Devices, and Interfaces. In *[45]*, 1996.
37. H. Montanari and F. Rossi. Concurrency and Concurrent Constraint Programming. In Andreas Podelski, editor, *Constraint Programming : Basics and Trends, Proceedings of the 1994 Châtillon Spring School, Châtillon-sur-Seine, France, May 1994*, number 910 in Lecture Notes in Computer Science, pages 171–192. Springer-Verlag, March 1995.
38. A. Mycroft. *Abstract Interpretation and Optimising Transformations for Applicative Programs.* PhD thesis, University of Edinburgh, England, 1981.
39. A. Mycroft (Ed.). *Static Analysis: Proceedings of the Second International Static Analysis Symposium.* Number 983 in Lecture Notes in Computer Science. Springer-Verlag, Glasgow, UK, September 1995.
40. R.A. O'Keefe. Finite fixed-point problems. In J-L. Lassez, editor, *Proceedings of the Fourth International Conference on Logic Programming (ICLP'87)*, pages 729–743, Melbourne, Australia, May 1987. MIT Press.
41. P. Palanque and R. Bastide. Petri net based Design of User-driven Interfaces Using the Interactive Cooperative Objects Formalism. In *[42]*, 1994.

42. F. Paternó (Ed.). *Proceedings of the First Eurographics Workshop on Design, Specification, Verification of Interactive Systems.* Bocca di Magra (La Spezia), Italy, June 1994. (Informal proceedings).
43. A. Podelski. *Constraint Programming : Basics and Trends, Proceedings of the 1994 Châtillon Spring School, Châtillon-sur-Seine, France, May 1994.* Number 910. Springer-Verlag, March 1995.
44. J. Ullman. *Principles of Database and Knowledge-Base Systems.* Principles of Computer Science. Computer Science Press, 1989.
45. J. Vanderdonckt and F. Bodart (Ed.). *Proceedings of the Third Eurographics Workshop on Design, Specification, Verification of Interactive Systems.* Namur, Belgium, June 1996. (Informal proceedings).
46. K. Verschaetse. *Termination Analysis of Logic Programs.* PhD thesis, Department of Computer Science, Katholieke Universiteit Leuven, Belgium, 1993.
47. P. Wolper (Ed.). *Computer Aided Verification.* Number 939 in Lecture Notes in Computer Science. Springer-Verlag, Liège, Belgium, July 1995.

This article was processed using the LaTeX macro package with LLNCS style

Device Models

G.P. Faconti[1] and D.J. Duke[2]

[1]CNUCE Institute, National Research Council of Italy, 56126 Pisa, Italy. email: faconti@cnuce.cnr.it
[2]Department of Computer Science, University of York, Heslington, York, YO1 5DD, UK. email: duke@minster.york.ac.uk

Abstract. Previous work on characterising the variety of interaction devices has focused either on physical properties of the devices or the range of behaviours that they can invoke. This work sets out a new approach to evaluating the usability of devices, one that accounts for the cognitive resources needed to use the device to perform particular tasks. The framework draws its expressive power from a technique called syndetic modelling that allows the description of both the device and cognitive resources to be captured in a common representation. In this paper syndesis provides a foundation for examining the coordinate spaces and transformations that are needed both by the operator and the computer system in performing tasks with a given device.

Keywords: Syndetic Modelling, Usability, Cognitive Resources, Formal Methods, Interactors, Graphical Devices.

1 Introduction

After an initial period of exploration, the interface to most personal computing systems and workstations has settled into a 'standard' model of a keyboard and mouse. Recently, there have been moves away from this, prompted by two socio-technical changes. One of these is the development of 'nomadic' computing devices, such as small portable notepads, for which a keyboard and mouse is inappropriate. The other is the development of immersive systems, where the operator has more degrees of freedom than can be adequately controlled via the usual input devices. These two developments have seen the development of a range of new interface techniques, including use of pens, the batfly (2.5D mouse), space-ball, and various glove-like devices for gesture.

Underlying practical use of these new technologies is the question of their suitability: are they appropriate for the tasks users need to perform, and what is their comparative ease of use? This paper sets out a framework for comparing what we believe are important characteristics of input devices that have so far been largely neglected: the coordinate spaces in which the operator manipulates the device. More importantly, we consider not just the physical coordinate space, but also the space in which input is rendered by the computer system, and the space in which it is interpreted within the user's cognitive processing. It is the relationship between these spaces, and the transformations that are necessary to move from one to another, that provides novel insight into usability.

2 Approach

We consider an abstract view of the flow of information between devices, users and system. To facilitate precise description and modelling at this level, we make use of a specification notation in which the various components (device, system and user) are modelled as interactors. The concept of an interactor has been described in detail elsewhere, for example [15, 17]. Briefly, an interactor is an object-like entity with an internal state, a presentation through which parts of the state (called percepts) can be perceived by a user, and actions - either user or system initiated - that bring about changes to the state. Interactors have been described using a number of formal notations including Z, LOTOS and MAL (Modal Action Logic), and it is the last of these that is used here. Briefly, MAL [30] is a typed first-order logic that extends the predicate logic with an additional operator. For any action 'A' and predicate 'P', the predicate '[A]P' means that after the action A is performed, P must hold.

Interactors can describe the logical and physical components of an interactive system, but by themselves give little direct insight into how a user might or might not be able to use the system. This is a problem, as many of the developments in interactive systems that can benefit from use of abstract models also depend critically on human abilities to process information. Syndetic models [12, 13] address this problem by expressing the behaviour of computing and cognitive systems within a common framework that supports reasoning about the conjoint system. Clearly, the 'computer' component of a syndetic model is determined by the system being represented, but for the cognitive side there is a range of models to choose from, each emphasising different aspects of human information processing. The approach that we have adopted for syndetic modelling is called Interacting Cognitive Subsystems, or ICS, and is summarised in Section 3. Importantly, ICS operates in terms of resources and information flow at a level of abstraction that is commensurate with that used to describe interactors.

The remainder of the paper is structured as follows. In Section 4 interactors are used to model a number of input devices. After extending the models in Section 5 to incorporate some basic tasks involved in graphical interaction, Section 6 defines the corresponding syndetic models that provide for the foundation for the comparative analysis in Section 7. Finally, the syndetic approach is related to previous works on input devices in Section 8.

3 ICS

ICS [3, 4] is a comprehensive model of human information processing that describes cognition in terms a collection of sub-systems that operate on specific mental codes. Although specialised to deal with specific codes, all sub-systems have a common architecture, shown in Figure 1. Incoming data streams arrive at an input array, from which they are copied into an image record representing an unbounded episodic store of all data received by that sub-system. In parallel with the basic copy process, each sub-system also contains transformation processes that convert incoming data into certain other mental codes. This output is passed through a data network to other sub-systems. If the incoming data stream is incomplete or unstable, a process can augment it by accessing or buffering the data stream via the image record. However, only one transformation in a given processing configuration can be buffered at any moment. Coherent data streams (see [4]) may be blended at the input array of a sub-system, with the result that a process can 'engage' and transform data streams derived from multiple input sources of the type supported by this sub-system.

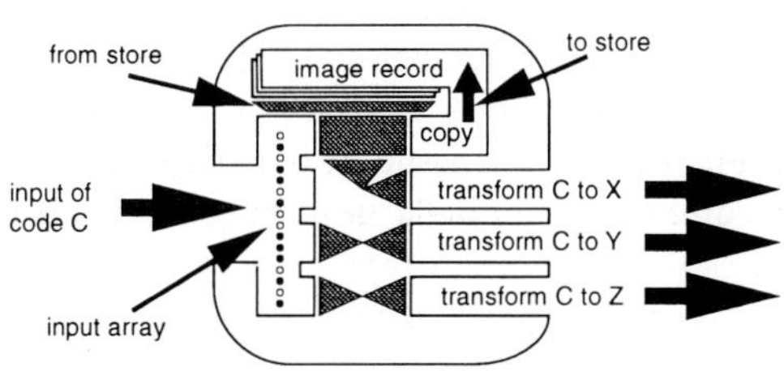

Fig. 1. *Generic structure of an ICS sub-system operating on code C.*

ICS assumes the existence of 9 distinct sub-systems, each based on the common architecture described above:

Sensory sub-systems
vis visual: hue, contour etc. from the eyes
ac acoustic: pitch, rhythm etc. from the ears
bs body-state: proprioceptive feedback

Structural sub-systems
mpl morphonolexical: words, lexical forms
obj object: mental imagery, shapes, etc.

Meaning sub-systems
prop propositional: semantic relationships
implic implicational: holistic meaning

Effector sub-systems
art articulatory: subvocal rehearsal and speech
lim limb: motion of limbs, eyes, etc

Overall behaviour of the cognitive system is constrained by the possible transformations and by several principles of processing. Visual information for instance cannot be translated directly into propositional code, but must be processed via the object system that addresses spatial structure. Although in principle all processes are continuously trying to generate code, only some of the processes will generate stable output that is relevant to a given task. This collection of processes is called a *configuration*. The thick lines in Figure 2 show the configuration of resources deployed while using a hand-controlled input device to operate on some object within a visual scene. The configuration has been developed jointly by the user and system modellers within the Amodeus project. The propositional sub-system (1) is buffering information about the required actions through its image record and using a transformation (written :prop-obj:) to convert propositional information into an object-level representation. This is passed over the data network (2), and used to control the hand through :obj-lim: (3) and :lim-hand: (4) transformations. However, both obj and lim are also receiving information from other systems. The users' view of the rendered scene arriving at the visual system (5) is translated into object code that gives a structural description of the scene; if this is to be blended at obj with the users' propositional awareness of their hand position (from 2) the two descriptions must be coherent. A propositional representation of the scene is generated by :obj-prop: and passed to prop (6) where it can be used to make decisions about the actions that are appropriate in the current situation. In parallel with this 'primary' configuration, proprioceptive feedback from the hand is converted by the body-state system (7) into 'lim' code (8) in a secondary configuration. The issue of blending between the :bs-lim: and the :obj-lim: stream (3) is discussed later in the context of the syndetic model.

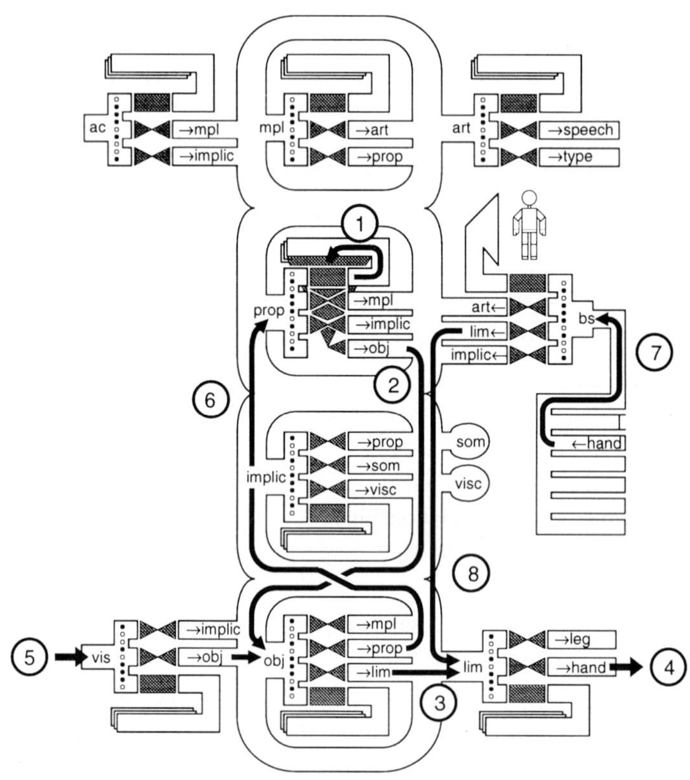

Fig. 2. *ICS (configured for graphical interaction).*

The key observation underlying syndetic modelling is that the structures and principles embodied within ICS can be formulated as an axiomatic model in the same way as any other information processing system. This means that the cognitive resources of a user can be expressed in the same framework as the behaviour of computer-based interface, allowing the models to be integrated directly. To begin this process, we define some sets to represent those concepts of ICS that will be used here. Here and elsewhere in this document we will make use of the Z notation [31] to define data types; much of this is based on common mathematical conventions for sets and relations, for example '$\times$' for cartesian product and '$\mathbb{P}$' for power set.

[sys]	- ICS sub-systems, e.g. vis, prop, obj etc.
[repr]	- Mental represententations
$\mathrm{tr} == \mathrm{sys} \times \mathrm{sys}$	- transformation processes, e.g. :vis-obj:.

Representations consist of basic units of information organised into superordinate structures. Coherence of units depends on several issues, including the timing of data streams, that will not be addressed here. Instead, coherence is captured abstractly in the form of an equivalence relation over representations:

$$\_ \approx \_ : \mathrm{repr} \leftrightarrow \mathrm{repr}$$

In describing ICS it is also useful to discuss the representations that are being delivered as part of a particular data stream. We therefore introduce a further set, code, whose elements are representations that have been labelled by the sub-system in which they were generated. Representations from or to the outside world are tagged with '*':

$\mathrm{code} == \mathrm{repr} \times (\mathrm{sys} \cup \{*\})$ - located representations

In general we will write $\mathrm{R}_{\mathrm{sys}}$ for the code $(\mathrm{R}, \mathrm{sys})$, and ':src-dst:' for the transformation $(\mathrm{src}, \mathrm{dst})$.

The state of the ICS interactor captures the data streams involved in processing activities and the properties of the streams such as stability and coherence which define the quality of processing, or in other words, user competence at particular tasks. The set of data streams onto which each transformation is 'locked' at any point in time is represented by a function 'sources' that maps each stream to the set of streams that it drawing input from. As an example, the sources of the ':obj-prop:' transformation are {:vis-obj:, :prop-obj:} as it is reflected in Figure 2. In general only a subset of transformations are producing stable output, and this set is defined by the attribute 'stable'. The codes that are available for processing at a sub-system are identified by a relation _@_, where 'c@s' means that code 'c' is available at sub-system 's'.

interactor ICS
attributes

sources	:	$\mathrm{tr} \rightarrow \mathbb{P}\,\mathrm{tr}$
stable	:	$\mathbb{P}\,\mathrm{tr}$
_@_	:	$\mathrm{code} \leftrightarrow \mathrm{sys}$

As not all representations are coherent, only certain subsets of the data streams arriving at a system can be employed by a process to generate stable output. The set 'coherent' contains those groups of transformations which, in the current state, are producing output that can be blended at the common destination. If the inputs to a process are coherent but unstable, the process can still generate a stable output by buffering the input flow via the image record and thereby operating on an extended representation. However, only one process in the configuration can be buffered at any time[1], and this process is identified by the attribute 'buffered'. The configuration itself is defined to be those processes whose output is stable and which are contributing to the current processing activity.

$$\begin{aligned} \text{coherent} &: \mathbb{P}\,\mathbb{P}\,\text{tr} \\ \text{buffered} &: \text{tr} \\ \text{config} &: \mathbb{P}\,\text{tr} \end{aligned}$$

Four actions are addressed in this model. The first two, 'engage' and 'disengage', allow a process to modify the set of streams from which they are taking information, by adding or removing a stream. A process can enter buffered mode via the 'buffer' action. Actual processing of information is represented by 'trans'. This allows representations at one sub-system to be transferred by processing activity to another sub-system.

actions

$$\begin{aligned} \text{engage} &: \text{tr} \times \text{tr} \\ \text{disengage} &: \text{tr} \times \text{tr} \\ \text{buffer} & \\ \text{trans} & \end{aligned}$$

The principles of information processing embodied by ICS are expressed as axioms over the model defined above. Axiom 1 concerns coherence, and states that a group of processes are coherent if and only if they have the same kind of output (in the code of the system 'dest') and that the representations produced by the processes and therefore available at 'dest' are themselves coherent.

axioms

1 $\forall \text{trs} : \mathbb{P}\,\text{tr} \bullet \text{trs} \in \text{coherent} \Leftrightarrow \exists \text{dest} : \text{sys} \bullet$

$$\forall s, t : \text{sys} \bullet \text{:s-t:} \in \text{trs} \Rightarrow t = \text{dest}$$
$$\wedge$$
$$\forall s, t : \text{sys};\ p, q : \text{repr} \bullet \begin{pmatrix} \text{:s-dest:} \in \text{trs} \wedge p_s@\text{dest} \\ \wedge \\ \text{:t-dest:} \in \text{trs} \wedge q_t@\text{dest} \end{pmatrix} \Rightarrow p \approx q$$

The scenario in Figure 3 illustrates two coherent data streams (these would be generated by :obj-prop: and :implic-prop:); the specific example is linked to the variables in the axiom through the annotations in parentheses.

The second axiom is that a transformation is stable if and only if its sources are coherent, and either it is buffered or the sources are themselves stable. A configuration then consists of those processes that are generating stable output that is used elsewhere in the overall processing cycle.

2 $t \in \text{stable} \Leftrightarrow \text{sources}(t) \in \text{coherent} \wedge (t = \text{buffered} \vee \text{sources}(t) \subseteq \text{stable})$

3 $t \in \text{config} \Leftrightarrow (t \in \text{stable} \wedge \exists s \bullet t \in \text{sources}(s))$

[1] This is actually a simplification for the purposes of the paper.

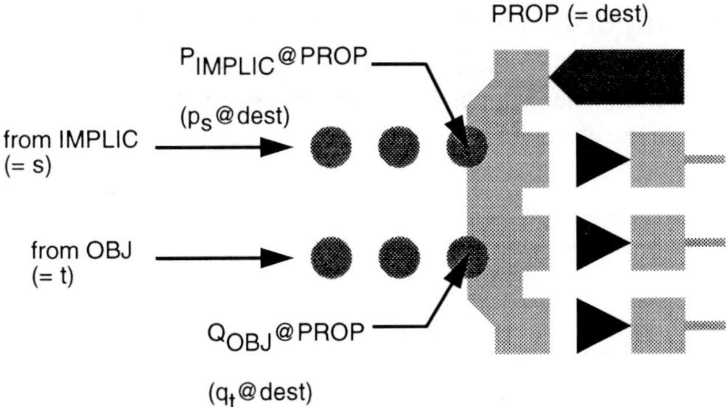

Fig. 3. *Coherent Data Streams.*

A process will not engage an unstable stream (axiom 4). If its own output is unstable, it will either engage a stable stream, disengage an unstable stream, or try to enter buffered mode (axiom 5).

4 $\textbf{per}(\text{engage}(t, src)) \Rightarrow src \in \text{stable}$

5 $$t \notin \text{stable} \Rightarrow \begin{pmatrix} \exists s \bullet s \in \text{stable} \wedge s \notin \text{sources}(t) \wedge \textbf{obl}(\text{engage}(t, s)) \\ \vee \\ \exists s \bullet s \notin \text{stable} \wedge s \in \text{sources}(t) \wedge \textbf{obl}(\text{disengage}(t, s)) \\ \vee \\ \textbf{obl}(\text{buffer}(t)) \end{pmatrix}$$

The effects of the buffer, engage, and disengage actions are straightforward and are given by axioms 6-8.

6 $[\text{buffer}(t)]\ \text{buffered} = t$

7 $\text{sources}(t) = S \Rightarrow [\text{engage}(t, s)]\ \text{sources}(t) = S \cup \{s\}$

8 $\text{sources}(t) = S \Rightarrow [\text{disengage}(t, s)]\ \text{sources}(t) = S - \{s\}$

The remaining two axioms define the effect of information transfer. Axiom 9 is the ‘forward’ rule: if a representation is available at a sub-system, then after trans a suitable representation will be available at any other sub-system for which the corresponding process is stable. Conversely, if after trans some information were to become available at a sub-system (dest), then there must exist some source system such that the information is available at the source, and the corresponding transformation is stable.

9 $p_x@src \wedge :src\text{-}dst: \in \text{stable} \Rightarrow [\text{trans}]\ p_{src}@dst$

10 $(\exists p : \text{repr};\ src, dst : \text{sys} \bullet [\text{trans}]\ p_{src}@dst)$

$\Rightarrow \exists x : \text{sys} \bullet p_x@src \wedge :src\text{-}dst: \in \text{stable}$

4 Graphical Input Devices

The availability of multi-media technology signals a departure from the methodology of interaction based on the use of keyboard, mouse and display 'standardized' so far. An increasing number of devices are employed within interactive systems with the aim of exploiting the 'natural' communication abilities of humans. Such devices range from traditional tablets to data gloves, from cameras to video recorders and players, from speakers to microphones, from flat to head mounted displays with stereoscopic views, and many others. The proliferation of devices explicitly calls for a strong concern about the usability of interactive systems in an interactionally rich environment.

In this paper we are explicitly addressing the specific class of 'graphical' input devices: that is devices that provide the system with input data interpreted in some geometric coordinate space. They include a broad range of physical and virtual equipment built around the basic operation of locating a position in space. This position can be entered directly into the system (positioning), be used to identify objects (selecting), or be accumulated with other positions to define shapes, trajectories, and, more recently, hand postures and face/head movements.

The fundamental component of any graphical input device is the 'positional measurement system': a mechanism that is able to determine the position of one or more sensors with respect to a source by means of different technologies from electro-magnetic to ultra-sound and mechanics. The more sophisticated technologies are able to determine both the position and the orientation of the sensor within a 3D space, thus enabling the definition of systems with 6 degrees of freedom.

As the focus of this paper is on the use of syndesis to model and reason about cognitive ergonomics, we have chosen to use simple but well-understood examples from 2D interaction. Our hypothesis is that coordinate spaces provide a useful framework for analysis, independant of whether interaction is in 2 dimensions or higher. The contribution of this paper is thus a framework and method by which these and other cognitive aspects of interaction can be addressed.

4.1 2D Graphical Input Devices

The most common and widespread graphical device is the 2D mouse, a physical device equipped with two transducers able to measure the distance between a current position and a next point along two axes and with a number of buttons (usually from one to three). The buttons have little value for the purposes of this paper, and are disregarded. The mouse can be described by a very simple interactor [14, 17, 27], where the type 'RelPos' represents *relative* positions, i.e. offsets.

interactor Mouse
attributes
    mouse : RelPos
actions
[bs] operate : RelPos

axioms

1 $\quad$ $[\text{operate}(\delta)]\ \text{mouse} = \delta$

2 $\quad$ $[\![\text{operate}]\!]$ in $[\![\text{Mouse}]\!]$

The Mouse interactor describes the state space of the device as a coordinate defining the distance of the current position from the previous one along two coordinate axis (RelPos == delta-xMouse × delta-yMouse). The **bs** decoration of the 'operate' action means that the device is sensed by the body-state sub-system when it is used, and the notation $[\![\ldots]\!]$ is used to refer to the perceivable aspect of an attribute, interactor or action. In this case, $[\![\text{operate}]\!]$ in $[\![\text{Mouse}]\!]$ says that the 'operate' action is perceived by the body-state sub-system (from the **bs** decoration) at any time the 'Mouse' interactor is also perceived.

While the mouse can be used as a pure input device, it is usually coupled with a cursor that provides the feedback of the current position in a reference space (usually a display). The cursor can be thought as a virtual device also described by means of an interactor

interactor Cursor

attributes

vis cursor : DispCoord

actions

render

axioms

1 $\quad$ $[\![\text{cursor}]\!]$ in $[\![\text{Cursor}]\!]$

The cursor attribute describes the state space of the cursor device represented by a pair of coordinates within the display coordinate system (DispCoord == xDisplay × yDisplay). The **vis** decoration indicates that the cursor position is visually perceivable, while the axiom states that the cursor must be visible whenever the interactor is activated. Mouse and cursor are linked by composing the two interactors and adding further information describing their mutual relations:

interactor Mouse-Cursor

Mouse, Cursor

attributes

mouseLocation : DispCoord × RelPos → DispCoord

axioms

1 $\quad$ $\text{mouse} = \delta \wedge \text{cursor} = \text{P}$

$\Rightarrow [\text{render}]\ \text{cursor} = \text{mouseLocation}(\text{P}, \delta) \wedge \text{mouse} = (0, 0)$

The attribute named 'mouseLocation' computes a new cursor position from the previous cursor position and the relative position of the mouse. The axiom states that after the rendering has taken place the new cursor position is the one computed by mouseLocation and that the mouse position is reset so that a new relative position is computed with respect to the current one. With this simple mechanism, relative positions in the mouse space can be reflected to the user in terms of absolute positions in the display space. As we will see this has a number of implications when considering the cognitive resources necessary to operate the device effectively.

In principle, a number of devices can be used in alternative to the 2D mouse; these includes, amongst others, various kind of tablets, pens, and touch panels. They range from dumb digitizers attached to a serial port to models with their own device drivers that place input streams in a shared memory buffer. Unlike the mouse, these devices all operate in absolute pointing mode. Some of them, for example the tablets, may need to be coupled with a cursor. Others, such as the notepads, allow an user to write directly on the face of a display, seeing a trail of 'ink' where the pen contacts the display surface. For brevity, we will discuss only tablets and notepads here. Again, similar reasoning can be applied to any device.

As in the case of the mouse, we describe tablets and pens by means of an interactor.

interactor Digitizer
attributes
[vis] digitizer : AbsPos
actions
[vis] [bs] operate : AbsPos

axioms
1 [operate(p)] digitizer = p
2 ⟦digitizer⟧ in ⟦Digitizer⟧
3 ⟦operate⟧ in ⟦Digitizer⟧

In contrast with the mouse, the state of the digitizer is made visible and the action of operating the device is made visually perceivable in addition of being sensed through the user's body-state. The digitizer may also be linked with the same cursor used for the mouse:

interactor Digitiser-Cursor
Digitizer, Cursor
attributes
digitizerLocation : AbsPos → DispCoord
axioms
1 [render] cursor = digitizerLocation(digitizer)

Clearly, the Digitizer-Cursor interactor has two visually perceivable attributes, the digitizer and the cursor. In the case of notepad devices the coordinate spaces are coincident since AbsPos = DispCoord, and thus digitizerLocation is the identity mapping. In the case of tablets, the two attributes are distinct since in general the coordinate space of the tablet differs from that of the display. When the cursor only is perceived, the tablet is 'cognitively equivalent' to the mouse.

5 Pointing vs Following Trajectories

Mice and digitizers are used for positioning and selecting, and for free-hand drawing and performing gestures.

- Positioning refers to the task of identifying a specific point in a coordinate system. As an example, positioning in a graphics editor refers to identifying the starting position from where an object is drawn. Similarly, in a text editor it refers to identifying the position at which text is written.

- Selecting refers to the task of identifying an object in a given context. This can be a graphical object, a menu entry, or a character in a text.

  Both positioning and selecting operations are influenced by the same kind of parameters:

  - current position: the point or the entity at which the device is currently located
  - target position: the point or the entity that must be identified
  - distance: the distance from the current to the target position
  - size: the area of the target position

  The device position is moved from the current to the target position. When the target is acquired an input token is sent to the interactive system, usually following the firing of a trigger. This trigger may either be implicit (i.e. the crossing of a window boundary) or explicit (i.e. the clicking of a swith). We already know from experiments [19, 25] that the movement performed with a specific device is influenced by the distance and the size of the target, assuming that no constraints are imposed over the trajectory. Similarly, performance varies across devices due to the different muscles involved in controlling the movement.

- Free hand drawing and gesturing refer to the task of creating a simple shape or a set of simple shapes linked by some relation. As an example, in hand writing, we have characters that can be written with a single stroke (i.e. c, e, ...) as opposed to characters that requires multiple strokes (i.e. t, i. ...).

  Shapes defined by multiple strokes are generated by single stroke operations with a positioning operation in between each two strokes. These operations are influenced primarily by the trajectory defining the shape and by the size of the shape itself. As in the case of positioning and selecting, the performance of operations may vary depending on the muscles involved in controlling the device movement.

In the following we will examine the potential of syndetic modelling to give insights on the requirements of cognitive resources to perform positioning versus gesturing. We will also compare the two classes of operations when performed with mouse versus notepads or tablets.

The interactors defined in the previous section adequately represent the system requirements and are suitable to be reused in a syndetic specification for the positioning and selecting operations. For what is concerning free-hand drawing and gesturing, a further interactor is required from the system side to express the rendering of the trajectory actually performed.

interactor Gesture
    Mouse-Cursor - (alternatively Digitizer-Cursor)
attributes
[vis] history : DispCoord∗
axioms
1 $[]\,\text{history} = \langle\rangle$
2 $\text{history} = H \Rightarrow [\text{render}]\,\text{history} = H \frown \langle \text{cursor} \rangle$
3 $[\![\text{history}]\!]$ in $[\![\text{Gesture}]\!]$

Initially (axiom 1) the history (trajectory) of points is empty, but can be extended by the 'render' action (axiom 2). The history of points is assumed always to be perceivable.

6 A Syndetic Model

The syndetic model of device interaction is created by introducing both the user and system models into a new interactor and then defining the axioms that govern the conjoint behaviour of the two agents. A new attribute (*goals*) is used to 'contextualise' the generic ICS model to the task of selection or trajectory formation by representing the sequence of points that the user wants to follow in the display space. Of course, it is highly unlikely that users will have such a precise mental model of their goals, and a more realistic approach might be to describe a class of desired or acceptable displays. However, it would add little to the analysis.

interactor Gesture-User
    Mouse-Cursor - the system: the device interactor
    ICS - the user: ICS resources and constraints
attributes
    goals : DispCoord∗ - user goals: to select/follow a series of points

The 'operate' actions defined in the device interactors are driven by the user's limb sub-system, and in order for the user to operate the device, the configuration must be set to transform a propositional representation of the desired display coordinate into musculature control, using the processes illustrated in Figure 2.

axioms

1 $$\textbf{per}(\text{operate}(D)) \Rightarrow \begin{pmatrix} [\![\text{cursor}]\!] \text{ in } [\![\text{Gest-User}]\!] \\ \wedge \\ \text{Gest-Config} \subseteq \text{config} \wedge \text{buffered} = \text{:prop-obj:} \end{pmatrix}$$

2 $\text{goals} = \langle C_i \rangle \frown G \Rightarrow [\text{operate}(P_i)]\ \text{goals} = G$

In order to operate the device and select a display coordinate 'D', the operator must be able to perceive the "current" position of the cursor. Whether or not the position of the cursor reflects all of the actions that the user has carried out is an issue that we will return to later. In addition, the set 'Gest-Config' of transformations which is assumed to contain the processes deployed in Figure 2, must be part of the configuration. These two requirements are captured in axiom 1. Axiom 2 simply states that a user works sequentially through the sequence of points that make up their current goal.

7 Analysis

In this section, we give an informal account of the analyis of the model. It would be a mistake to evaluate syndetic modelling only on the basis of its support for formal reasoning about interaction. This is an ambitious goal to which a community of researcher is currently contributing in several aspects and it is seen as a long term research activity. However, the representation of user and system models in a common framework already provides in the current stage a starting point for rigorous and informal reasoning about design issues that involves both entities.

Starting from these premises, we compare the use of mouse and digitizer devices for positioning and selecting, and for following trajectories. The comparison reveals that there exists a different cognitive load of the user depending on whether she performs the related tasks with one device or with the other.

7.1 Positioning and Selecting

As mentioned in section 5, pointing and selecting refer to the task of identifying respectively a position or an object (actually an area) within the display space.

At the propositional system, a corresponding goal is formulated that consists of only one coordinate, namely the propositional representation of the target point or area location(s) that the cursor is to be moved to. In terms of the Gesture-User interactor, this is an assumption that in the 'initial' state goals $= \langle \text{location}_{\text{PROP}} \rangle$. The goal is satisfied when the distance of the cursor or of the digitizer from the $\text{location}_{\text{PROP}}$ is perceived to be zero (or within some threshold).

According to axiom 1 of Gesture-User (henceforth axioms are abbreviated as GU.1 etc), we also have that

$$\{\text{:prop-obj:}, \text{:vis-obj:}, \text{:obj-prop:}, \text{:obj-lim:}, \text{:bs-lim:}, \text{:lim-hand:}\} \subseteq \text{config},$$

and buffered $=$:prop-obj:. The same axiom also states that the 'current' position of the cursor must be visible, a requirement that is in fact guaranteed by the display model (Dis.1). For the devices to be operated, the conditions '$[\![$operate$]\!]$ in $[\![$Mouse-Cursor$]\!]$' and '$[\![$operate$]\!]$ in $[\![$Digitizer$]\!]$' must also hold.

If :vis-obj: and :bs-lim: are to be part of the configuration, we must also assume that the input to these sensory systems is stable (from ICS.3), and thus after a processing cycle (modelled by the 'trans' action) we have that $\text{cursor}_* @\text{VIS}$ and $\text{operate}_* @\text{BS}$ (from axiom ICS.9).

Since :prop-obj: and :vis-obj: are part of the configuration and are both sources of :obj-prop: and :obj-lim:, they must also be coherent, to satisfy axiom ICS.2. Provided this condition is met, processes within the object system are permitted to engage the streams (ICS.4) and thus blending of the input data streams can occur. After a 'trans' action, the visual information becomes available at the object system, so that the following holds:

$$\mathrm{location}_{PROP}@OBJ \wedge \mathrm{cursor}_{VIS}@OBJ$$

In the case of the Digitizer, we will also have $\mathrm{digitizer}_{VIS}@OBJ \wedge \mathrm{operate}_{VIS}@OBJ$.

Using its own encoding of this information, object system processes are able to derive propositional information on the distances between objects, and limb-based code specifying the musculature control needed for the cursor or the digitizer to get closer to the target location within the display space. Let us use Δpos to refer to the difference between the intended position $\mathrm{location}_{PROP}@OBJ$ and current position of the cursor as determined visually (i.e. $\mathrm{cursor}_{VIS}@OBJ$). The two representations produced by OBJ processes will then be $\Delta\mathrm{pos}_{OBJ}@PROP$ (from :obj-prop:) and $\Delta\mathrm{pos}_{OBJ}@LIM$ (from :obj-lim:). These will be transferred over a stable stream to the destination systems as required by ICS.9.

Provided that the new representation $\Delta\mathrm{pos}_{OBJ}$ of the required device movement is coherent with respect to the current goal, PROP can sustain its output toward OBJ in a reciprical loop. In fact, if $\mathrm{location}_{PROP}$ and $\Delta\mathrm{pos}_{OBJ}$ are coherent, the reciprocal exchange of information will be stable and self-sustaining, and as a result it will not be necessary for :prop-obj: to be buffered. However, this aspect of processing is yet to be captured in the formal model of ICS.

The limb sub-system receives an input stream from the body-state system providing proprioceptive feedback, encoding dimensions such as skeletal muscle tensions. According to the definition of the Mouse and Digitizer interactors, the limb system produces an output stream that enables the hand to control the corresponding devices through the 'operate' actions. Consequently, the body-state system receives a proprioceptive feedback from these devices (M.2, D.3). From this information the system is able to derive information on the direction and on the velocity and the acceleration of the arm/hand movement as perceived in the device spaces, so that we can state that $\Delta\mathrm{dev}_{*}@BS$. This information is carried on a stream of data from the :bs-lim: transformation, and will be stable provided that the movement is within expected bounds. Finally, the limb system is permitted to engage the stable streams from the object and body-state systems (axiom ICS.4) and after a trans action we will have

$$\Delta\mathrm{pos}_{OBJ}@LIM \wedge \Delta\mathrm{dev}_{BS}@LIM.$$

Now, for the output of the :lim-hand: transformation to be stable we know from axiom ICS.2 that its sources (:obj-lim: and :bs-lim:) must also be both stable and coherent. Since the representation of $\Delta\mathrm{pos}_{OBJ}@LIM$ and $\Delta\mathrm{dev}_{BS}@LIM$ are actually based in different coordinate spaces (AbsPos and RelPos) in the case of a mouse-like device, we argue that they are not coherent. Under this assumption, axiom ICS.5 indicates that the limb system will either (a) try to engage a new stable stream, or (b) try to disengage from one of the two streams, or (c) enter buffered mode. There is no other available stream in ICS for LIM code, so option (a) is eliminated.

Option (b) captures the situation where a user consciously tries to ignore either visual or proprioreceptive feedback, possibly by watching the motion of the mouse rather than the cursor. In the case of option (c), axiom ICS.6 requires that buffering is transferred from :prop-obj: to :limb-hand:. This cognitive configuration is different from that needed to operate a notepad/digitizer. However, it can be achieved, as once a stable representation of the desired location has formed in the :prop-obj::obj-prop: loop, the buffer can be moved. In the notepad case, the device space is the same as the display space so that Δpos_{OBJ}@LIM and Δdev_{BS}@LIM are coherent, and thus there is no need for the limb system to enter buffered mode.

7.2 Following trajectories

The task of following a trajectories is used in creating shapes, for example in hand writing of drawing. The cognitive configuration necessary to perform the task doesn't differ from the one required for positioning and selecting. However, the goal formulated at propositional level is rather different since we have to select a series of coordinates. Consequently, the goal set up at the propositional level will be a sequence of required locations. In contrast to straight selection, the buffer *cannot* be transferred from :prop-obj: to :lim-hand: without disrupting task performance, since the subsequent positions to be reached by the cursor can be generated only by reading the image record of the propositional sub-system. This requires access to the buffer. Thus the operation of a mouse-like device will cause an oscillation of the cognitive configuration by continuously transferring the buffer between :prop-obj: and :lim-hand:. This explains why it is difficult to perform even simple gestures with the mouse, unless the movement becomes 'proceduralised' into the :lim-hand: process by repeated rehearsal. In contrast to the mouse, the notepad/digitizer doesn't require any buffer transfer and can be used effectively to follow a trajectory.

7.3 System feedback

A further problem that can arise with either input device is due to potential delays in the system providing with cursor feedback. Axioms M.1 and M-D.1 link the 'operate' action with the cursor attribute (similarly with axioms D.1 and D-C.1). When the rendering of the cursor is delayed, the input to the visual system will be a cursor position that diverges from the next location that the user wants to select. Once this divergence reaches a threshold, the streams of locations derived from :vis-obj: and :prop-obj: will become incoherent. Consequently, the streams generated from :obj-prop: and :obj-lim: will not be coherent, and so buffering will also be required at :prop-obj: and :lim-hand: resulting in the impossiblity of achieving the formulated goal. This explains how a device that responds too slowly to user actions becomes very difficult to operate interactively. An all too common example of this is where a mouse becomes 'sticky', that is, the hardware does not properly register movement. The resulting need for focal awareness to oscillate between problem solving at a propositional level, and control over a wayward device, manifests itself as a sense of frustration as changes anticipated via feedback from body-state are not reflected in the display.

8 Related Work

Input devices have received considerable attention in the past and a number of early and fundamental results in the subject are available in the literature [20, 21, 1, 29, 28]. By the late 80's the technology underlying input devices was developed to the point where it was thought to be appropriate to systematize the available knowledge.

Cardelli and Pike [9] have started exploring the concurrent use of multiple devices and the potential of multi-thread interactions. The Squeak specification language they have developed is directly derived from Hoare's CSP [22], Milner's CCS [26] and ESTEREL [5]. It is given a formal semantics by using the methods of operational semantics and represents one of the very first works on interaction grounded on a rigorous formal approach.

Duce et al. [11] have formally described in CSP the input model of standard graphics systems and have extended the model to describe hierarchies of input devices [10]. Similarly, Faconti et al. [16] have developed a complete specification of the standard device model by using the LOTOS [23] notation. They have formally described also an enhanced model that allow to extend the device operating modes by means of constraints and composition operations [18]. The same method has been applied to the broader framework of human computer interfaces to describe the composition of Interactor [17], basic components of an interaction described in terms of LOTOS processes.

The work of Cardelli, Duce and Faconti is mainly devoted to describe the behaviour of devices within interactive systems and to gain insights on properties that are intrinsic to these systems. No attention is paid to the user component of an interactive system. When the user is explicitly considered [10, 16], the corresponding behaviour is modelled with the same formalism as the system through a process definition that is prescriptive with respect to user behaviour. This observation isn't a criticism of these papers; rather it acknowledges that user processes are introduced to verify the robusteness of the system. In fact the aims clearly declared by the authors are to describe computer and device models rather than to investigate user-oriented properties.

A different approach is found in the work of Card et al, in [7, 8]. They have built a taxonomy of input devices that extends previous approaches such as Baecker and Buxton [2] and Foley et al [24]. The key concept underlying the Card's work is based on a rather simple model: the interaction occurring between a human being and an embedded computer is modelled as the interaction in an artificial language among three agents (i) a human, (ii) a user dialogue machine, and (iii) an application. The semantics of an input device is traced by mapping human actions into device events and finally device events into changes in the application state. Following the semantic analysis, a design space of input devices is generated and, subsequently, human performance theories and data are used for the evaluation of points in this design space. Input devices are positioned in a 6-dimensional space where each point (i.e. each device) can be evaluated according to its expressiveness and effectiveness by a number of figures of merit such as device bandwith, pointing precision, errors, and desk footprint.

The taxonomy implicitly addresses (some) properties of users since the metrics of the device space is built directly from field experiments, from the observation of the use of existing systems and devices, and from human performance literature. Consequently, it is effective in highlighting the different expressiveness and effectiveness in performing specific tasks by means of devices that would have resulted to be equivalent with the previous approaches.

However, Card's work doesn't take into consideration, neither explicitly nor implicitly, issues on cognition that are at the core of our approach. In fact, the majority of the user properties embedded into the taxonomy refer either to physical or physiological ergonomics. Incidentally, the bandwidth of input devices, as defined in the taxonomy, might be used (loosely) to get a (qualitative) measure of the proprioceptive feedback from the body-state sub-system of ICS. This will be an indirect evaluation of the capability of blending body-state and object information at the limb sub-system that has a definite influence in the usability of a device in a given context.

9 Conclusions

Traditional approaches to evaluating or comparing input devices have focussed either on the logical behaviour of the device, or ergonomic aspects of its use. This paper has presented a framework that allows analysis of the *cognitive* ergonomics of interaction, in terms of the mental resources needed to utilise a particular device for a specific task. We have used the model to present a systematic account of the diffences between mouse and tablet for trajectory formation. The example was chosen for familarity, rather than for novelty, and although the conclusions are not in themselves profound, the approach is one that can be extended to rather more sophisticated and problematic techniques such as gesture and multi-modal input [13].

The account of ICS given in this paper is more detailed than in previous analyses, and in particular includes the concepts of stable and coherent data streams that were particularly useful in reasoning about device interaction. Our next objectives include a systematic account of the structure and relationship between representations, and the modelling of phases of cognitive activity [3].

One argument raised against the wider use of syndetic modelling for human factors evaluations is the level of formality involved. This is a reasonable concern, and there are two responses. The first is that the work on syndesis carried out so far has been primarily concerned with establishing its feasibility as a model for explaining interaction, rather than as a practical tool for industrial use. We are now beginning to explore means by which the level of formality can be tamed, both by supporting development of formal models with software tools, or by encapsulating the technique within a tool to support scenario-driven analysis of interaction.

The second response to concern about formalism is that the complexity of modern interfaces, and the subtle demands that they place on users' cognitive abilities, calls for an expressive and analytically powerful method for modelling and evaluation. Such a method needs to be able to span both the user and the system, in order to capture the interplay between the information available from the system, the actions that can be taken, and the tasks and knowledge of the user. We are not advocating syndetic models as a replacement for other design representations. There is an inherent trade-off between the power and generality of a notation [6], and there important issues, for example based on social factors or domain requirements, for which syndetic models are either inappropriate or inadequate. Likewise however, syndesis brings considerable analytical power and authority (in the form of the underlying cognitive theory) to bear on problems whose complexity makes the use of less formal design techniques problematic.

Acknowledgements

The work reported in this paper was funded by the European Union with grants through the Human Capital and Mobility Programme, Contract CHRX-CT93-0099, and ESPRIT Basic Research Action, Contract AMODEUS 7040

References

1. E. Anson. The device model of interaction. *Computer Graphics*, 16(3), 1982.
2. R.M. Baecker and W. Buxton, editors. *Readings in human-computer interaction: A multidisciplinary approach*. Morgan-Kaufmann, 1987.
3. P.J. Barnard and J. May. Cognitive modelling for user requirements. In P.F. Byerley, P.J. Barnard, and J. May, editors, *Computers, Communication and Usability: Design Issues, Research and Methods for Integrated Services*, North Holland Series in Telecommunication. Elsevier, 1993.
4. P.J. Barnard and J. May. Interactions with advanced graphical interfaces and the deployment of latent human knowledge. In *Eurographics Workshop on Design, Specification and Verification of Interactive Systems*, pages 15–49. Springer, June 1994.
5. G. Berry. The ESTEREL synchronous programming language and its mathematical semantics. In *NSF/SERC workshop on concurrency*. CMU, 1984.
6. A. Blandford and D.J. Duke. Integrating user and computer system concerns in the design of interactive systems. *IEEE Transactions on Software Engineering*, 1996. Submitted for publication.
7. S. Card, J. Mackinlay, and G. Robertson. The design space of input devices. In *Proc. of CHI'90*. ACM Press, 1990.
8. S.K. Card, J.D. Mackinlay, and G.G. Robertson. A semantics analysis of the design space of input devices. *Human-Computer Interaction*, 1990.
9. L. Cardelli and R. Pike. Squeak: a language for communicating with mice. *Computer Graphics*, 19(3), 1985.
10. D. Duce, P. ten Hagen, and R. van Liere. An approach to hierarchical input devices. *Computer Graphics Forum*, 9(1):15–26, 1990.
11. D.A. Duce, R. Van Liere, and P.J.W. ten Hagen. Components, framework and GKS input. In *Proceedings of Eurographics'89*. North-Holland, 1989.
12. D.J. Duke. Reasoning about gestural interaction. *Computer Graphics Forum*, 14(3):55–66, 1995. Conference Issue: Proc. Eurographics'95, Maastricht, The Netherlands.

13. D.J. Duke, P.J. Barnard, D.A. Duce, and J. May. Systematic development of the human interface. In *APSEC'95: Second Asia-Pacific Software Engineering Conference*, pages 313–321. IEEE Computer Society Press, 1995.
14. D.J. Duke and M.D. Harrison. Abstract interaction objects. *Computer Graphics Forum*, 12(3):25–36, 1993. Conference Issue: Proc. Eurographics'93.
15. D.J. Duke and M.D. Harrison. Interaction and task requirements. In P. Palanque and R. Bastide, editors, *DSV-IS'95: Eurographics Workshop on Design, Specification and Verification of Interactive Systems*, pages 54–75. Springer-Verlag, 1995.
16. G. Faconti, M. Caneve, E. Salvatori, and N. Zani. A LOTOS view of the input model of standard graphics systems. In *Proceedings of Eurographics Workshop on Formal Methods in Computer Graphics*, 1991.
17. G. Faconti and F. Paterno'. An approach to the formal specification of the components of an interaction. In C. Vandoni and D. Duce, editors, *Eurographics 90*, pages 481–494. North-Holland, 1990.
18. G.P. Faconti, N. Zani, and F. Paterno'. The input model of standard graphics systems revisited by formal specification. *Computer Graphics Forum*, 11(3), 1992. Proceedings of Eurographics'92.
19. P.M. Fitts. The information capacity of the human motor system in controlling amplitude of movement. *Journal of Experimental Psychology*, 47:381–391, 1954.
20. J.D. Foley and V.L. Wallace. The art of natural graphic man-machine conversation. In *Proceedings of IEEE 62*, 1974.
21. R.A. Guedj, editor. *Proceedings of IFIP Workshop on methodology of Interaction*. North-Holland, 1980.
22. C.A.R. Hoare. *Communicating Sequential Processes*. Series in Computer Science. Prentice Hall International, 1985.
23. ISO Central Secretariat, Geneva. A formal description technique based on temporal ordering of observational behaviour, 1988. ISO/IS 8807.
24. P. Chan J.D. Foley, V.L. Wallace. The human factors of computer graphics interaction techniques. *Computer Graphics and Applications*, 4(11), 1984.
25. G.D. Langolf. *Human motor performance in precise microscopic work*. PhD thesis, University of Michigan, 1973.
26. R. Milner. *Communication and Concurrency*. Series in Computer Science. Prentice Hall International, 1989.
27. F. Paterno' and G. Faconti. On the use of LOTOS to describe graphical interaction. In A. Monk, D. Diaper, and M. Harrison, editors, *People and Computers VII: Proc. of the HCI'92 Conference*, Conference Series, pages 155–173. British Computer Society, 1992.
28. G. Pfaff, editor. *User Interface Management Systems*. Eurographics seminars, Springer-Verlag, 1985.
29. D.S.H. Rosenthal et al. The detailed semantics of graphical input devices. *Computer Graphics*, 16(3), 1982.
30. M. Ryan, J. Fiadeiro, and T. Maibaum. Sharing actions and attributes in modal action logic. In T. Ito and A.R. Meyer, editors, *Theoretical Aspects of Computer Software*, volume 526 of *Lecture Notes in Computer Science*, pages 569–593. Springer-Verlag, 1991.
31. J.M. Spivey. *The Z Notation: A Reference Manual*. Prentice Hall International, second edition, 1992.

This article was processed using the LaTeX macro package with LLNCS style

A Formal Description of Low Level Interaction and its Application to Multimodal Interactive Systems

Johnny Accot [1,2], Stéphane Chatty [1], Philippe Palanque [1,3]

[1] CENA
7 av. Edouard Belin
31055 Toulouse cedex,
France
chatty@cena.dgac.fr

[2] D.G.P.
University of Toronto
Toronto, ON M5S 1A4
Canada
accot@ cena.dgac.fr

[3] LIS - IHM
Université Toulouse I
31042 Toulouse cedex,
France
palanque@cict.fr

Abstract The lack of formal models for describing low-level interaction restricts programmers to interactors provided by toolkits. It impedes the construction of highly interactive systems and the design of new interaction styles, such as multimodal interaction. This article reports on our experience with formalising low-level graphical interaction. We propose primitives for event specification and handling that can be used along with Petri nets to model such interactions. We then show how multimodal interactions can be built from monomodal ones by combining those models. This is exemplified by an experimental two-handed graphical editor that has been built using the proposed model.

1. Introduction

Though this is an active field of research [4,11,12], most interactive systems designed today do not rely on formal development or specification methods. This is especially true as far as low-level interaction in user interfaces is concerned. This leads interface programmers to stick to widespread interaction styles and to rely on the experience of toolkit designers. Those limitations cause serious problems to the designers of complex and critical systems such as air traffic control displays: either they use well-known interaction styles, with possible consequences on the usability of the system, or they are faced with the problem of specifying low level interaction without any help for that purpose. Another consequence of that absence of formal models is the difficulty to teach graphical interface construction. Whereas basic algorithms such as list sorting or rendezvous are described in numerous books, no formalism is available to describe even a double-click or the behaviour of a menu in a precise way. Building widgets is thus closer to craft knowledge than to a reliable and structured engineering process. This leads to unacceptable situations such as simple widgets of widespread commercial toolkits exhibiting incoherent behaviours.

This lack of models for graphical interaction has obvious consequences on the design of more complex systems. In multimodal interfaces, for instance, parallel event flows can lead to unpredictable incoherent situations in the same way as in multitasking

concurrent programming systems. What is a nuisance with graphical interaction can become an obstacle with multimodal interaction. Our previous work on two-handed interaction [7] confronted us with this problem, for instance when trying to build a two-handed graphical editor, using techniques such as Toolglass and Magic Lenses [3]. In order to be able to specify precisely and without ambiguities the expected behaviour of the system, we were faced with the need to use formal methods. This paper reports on the solutions we proposed and implemented to describe discrete aspects of traditional and multimodal interaction at a low level of interaction.
The next section proposes a flexible event specification scheme proposed as a replacement for callbacks. Section 3 shows how this specification scheme can be fruitfully integrated with Petri nets to fully describe event-based interactions. This leads to a generic framework for modelling low level interactions. Section 4 shows how that framework can be easily exploited to deal with multiple threads of input for multimodal interaction. Section 5 gives an example of how a model designed with that framework can be implemented in a real application. The example proposed here is a graphical editor with two-handed interaction capabilities. In the last section the work described in this paper is positioned relatively to previous work in the field of dialogue modelling.

2. Primitives for event description

When modelling interactive systems, the interaction primitives are at least as important as the formalism used to manipulate them, because they determine the level of control that programmers have over the behaviour of the system. In an event-based model, the most important primitives are the event selection scheme and the method of linking events to behaviours. Most toolkits select events by their types, the window in which they occur, and sometimes the graphical object located under the cursor. They use callbacks to associate events to behaviours: when selecting a class of events, the procedure to be called is specified. However, callbacks are known to be a problem, because of their low level of abstraction [19]. In addition, as shown in [7], event selection sometimes needs to use notions more precise than event types. For instance, one often needs to specify which button of the mouse or which key of the keyboard was pressed. We thus considered an association of two primitives: *reactions* and *criteria*.

- A *reaction* links an object's method to a set of possible events. When an event matching the specification of the reaction occurs, the method is invoked. The object that reacts to the event does not have to be the graphical object on which the event occurred. It does not even have to be a graphical object. This allows us to have icons react to mouse movements on the background of a window (when dragging), or to have functional core objects react to key presses (when pressing ‘q’ to quit an application).
- A *criterion* is the basic entity used in the specification of reactions. Criteria check properties of events, and a reaction is triggered only if all its criteria are satisfied. With current callback-based models, event selection can be expressed as "bind this function to this type of events". With reactions and criteria, it can be expressed as "this object is interested in events having such and such properties". This allows

designers to explicitly express constraints, instead of implementing them by hand in the callbacks. The most commonly used criteria are event types, event targets, and devices. Others can refer to the attributes of events (amount of movement, for instance) or to the state of any object. A classical example of amount of movement is the threshold parameter in the options of a mouse. The threshold helps users in performing clicks and double-clicks by ignoring small mouse movements between clicks.

Using criteria and reactions provides a flexible way of expressing simple behaviours of objects in an event-based application. However, more complex behaviours triggered by sequences of events need formalisms such as finite state machines or Petri nets to be expressed. This is the topic of the next section, where criteria are used to label the transitions of Petri nets.

3. Interaction level

This section aims at showing how it is possible (using the primitives for event description described in the previous section) to model the production of high level events from the physical model of an input device. We will show the model for a mouse and we will prove the consistency of the physical model with the one responsible for the generation of high level events (called interaction level events).

Interaction level correspond to the user's actions while manipulating physical devices such as a mouse or a keyboard. For example, such actions can be: press the right button on the mouse, release the left button of the mouse, press a key, etc. It can be easily seen that the user's behaviour heavily depends on the physical behaviour of the device, e.g. a button can only be released if it has been pressed before. Describing this behaviour is quite straightforward when dealing with classical devices such as mice and keyboards. The behaviour of a mouse featuring only one button is easily described with an automaton, or with a Petri net as shown in Figure 1. Petri nets are a formalism devoted to the modelling of discrete event systems in which parallelism plays an important role. When modelling with Petri nets, a system is described in terms of state variables (called places and depicted as ellipses) and by state changing operators (called transitions and depicted as rectangles), connected by arcs. The state of the system is given by the marking of the net, which is a distribution of tokens in the net's places. This use of tokens allows designers to describe states in a very concise way.

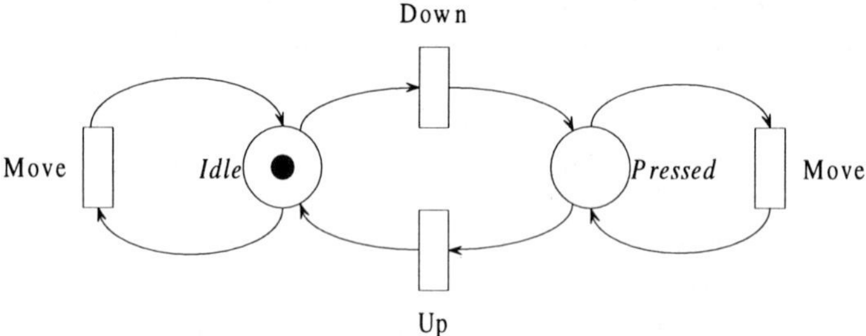

Fig. 1 The physical behaviour of a one-button mouse

In Figure 1 the Petri net describes that, in the initial state, the device is idle: it waits for an event to occur. While in this state, two kinds of events may occur; this is represented by the two different outgoing transitions. The transition Move removes the token from the place and puts it back into it, which means that the system remains in the same state. If the event Down occurs, the token in the place *Idle* is removed and put in the place *Pressed* thus describing a change of state in the system. From that state, both Move and Up events may occur: Move keeps the system in the Pressed state, and Up puts it back to its initial state.

This model describes precisely the basic behaviour of the device. However, in order to fully exploit the benefits of this device, it is important to take into account a higher level of interaction represented by the events associated to sequences of physical actions. For example, it is important for the designer to be able to process the sequence (Down, Up) either as two different events or as a single event: Click. The management of those higher level events is less simple than that of the physical ones. For instance, double-clicks events involve a time constraint: the sequence (Down, Up, Down, Up) has to be performed within a given amount of time, else it is interpreted as two isolated clicks. This is usually implemented by activating a timer, and taking into account time-out events. In addition, if the mouse is moved during the sequence, this is not a double-click either. Figure 2 presents the different events associated to a mouse. The physical events are described in the left column, the interaction level ones in the right column.

d: Button Down	C: Single Click
u: Button Up	DC: Double Click
m: Mouse Move	B: Begin Drag
t: Time Out	D: Drag
	E: End Drag

Fig. 2 Physical (left column) and interaction level events

In the case of mouse events, most applications are at least as much interested in higher level events as in the low level ones produced by the graphical layer. Clicks, drags and double-clicks are the results of sequences of physical events. Usually, these higher level events are produced either by dedicated algorithms, or through models such as Garnet's interactors [19]. We call them interaction level events, as they may be used as input to other processes that would use them as basic events. The creation of those interaction level events depends on the state of the interaction, which in turn depends on the physical events that previously occurred. This type of behaviour is easily described with automata or Petri nets. We chose to use Petri nets, so as to be able to express parallelism and synchronisation, as we will see in the next sections. Using this formalism we have been able to test different designs for double clicks, and provided a good basis for thinking about interaction styles involving more than one device. It also proved useful when integrating exotic devices such as a telephone into graphical applications [8].

However, the use of basic Petri nets does not permit the description of the types of events and to change the behaviour of the model according to the type of these events. Thus we upgraded the basic model by adding the criteria and reactions presented in section 2. The next section shows how this integration can be used for describing high level event production.

3.1 An example of generation of interaction level events

Figure 3 shows the physical events consumed by the Petri net and the interaction level events it produces. This is one of the possible designs for clicks, double-clicks and drags according to both physical and higher level events. Labels associated with transitions have the following shape:

$$\frac{\textit{Event, criteria / Reaction}}{\textit{Production}} \qquad (1)$$

where

- *Event* is the low-level event consumed
- *Criteria* is the set of additional criteria that the event must satisfy for the state change to occur
- *Production* is a set of interaction level events that are produced during the state changing
- *Action* is the additional action performed by the model during the state change

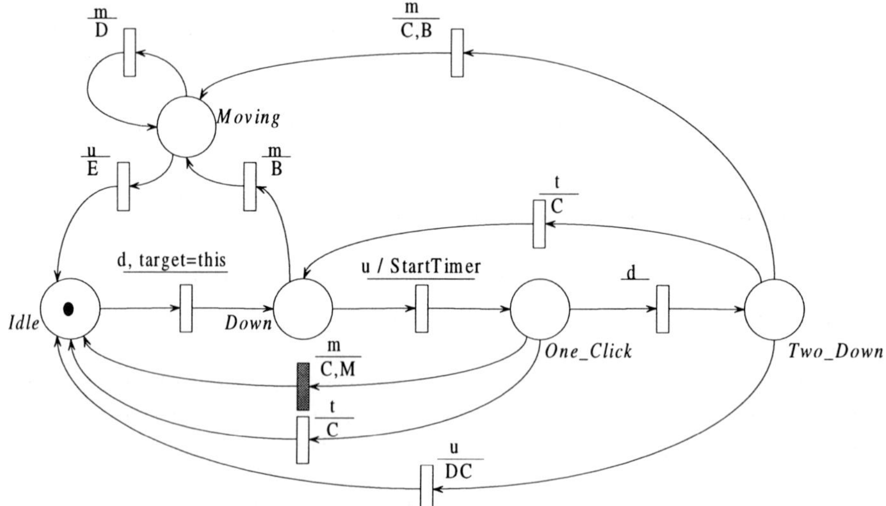

Fig. 3 A design for clicks, double-clicks and drags

This Petri net describes the interaction level events policy for a given interaction style. It tells when and how those events are produced according to the user's actions on the device. In this Petri net, the policy works like a transducer: each time a physical event is accepted, the Petri net fires a transition and creates higher level events. For example, in the policy represented in Figure 3, the physical mouse-move (`m`) is transformed into a higher level mouse-move (`M`), e.g. the transition between places *One_Click* and *Idle* reacts to the event `m` by generating another event `M` plus an event click (`C`). However; all physical events are not immediately translated into interaction level events. For example, each event `d` received while the system is in the initial state, is consumed without any production.

In the simple model described in Figure 3, criteria are only event types, except for the first down event (the transition between places *Idle* and *Down*). More sophisticated designs could be obtained, for instance by attaching a criterion to the arc from place *Down* to place *Moving* so that small moves are ignored. The reaction StartTimer on the transition between place *Down* and place *Click* states that a timer is started as soon as the physical event up (u) is received.

The Petri net in Figure 3 was built as follows. The four states *Idle, Down, One_Click* and *Two_Down* and the transitions between them represent the sequence of physical events (d, u, d, u) that corresponds to a double click (DC). The timing constraint is represented by the action StartTimer on the transition between *Down* and *One_Click*, and by the transitions labelled by t on other states. The most significant such transition is the one that links *One_Click* to *Idle:* after a certain time; the sequence (u, d) is considered as a C. Similarly, the no-move constraint is expressed by the transitions that lead to place *Moving*. When in that state, all moves (m) are parts of a drag (D). A m received when in state *One_Click* also forces the emission of a C, and the return to *Idle*, though the greyed[1] out transition in Figure 3.

3.2 Analysis of conformance between physical and interaction level events

An important point while designing a model for interaction level events is to make sure that the model is consistent with the underlying physical behaviour of the device. One possible way to prove this kind of property is to consider the two models as a client and a server cooperating together. In our case, the model describing higher level events (called MHLE) is the client and the one describing the physical events (called MPLE) the server. Thus, the low level events in the MHLE describe the demand of the model towards the MPLE. Conversely, the events modelled in the MPLE describe its offer.

Proving that the models are consistent is equivalent to proving that the demand of the client is included in the offer of the server. In previous work on the verification of the consistency of models in CSCW applications [22], we have shown that, for Petri nets, the demand and offer of the models are the same as the demand and offer of the automata corresponding to the marking graph of the Petri nets. Here, the Petri nets in Figure 1 and Figure 3 are state machines, hence their marking graphs are the models themselves. We thus only have to prove that the language of the automaton corresponding to MPLE is included in the language of the automaton corresponding to the MHLE.

Using regular expressions to represent the languages, the language accepted by the Petri net in Figure 1 is: (m*dm*u)* and the language accepted by the Petri net in Figure 3 is: (dm*u | dum | dudmm*u | dudu)*. The time-out events have not been taken into account in the calculation of the language because they can be considered as internal events of the MHLE. A more precise calculation could have been made by considering that those events are produced by another server responsible for handling temporal aspects.

---

[1] There is no special meaning for the greyed out transitions. It is only for sake of readability that it has been greyed out

It can easily be proved that the language generated by each of the terms of the second regular expression are included in the language generated by the first regular expression. Hence, the language generated by the second regular expression is included in the language generated by the first one. That proves that the MHLE only describes actions that can result from sequences of events emitted by the MPLE. In other terms, this means that the model of interaction level production only reacts to sequences of physical events that can be provided by the physical device. It is important to notice that the model is not supposed to react to all the possible sequences of the physical device as this is a matter of design.

The next section is devoted to show how the work presented above can be reused in order to model the production of multimodal events in an interactive system. The emphasis is not on the design process which will be the focus of a future paper but on the solution of the precise problem of generating multimodal events in two-handed multimodal systems.

4. Multimodal events

The technique described above can be used to describe multimodal interaction, as we will show here for the case of two-handed interaction. As shown in [7], two-handed interaction is - at least technically speaking - a form of multimodal interaction, and exhibits the same variations. There are many possible two-handed interactions styles, and thus tools are needed to design, implement and evaluate them. Examples of such interaction styles are those that would involve touch screens manipulated [7] with two or more fingers[2]. With such devices, one can perform combined clicks (i.e. two fingers 'clicking' at the same time) and many other combinations. For instance, air traffic controllers currently use their finger tips to evaluate and compare distances on their radar screens. With touch-screen based displays such as those explored at CENA [9], such comparisons could be improved by detecting combined clicks and displaying the appropriate information. But such combinations are complex to implement, and we soon found that a formal method was necessary in order to handle all possible cases, and even to communicate among ourselves. We thus generalised the use of the framework described in the previous section. Here the set of multimodal events is limited to two: a combined click (a classical click done with both hands simultaneously) and a double combined click (a classical double click done with both hands simultaneously). However, the approach presented aims at being much more generic as we are currently working on composition techniques in order to describe other multimodal interaction level events, and to build them for other input devices.

---

[2] Such touch screens are not commonly available yet, but their behaviour can be simulated with multiple pointing devices

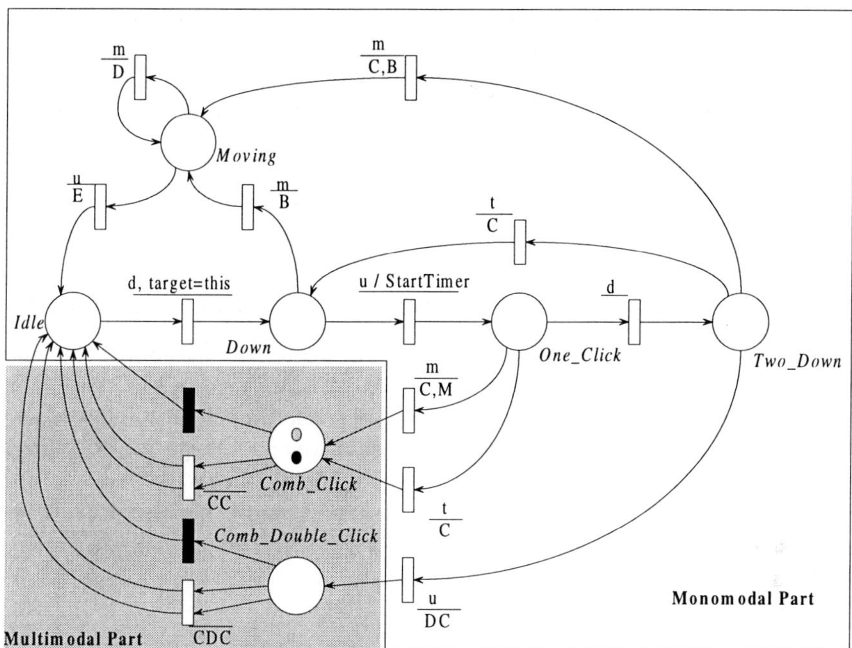

Fig. 4 One possible Petri net for managing two-handed interaction

The use of Petri nets allows us to describe two-handed interaction as an extension to traditional interaction. The presence of two similar devices is represented by the presence of two tokens in the same Petri net: this technique is known as folding and is the basis of an early extension to basic Petri nets called coloured Petri nets [17]. For that reason, in Figure 4 the two pointers (possibly fingers) are represented by a grey and a black token. Places and transitions are added to model the interaction level events produced in reaction to combined actions on the devices. In the case of combined clicks, the extension to the model of Figure 3 is represented in the greyed out region of Figure 4. Two places were added and the transitions CC and CDC (which feature double incoming arcs and double outgoing arcs) detect the occurrences of combined clicks and combined double-clicks. For instance, a combined click occurs when two tokens reach place *Comb_Click* at nearly the same time.

The Petri net in Figure 4 is currently in the state when the user has pressed and released each pointer. This is described by two tokens in the place *Comb_Click*, thus enabling the transition with two arcs that will produce the CC event. The need for the click events to be temporally close is represented by the other output transition of the place *Comb_Click*, which consumes tokens according to the duration of their stay in that place. These black transitions correspond to timed transitions, which means that after being enabled they wait for a given amount of time before being fired. This modelling has been made possible thanks to the easy quantitative time modelling in Petri nets. This kind of Petri net is known as timed Petri nets and further information about their functionalities can be found in [23].

As there is one token for each device it is important to be sure that an event occurring on a given device d is associated to the token corresponding to the device d in the model. This is done by associating criteria to transitions. For example, at the initial state the user can press the button on each device. The criterion associated to the only available transition will select the token corresponding to the device currently used by the user and then the transition will remove the selected token from the place *Idle* and put it in the place *Down*. An example of easy representation of this kind of criteria is to use colours as the tokens have to be differentiated. However, as the production of CC and CDC events is related to quantitative temporal evolution of the model the expressive power of Coloured Petri nets is not sufficient.

5. Implementing an example: an editor for building automata

The aim of this section is to show on a simple example how the formal framework presented above can be used in real applications. The example we use is graphical editor for drawing automata, extended with two-handed input capabilities, that we developed as a test-bed for two-handed interaction styles. Though we are currently working on two-handed interaction from a usability point of view, we will here focus on very simple interaction techniques for the sake of clarity. We also provide more details on the implementation of our framework.

5.1 The example

The first step when designing an automaton consists in drawing the skeleton of the automaton. To do so, the editor provides users with two basic objects: states (represented by circles) and transitions (depicted by labelled arcs). It also features functions to manipulate these objects: add, move, delete, and edit.
The sample editor we have built is shown in Figure 5. It features a working area, where the automaton is built and displayed, and a set of buttons for three of the functions: create, move, and delete. Editing the contents of a transition is performed by double-clicking on its graphical representation, which opens a transition editor (not presented here). As an example of two-handed interaction, let us consider a cooperation of two hands (or rather two fingers). An example of such cooperation consists in the creation of a transition between two states by clicking on a state, then nearly simultaneously on another. In that case the transition is automatically created from the state that has been clicked first to the other one. The double click and the two-handed manipulations in this editor have been implemented using the Petri net model for the management of multimodal aspects presented in section 4 of this paper.

5.2 Implementation

Our implementation of the model is like traditional interpreted implementations of Petri nets [15] where arcs, places and transitions are stored in lists. Depending on the current marking, the only active criteria are those corresponding to transitions that might potentially be fired. Criteria are implemented through an event subscription mechanism, so this means that the model will only receive events that are meaningful according to the current state.
Referring to figure 4 for instance, if both tokens are present in the *Idle* place, the net is only sensitive to button-down events and does not receive any information regarding other kinds of events. Then, when one of the transition fires and the token moves from

the *Idle* place to the *Down* place, the net unsubscribes to button-down events from the device corresponding to the token that moved, and subscribes to button-up and move events.

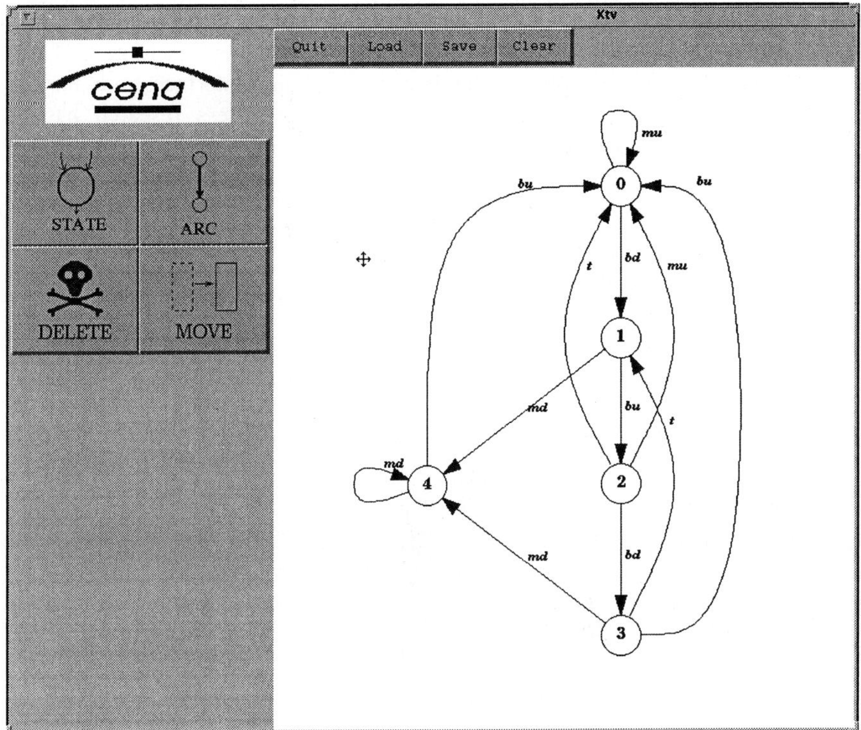

Fig. 5 The editor for designing automata

This approach based on an interpretation of the Petri net a run-time has two important consequences on the characteristics of the net, regarding the choice and specialisation of behaviours:

- a graphical object is able to 'choose' a behaviour (by inheritance in our approach). In user interfaces for instance, a window does not have the same behaviour as a menu button: when a user grabs a window, this window should always be attached to the mouse pointer and all mouse drag events should be redirected to it, whereas a menu button should not grab all events event if the button-down was performed on it. A set of different "standard" behaviours would be proposed by graphical toolkits so that developers would just have to pick a behaviour and, if needed, modify it as suggested below.
- the Petri net can be 'customised', which means that developers or even end-users can easily change the interpretation of events from devices according to their needs. These modifications primarily concern the parameters of the net, such as time delays or movement tolerance if any, but also the net structure itself. Even if these dynamic changes are not fully understood, we believe that this may be possible and useful if assisting users in this task.

6. Related Work

A lot of work has been devoted to the formal specification of interactive systems. This work has been done at different levels of abstraction from higher level such as sequencing of windows, sequencing of interaction objects triggering within a dialogue window, to elementary behaviour of a single interaction object. A large part of the results obtained have been applied to 'indirect manipulation interfaces', i.e. interfaces controlled through menus, buttons and dialogue boxes. Less research has been devoted to highly interactive interfaces featuring direct manipulation or animation.

Figure 6 presents a summary of the related work in the formal specification of interaction. This table does not represent more theoretical work on interactive systems such as [13, 16, 10] in which the main point is to address properties of interactive systems rather than their inner behaviour.

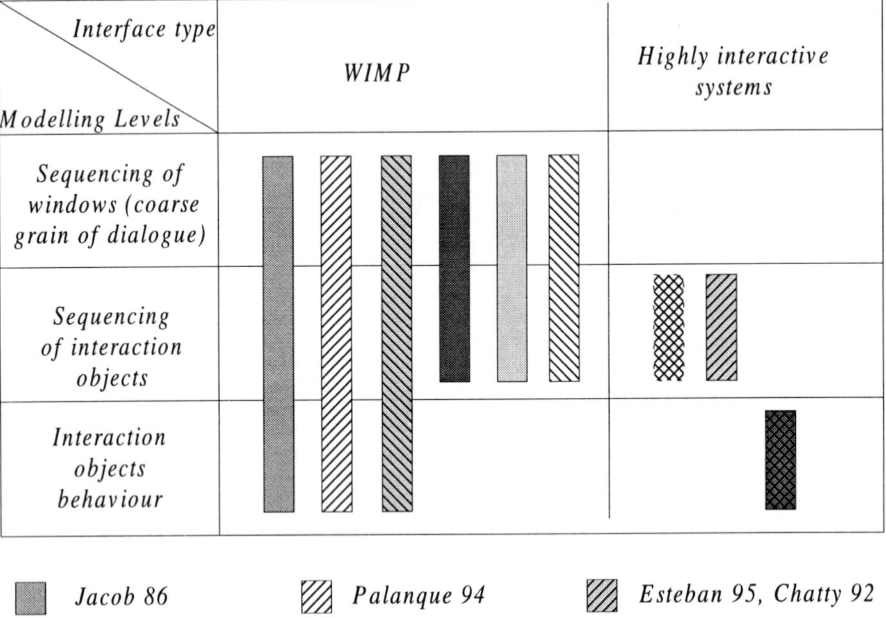

Fig. 6 Relative position of related work on low level behaviour modelling

The figure is organised as follows. Work located in the left column corresponds to work on indirect manipulation systems. Work located in the right column concerns highly interactive systems. The three rows correspond to the grain of dialogue taken into account. The lower row is related to the finest grain of dialogue (the behaviour of interaction objects themselves) while the upper one concerns higher grain of dialogue (such as the sequencing of local dialogues). The work presented in [25] ( ) is in dashed lines because it only concerns animation and not interaction in applications. The work presented in this paper is located in the lower right box of the Figure 6 as it concerns multimodal interactive systems with direct manipulation facilities.

7. Conclusion

We have shown in this paper how novel primitives for event description can be used as the basis of a formal model of low level interaction in direct manipulation interfaces. We have also shown how the same formalism, based on Petri nets, can be used to describe multimodal interaction in a way that allows the reuse of well known interaction styles to design new ones. This work has been integrated to the X_{TV} toolkit [1], and is currently used to experiment with the design and evaluation of two-handed interaction styles at the University of Toronto.

This works opens a whole set of new questions and perspectives. It suggests a more general event-based model of interactive applications, where physical events would successively be translated into more and more abstract events until they are caught by the functional core. In such a model, even the internal behaviour of widgets would be specified formally, thus allowing formal verification of the whole behaviour of an interface, and possibly performance evaluations within the GOMS methodology. This works also calls for its integration into a graphical direct manipulation interface editor such as Whizz'Ed [14]. We are currently working on methods that would help the construction of complex behaviours from basic ones in such an editor. We are also working on the merging of this model with the data-flow model of Whizz [6], so as to span as widely as possible the spectrum of user interfaces behaviours, from simple callbacks to continuous phenomena such as animation.

References

1. M. Beaudouin-Lafon, Y. Berteaud, S. Chatty. Creating direct manipulation interfaces with X_{TV}. EX'90. European conference on the X Window System. London 1990.
2. A. Beck, C. Janssen, A. Weisbecker, J. Ziegler. Integrating object-oriented analysis and graphical user interface design. In Coutaz J. & Taylor R. (Eds) LNCS Springer Verlag 1995.
3. E. Bier, M. Stone, K. Fishkin, W. Buxton, T. Baudel. A taxonomy of see-through tools. In proceedings of the CHI'94 conference. ACM Press, 1994, p. 358-364.
4. P. Brun & M. Beaudouin-Lafon, A taxonomy and evaluation of formalisms for the specification of interactive systems. HCI'95, 1995, p. 197-212.
5. W. Buxton. A three state model of graphical input. In proceedings of the Interact'90 conference, p.449-456, North Holland 1990.
6. S. Chatty. Defining the behaviour of animated interfaces. Engineering for Human Computer Interfaces conference 1992. p. 95-109. North-Holland.
7. S. Chatty. Extending a graphical toolkit for two-handed interaction. In ACM UIST'94, pages 195-204. ACM Press, 1994.
8. S. Chatty, P. Lecoanet. Pen computing for air traffic control. CHI'96 conference proccedings ACM Press, 1996, p. 87-94.
9. S. Chatty, P. Girrad, S. Sire. Vers un support multimedia aux collecticiels synchrones. Techniques et Sciences Informatiques, vol. 15, n°9, 1996, 28p.
10. A. Dix. Formal Methods for Interactive Systems. Academic Press, 1991.

11. Proceedings of the First Eurographics workshop on Design, Specification and Verification of Interactive Systems, F. Pernó Ed. Springer Verlag 1995.
12. Proceedings of the Second Eurographics workshop on Design, Specification and Verification of Interactive Systems, P. Palanque & R. Bastide Eds. Springer Verlag 1995.
13. D. Duke, M. Harrison. Abstract Interaction Objects. Computer Graphics Forum 12(3), p. 25-36 1993. Eurographics 93.
14. O. Esteban, S. Chatty, P. Palanque. Whizz'Ed: a visual environment for building highly interactive interfaces. Proceedings of the Interact'95 conference, p. 121-126.
15. F. Feldbrudge. Petri net tool overview 1992. Advances in Petri nets 1993. In G. Rozenberg (Ed.), Lecture Notes in Computer Science n° 674, p. 169-209. Springer Verlag 1993.
16. M. Harrison, H. Thimbleby. Formal Methods in Human Computer Interaction. Cambridge University Press, 1990.
17. K. Jensen. Coloured Petri nets and the invariant method. Theoretical Computer Science 14, 1981, North Holland, p. 317-336.
18. K. Jensen. Coloured Petri nets. Vol. 1 (Basic concepts) and Vol. 2 (Analysis methods and practical use) Springer Verlag, 1995.
19. B. Myers. Comprehensive support for graphical, highly interactive user interfaces. IEEE Computer, p. 71-85, Nov. 1990.
20. L. Nigay & J. Coutaz, A generic platform for addressing the multimodal challenge. Proceedings of CHI'95, 1995, p. 98-105.
21. P. Palanque & R. Bastide. Petri net based design of user-driven interfaces using the interactive cooperative object formalism. In [11], p. 383-401.
22. P. Palanque & R. Bastide. Formal specification and verification of CSCW using the Interactive Cooperative Object formalism. In HCI'95 conference, People and Computers X, p. 213-231. Cambridge University Press 1995.
23. P. Palanque & R. Bastide. Time modelling in Petri nets for the design of Interactive Systems. GIST workshop on Time in Interactive Systems. Glasgow, July 1995, and also SIGCHI bulletin vol 28 n°2, p. 43-46.
24. F. Paternó & A. Leonardi. A semantic-based approach for the design and implementation of interaction objects. Computer Graphics Forum 13(3) p. 195-204, 1994.
25. J. T. Stasko. TANGO: A framework and System for Algorithm Animation, PhD thesis, Brown University, 1989.
26. W. Van Biljon. Extending Petri nets for specifying man-machine dialogues. Int. J. Man-Machine Studies (1988) 28, p. 437-455.
27. A. Wasserman. Extending state transition diagrams for the specification of human-computer interaction. IEEE Transactions on Software Engineering, 11(8), August 1985.

Deriving a formal model of an interactive system from its UIL description in order to verify and to test its behaviour

Bruno d'AUSBOURG, Guy DURRIEU, Pierre ROCHE

Centre d'Etudes et de Recherches de Toulouse - ONERA
2 Avenue E. Belin - B.P. 4025 - 31055 TOULOUSE, France
email:{ausbourg,durrieu,roche}@cert.fr

Abstract This paper focuses on verifying and testing the interaction or dialogue between a user and an interactive system, especially in case of safety critical systems. In order to verify that the interface of a system behaves as intended by the user, we based our ongoing research on a compromise by allowing the use of informal (but practical) and formal methods. In fact, a formal description of the user's interests and activities through the Interface seems very hard to produce by common designers. A more realistic attitude consists in deriving a formal model from the description of the intended interface as it was informally designed. Practically, a tool generates models in the language Lustre from a user's UIL description and these models are used for verification or test purposes.

1 Introduction

User interfaces systems are becoming a quite important part in applications. Indeed, they ensure the well understanding of the internal state of the application by the user and they make easier communications between user and application. In consequnence, users are looking for a more efficient use of their environment[6].

In accordance with the increasing need of user's considerations in the Human Computer Interface (HCI) area, even from the specification step, an approach based on prototyping is often proposed against the usual technics of Software Engineering. A wide variety of tools have been developed in order to rapidly prototype graphical user interface systems. However, this approach leads to some problems related to misunderstanding, disability and non-fulfilment of requirements. Moreover, in some contexts, especially in safety or security critical applications, it may be helpful to verify that systems built with these tools behave as intended. In other words, it may be crucial to verify that the interactive behaviour of theses systems possess certain formally expressed properties. It is interesting also to get some test scenarii in order to verify that the final product satisfies the specified properties.

Using formal specification methods is generally suggested as an alternative. But, in this case, they are not necessarily suitable for the full description of the system behaviour. Indeed the designer or the user are not familiar with such formalisms and will have some difficulty in writing or in reading such a description. Moreover, the existing formalisms cannot really capture some characteristics of this behaviour, as the user's opportunism: practically, they restrain the user from having a natural behaviour by limiting the order of actions to a sequential order[13].

So we feel that a problem lies in the following paradox. In one hand, an experimental and empirical approach, based on prototyping, is able to fulfil the user's requirements, but is able also to produce unsatisfying interfaces. In the other hand, more rigorous formal methods are able to produce satisfying interfaces provided that the full requirements of the users and their natural behaviour are formally expressed: and in practice it is not really the case.

The basic idea of our approach consists in keeping the final user in the process of specifications of the interface. In particular, it seems important to respect the modes that the user uses to express his requirements. And *in practice*, the final user or the customer does not generally use formal techniques for specifications, but rather informal techniques as prototyping. This is clearly an empirical approach. But it meets the objective: associating the final user to the specification process, and taking account of his opportunist behaviour when defining a prototype model.

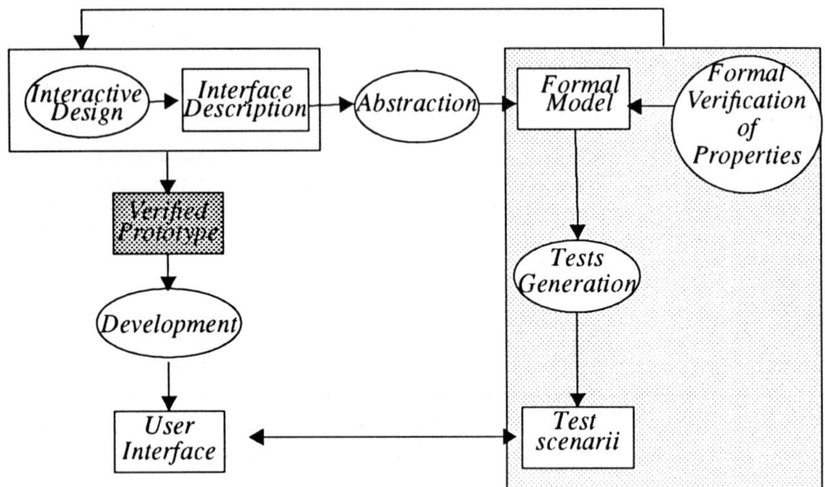

Fig. 1 The development and verification processes

We suggest to combine a formal validation step with this approach. The objective of our ongoing research project is to represent and to describe the interactive behaviour of a system by building an abstract formal model of it and to verify automatically that this behaviour possesses the required properties. This approach is quite close to the approach developed in [4]. In this case, the problem is not in specifying the behaviour of interactive systems in order to design them properly, but rather to model the behaviour of a prototype as it was generated by appropriate tools, as an interface generator for example.

A description of the user interface is generally produced as a result of prototyping tools. This description is expressed in a particular language (a standard one is UIL for instance). The idea is then to abstract a formal model from this UIL description and to use this model to perform formal verifications. Clearly, this verification step focuses on the Dialogue part of the Seeheim model [13]. Iterations between the process of interactive design and the abstraction process permit to converge on a verified prototype that fulfils most of the user requirements. The formal model is also used to generate some test scenarii that are used to verify that the resulting user interface satisfies the modelled requirements.

The paper is organized as follows. In section 2 we explain how an interactive system is modelled as a dataflow system and, in section 3, how it can be described using the language Lustre. Thus, in section 4, we show how to derive some Lustre code from an UIL description in order to build a model of a prototyped interactive system. In section 5, we describe how some properties of the aimed interface system are verified and how to get some relevant test cases.

2 Modelling Interactors with flows

The difficulty is in finding a model and a formalism on which automatic formal verifications can be processed. Several models ([14], [2]) exist, that use various formalisms. Our work is inspired by previous works at York ([9],[10]) and CNUCE([19]). We also use the notion of interactor. An Interactor abstracts some inputs (some user actions or some application signals) and produce some outputs: the interactor collects the input data of the Dialogue component of the system and computes activating signals for the application and a rendering state that is observed by the user.

Moreover, [1] makes a difference between events that can be viewed as discrete signals and status that can be viewed as continuous signals that last a certain time. Distinguishing them is important in order to build an expressive model. In [9], Duke and Harrison argued that the display observed by the user is in fact the translation of the internal state of the interactor in terms of rendering. This leads to devise interactors as made up by two components: the first is in charge of computing internal states and output signals addressed to the application while the second is in charge of computing renderings.

To sum up, an interactor receives input signals (event or status signals) and computes output (event or status) signals: some of these signals are rendering signals and reflect the internal state of the interactor. The Fig.2 depicts a structural model of interactor. It is important to describe also the behaviour of this structure of interactor. Indeed, the behaviour is typically what has to be verified.

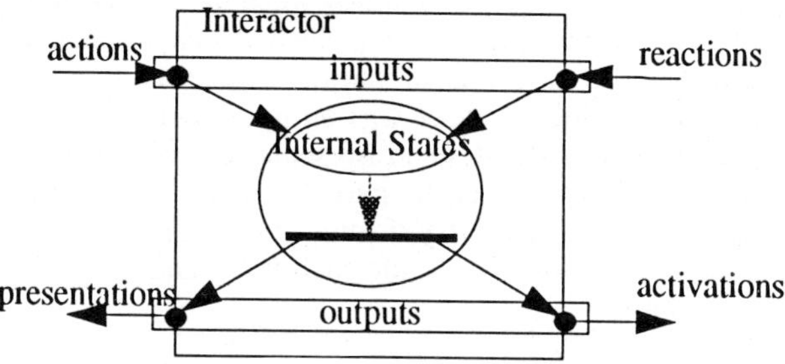

Fig. 2 Structure of an interactor

The idea is to consider a user interface system as a *dataflow* system. In accordance with this structural model, the system is viewed as a network of operators (interactors)[7][8] transforming flows of data. These flows are input data and interaction events from the user, signals from the application, and also output data. The behaviour of the network of interactors is captured by the mean of a system of equations in accordance with the dataflow paradigm ([12],[18]). The main advantages of this model are:

- it is a functional model, with its subsequent mathematical cleanness. This makes it well adapted to formal verification since functional relations over data flows may be seen as time invariant properties;
- it is a parallel model, where any sequencing and synchronization constraints arise from data and events dependencies. This is a nice feature which allows the description of inherently parallel processes as user interactions.

Informally, flows are sequences of values. Flow values can reflect both events, states and renderings. Then, they permit to describe classical interactors and to compose them in order to get more complex interactive systems. As an example, the Fig.3 describes the behaviour of a button interactor with boolean flows. This interactor has three input action signals from the user: *Push* and *Release* that indicate, when *true*, that the user has pushed or released the button of the mouse and *On* that indicates, when this flow has a value *true*, that the cursor is on the interactor. It uses two output signals: the *Activation* signal emits (when the flow has a value *true*) a signal to the application in order to activate a particular task, and *Revvideo* is a *presentation* flow that indicates, when true, that the rendering on display uses a reverse video mode.

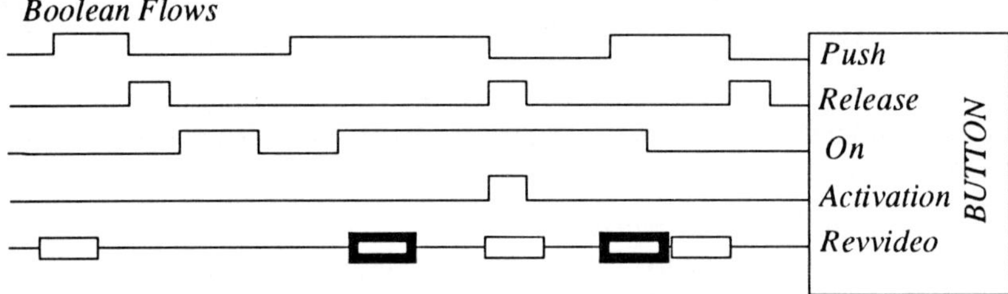

Fig. 3 Describing the behaviour of a button with flows.

The structure of flows shows how the interactor reacts when events occur. For instance, on Fig.3, an *activation* signal is emitted when the user is on the interactor and releases the button of the mouse (that is *not* pushed at this instant). The rendering of this interactor changes at the same instant and the display is in normal video after it had been in reverse video while the user was on the interactor and was pushing the button of the mouse. So, the order defined by sequences of values on flows gives a partial order on events, and therefore, the partial ordered structure of events can be examined and analysed in order to verify if interaction properties are satisfied or not. We use the formal language Lustre to denote such flows and to specify the behaviour of such interactive systems.

3 Modelling with Lustre

3.1 The language Lustre

We won't give here a detailed presentation of the language Lustre, which can be found elsewhere (in [5], [15] and [16] for example). Lustre is a declarative language devoted to the description and the programming of reactive synchronous systems. It was designed in order to program real-time systems and to describe circuit behaviour. We propose to take advantage of the formal nature of the language in order to describe (specify) and verify interactive systems. A Lustre program describes a functional relation between input and output variables. A variable is intended to be a function of time. And time is considered as isomorphic to the set of natural numbers. Relations between variables are defined by equations and assertions. An equation $X=E$, where E is a Lustre expression, specifies that the variable X is always equal to E, and an assertion *assert(E)*, where E is a boolean Lustre expression, specifies that the expression E is always *true*.

In Lustre, variables and expressions denote flows. A flow is a pair made of a sequence of values of a given type and a clock representing a sequence of instants. A flow takes the n-th value of its sequence of values at the n-th instant of its clock. Expressions are made of variables identifiers, constants, usual arithmetic, boolean and conditional operators, and only two specific temporal operators: the *previous* and *followed_by* operators which act on flows in the following way.

Pre (for previous) acts as a memory: if $(e_1,e_2,...,e_n,...)$ is the sequence of values of expression E, the sequence of values of *pre(E)* is $(nil,e_1,e_2,...,e_{n-1},...)$ where nil is an undefined value. *Pre* acts as a delay operator.

If E and F are expressions, with respective sequences $(e_1,e_2,...,e_n,...)$ and $(f_1,f_2,...,f_n,...)$ then E→F (for E followed by F) is an expression whose sequence of values is $(e_1,f_2,f_3,...,f_n,...)$. In other words, E→F is always equal to F, but except at the first instant, where it is initialised with the first value of E. The operator → is then an initialisation operator.

So, taking back the system depicted in Fig.3, we define the following expressions in Lustre denote some particular flows.

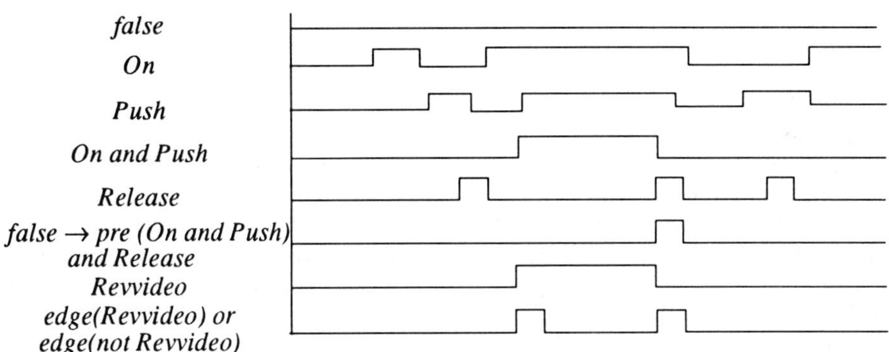

Fig. 4 Flows denoted by some Lustre expressions

The expression *On and Push* denotes a flow that is true when both flows *Push* and *On* are *true*. This expression can be used to denote the fact that the interactor is selected by the user. In this case, a particular flow can be defined as:

Selected = On and Push;

The flow *false → pre(on and Push) and Release* is *true* when *Release* is *true* and when the interactor was selected at the previous instant. This denotes the flow associated to the activation signal of the interactor:

Activation = false → pre Selected and Release;

Set of equations and of assertions are structured within *nodes* inside Lustre programs. A node can be viewed as a subprogram specifying a relation between input and output parameters (flows). This relation is expressed by the unordered set of parallel equations and assertions, possibly involving local flows (local declarations of variables). As an example, the following rising edge program accepts one boolean input flow *b* and returns one boolean output flow *edge* that is *true* at each rising edge of its input *b*:

```
node edge(b:bool) returns (edge:bool);
let edge = false → b and not pre b; tel
```

Another interesting temporal node can be built in the following way: this node takes two input signals, *begin* and *end*. The output of this node is a boolean flow that is *true* from the instant when the *begin* flow is *true* (and not the *end* flow) until to the instant when the *end* flow becomes *true*.

```
node from_to (begin,end:bool) returns (ft:bool);
let
  ft = begin and not end →
  if (pre ft or begin) and not end then true
  else if end then false else pre ft;
tel;
```

So, using these flows and these operators, a particular node *PushButton* can be written in Lustre: This node describes the behaviour of a *PushButton* interactor:

```
node PushButton (push,Release,On:bool)
   returns (Activation, Revvideo, Modif_Pres:bool);
  var
   Selected : bool;                          -- internal state flow
let
  Selected = Push and On;                    -- the interactor is selected
  Activation = false → pre Selected and Release;
  Revvideo = from_to(selected, not selected);  -- Presentation flows
  Modif_pres = edge(Revvideo) or edge(not Revvideo);
tel;
```

Fig. 5 PushButton interactor

It can be verified that the behaviour depicted in Fig.3 is compatible with the model of behaviour described by this node. This approach can be extended in order to model other interactors.

3.2 Modelling interactors with Lustre

We use the flows defined in Lustre to depict the occurrences of events and states. Events and states of an interactor are Lustre boolean variables. Each boolean variable is denoted with some additional subtypes. The first subtype depicts the kind of the variable: event or status. The second subtype makes the difference between global variables or flows (the flows *Push* and *Release* that are shared by all objects for example) and local variables or flows (the *on* status signal, that is in relation with the mouse pointing on an object, is local to each object). In fact, we use the syntax suggested in [1]: *in* for a discrete input, *in_status* for an input state, *global* or *local*. In the same way, we define *out* et *out_status* for the output. The internal states are named *states*:

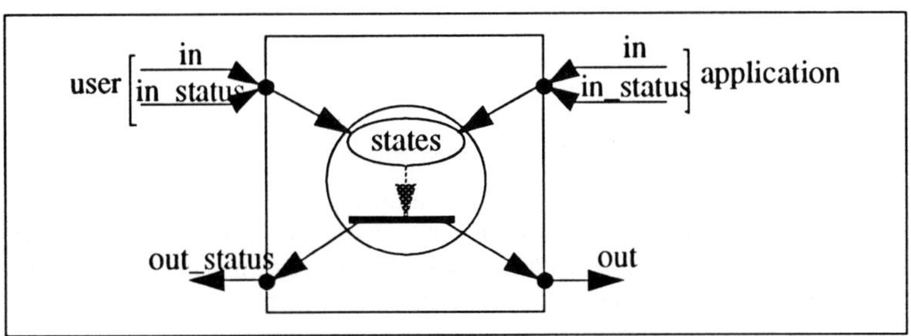

Fig. 6 A generic model of interactor

Thus, some variables represents internal states, renderings and activations that are built by using input flows. By applying the types and subtypes referenced above, we make up a general model of interactor including input, output and internal flows. However, Fig.7 shows that the output flows are local and specific to each interactor in the same way as the particular actions that characterize the dynamic behaviour of an interactor:

```
node gen_interactor(list_in_global,             ]  from the user
                    list_in_local,              ]  from the application
                    list_in_status_global,      ]
                    list_in_status_local)       ]
   returns (list_out_local,
            list_out_status_local );
  var list_internal_states;
let
   -- internal states
   -- output flows                              ]  to the user
tel;                                            ]  to the application
```

Fig. 7 A general interface of a Lustre interactor

4 Building Lustre interactors from UIL

We made the assumption that the user or the designer builds the interface by using an interface generator tool (in our case this tool is UIM/X). This tool generates an UIL description of the interface. The objective is to derive a formal model from this description in order to verify and to test the interface and its properties. Here, we explain the aimed automated translation.

4.1 Language UIL

All details exposed here can be found in [17]. UIL is a language that permits to describe the hierarchy of objects making up the interface in terms of a list of typed objects. Let an example on Fig.8.

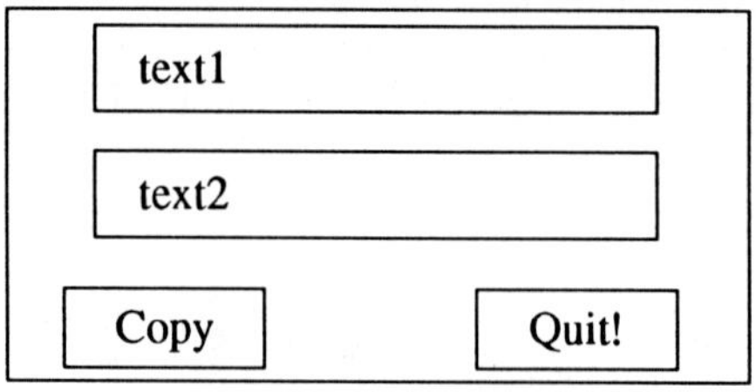

text1	
text2	
Copy	Quit!

Fig. 8 An example

It is composed of a list of basic objects such as buttons or textfields that are all included in a container object. Links between objects and dependencies are defined in the *controls* structure of the description. In Fig.9, the container is an *XmBulletinBoard* object: its name is *ApplicationCopy*. It includes two *XmPushButton* objects: *Quit* and *Copy*. It includes also two *XmTextField* objects: *Text1* and *Text2*.

```
object ApplicationCopy : XmBulletinBoard
{ arguments
      { XmNwidth = 280; ... };
  controls
      { XmPushButton Copy;
        XmPushButton Quit;
        XmTextField text1;
        XmTextField text2;
}};
```

Fig. 9 UIL code of the container object

In addition to this structure, two other structures may be included inside the description of each object: *arguments* section describes the initial presentation of the object (colour and fonts for example) and other attributes (resizing is allowed for example); the *callbacks* structure links some given events to some specific procedures that are described in an external C file. In Fig.10, the callback structure inside the *Copy* description shows that *activate_copy* is the procedure that is called when the button *Copy* is activated (an activation signal is emitted).

The C code that defines the *activate_copy* procedure performs the copy of the value contained in the *Text1* field into *the Text2* field. More generally, the C code of callbacks may contain some statements that modify the presentation of the interface. Thus, taking account of the C code when composing interactors is necessary to model exactly the behaviour of the interface.

```
object Copy : XmPushButton
{ controls
  { XmNactivateCallback =
      procedcure activate_copy ();
}};
```

```
static void activate_copy ()
{
  Widget w =
 XtNameToWidget(mainwindow,"text2");
  XmTextFieldSetString (w, string_text1);
```

Fig. 10 An example of a callback

4.2 Generic Lustre Interactors

Each UIL description uses basic or generic objects. The idea is to link each generic object of an UIL description to a Lustre node. This node abstracts the dynamic behaviour of the object and has the same structure as described in Fig.7. It can be viewed as a generic interactor. In order to build such a generic interactor, it is necessary to define:

- the exhaustive list of its input and output flows;
- the internal states and how they are functionally built using these flows;
- a relation between internal states and their rendering.

By *«exhaustive»* we mean that our goal is to take account of all signals and all states that participate in building the behaviour of the interactor. This can be done by reading the programmer's MOTIF guide and by listing all the events and procedures that are attached to a given typed object. These events and procedures are mapped to flows of the corresponding Lustre interactor model.

For instance, a *Textfield* has four implicit input signals: *push* et *release* events that bracket the click of a mouse button, the *on* status that indicates that the mouse is pointing the object and *char* that indicates the input of a character. These signals are global because they are the same for all the objects of the interface. The *XmTextFieldSetString* function can be called inside a callback procedure. This event is abstracted by the input signal *set*. The output signals may activate an application task and may produce a new presentation. So Fig.11 gives an abstract model of the node *TextField*. The input flows denote the four implicit input signals and the signal *set*. Its output flow denotes a change in the presentation of the object.

```
node TextField (push, release, char, on, set : bool) returns (modif_pres : bool);
    var active : bool;
let
   active = from_to (edge (push) and on, edge (push) and not on);
   modif_pres = ((active and char) or set) or edge (active) or edge (not active);
tel;
```

Fig. 11 The Lustre model of a TextField object

4.3 Translating and Composing Interactors

The definition of generic interactors is insufficient to achieve the full translation from an UIL description to Lustre models. Indeed, the global interaction is complex and results from combining some generic Lustre interactors into a composed node. Taking back the example, let us try to compose a *PushButton* node and a *TextField* node inside a *Compose* node.

In order to perform this composition of basic interactors, it is necessary to map the hierarchy defined by the UIL description into a hierarchy of the basic Lustre nodes. In fact, a node is associated to each object of the UIL description. When the object is a basic object, the mapped node is generic and corresponds to the predefined Lustre node.

When the object is not a basic object, its *controls* structure includes other objects. The input and output flows of this object combine the input and the output flows of the included objects by making connections between them. Indeed, some output flows of one of these included objects can be linked to the input flows of some other included objects.

In order to build these links automatically, a diagram of implicit dependencies can be used. It may contain three kinds of dependency links. A first kind of link is built between an output Lustre signal, that corresponds to an event that is handled by a callback in the UIL description, and the C procedure that is called. Taking back the same example, the *activation* signal of the *PushButton* node in the Lustre model is linked to the *activate_copy* procedure because this procedure is the callback procedure that is associated in the UIL description to the activation of the *PushButton* interactor.

The second kind of link is built between a callback procedure and a statement *stat* that is executed inside the C code attached to this callback and that modifies the presentation. In the example, the code of *activate_copy* performs a *XmTextFieldSetString* statement that changes the displayed content of the text field.

The third kind of relation is between a C function or procedure that changes the display of an object and the input Lustre signal that is devoted to reflect this change requirement. In the example, the execution of the *XmTextFieldSetString* function results in setting (*set*) the *Text2* field. By drawing these relations on the diagram of the Fig.12, we deduce a new induced relation, *activation* → *set*. So, in the Lustre model, the *activation* output signal of the *Copy* model is linked to the *set* input signal of the *Text2* model.

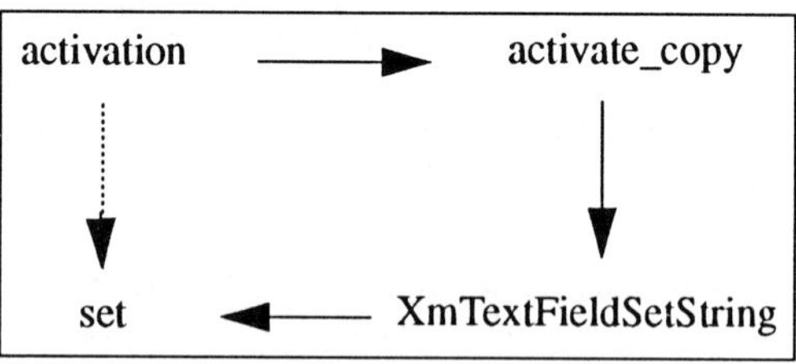

Fig. 12 Induced dependencies

Thus, the *Compose* object is built firstly by instantiating the nodes that model the objects under its control. These nodes are the *PushButton* node that is related to the button *Copy* and the *TextField* node that is related to the textfield *Text2*. We just mapped the *activation* flow to the *set* flow because they are connected and simultaneous, as previously shown.

However, we emphasize a technical aspect: the generic Lustre nodes that model the *Copy* PushButton and the *Text2* TextField have a local generic input flow, *on*, which is not the same on each object. In order to be sure that they are not the same effectively, this signal is suffixed by the node that uses it. All local flows are distinguished by using the same mean.

Technically, the translation of an UIL description (list of objects) consists in building a corresponding hierarchical tree in Lustre. The Fig.13 depicts such a tree. The root node has two children nodes: *Copy* and *Text2*. Data concerning input or output flows and dependencies are added to each node of the tree when it is built.

Leaves are tackled in first. They correspond to the generic basic interactors. These interactors contain some predefined data that are immediately attached to the leave in the tree: input and output flows as they are defined in the Lustre generic interactor and the implicit dependencies.

Then, these data are included in the parent node in the following way. The input and output flows of the parent node are the set union of the input and output flows of its children, and its dependencies arc given by the transitive closure of the dependencies contained in its children.

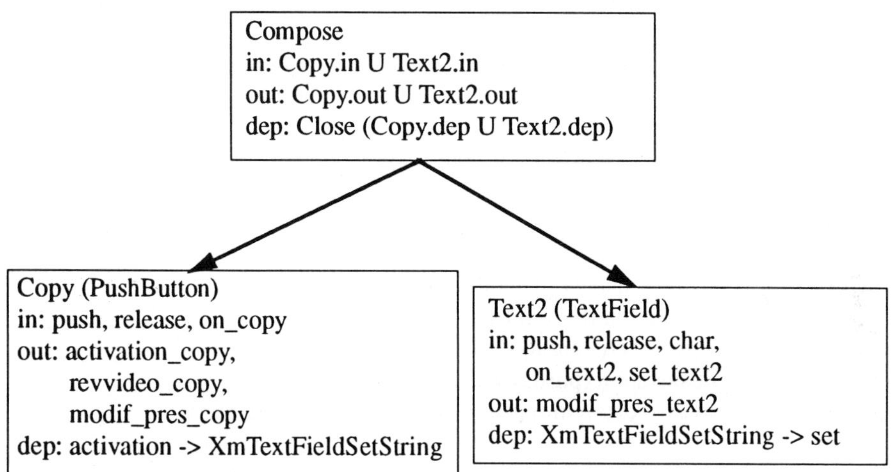

Fig. 13 Defining a parent node in the hierarchical tree

The final process translates each node of the tree into a Lustre node: leaves are connected to a generic Lustre nodes and the other nodes of the tree are defined as composite nodes including the nodes that model the objects under control. The composition illustrated in Fig.13 leads to the following Lustre node:

```
node Compose (push, release, char, on_copy, on_text2 : bool)
   returns (activation_copy, revvideo_copy,
   modif_pres_copy,modif_pres_text2: bool);
let
(activation_copy,revvideo_copy,modif_pres_copy) =
               PushButton(push,release,on_copy);
modif_pres_text2 = TextField (push,release,on_text2, activation_copy);
tel;
```

Fig. 14 Result of composing Text2 and Copy

We restricted the explanation to the composition of two objects of the description. The complete translation requires the nodes that model the *PushButton Quit*, the *TextField Text1*, and the mapping of the connections of flows between these nodes. The *Text1* interactor reacts when a char is input. Clicking on the button *Quit* generates the signal *activation_quit*. This signal stops the application and then deactivates all the interactors. So, this signal is viewed as a new input signal in each model node.

In fact, it suggests an enlarged set of input flows for each generic interactors in order to take account of this common event. So, by adding a signal *exit* in each generic node, a new object *Compose* is obtained.

But the node modelling the *PushButton Quit* includes *exit* in its input signals, which is in fact its output signal *activation_quit*. To avoid a cycle in the calculus of *activation_quit* and *exit*, we have to add an equation that connects the n-th value of *exit* to the *pre* value of *activation_quit* :

```
node Compose (push, release, char, on_copy, on_text2 : bool)
   returns (activation_copy, revvideo_copy, modif_pres_copy,
            activation_quit, revvideo_quit,modif_pres_quit,
            modif_pres_text1, modif_pres_text2: bool);
  var
   exit : bool;
let
(activation_copy,revvideo_copy,modif_pres_copy) =
               PushButton (push,release,on_copy, exit);
(activation_quit,revvideo_quit,modif_pres_quit )=
               PushButton (push,release,on_quit, exit);
modif_pres_text1 = TextField (push,release,on_text1, false, exit);
modif_pres_text2 = TextField (push,release,on_text2, activation_copy, exit);
exit = false → pre action_quit;
tel;
```

Fig. 15 Full Node modelling the example

5 Verifying properties on the model and using the model to generate test cases

5.1 Verification

The declarative aspect of the language Lustre may be used in order to express and verify properties of programs ([16]). Moreover, Lustre can be considered as a subset of temporal logic ([22]). The principle is to translate such a property P into a boolean expression E_p so that P is satisfied if and only if E_p denotes a flow that is always *true*. Expressing safety properties by the invariance of a boolean Lustre expression is easier if general temporal operators are available. We defined such a temporal operator in 3.1 :*from_to*. Some properties to verify can be found in [7] for example.

In order to prove that an expression E is an invariant of the program P, we build a new program P' made of the body of P and of the system of equations defining E, and whose only output is E. It is necessary then to prove that this output E is always *true*.The Lustre compiler performs an automaton generation from a Lustre program. Roughly speaking, the verification consists in traversing the produced automaton and checking that the output E is never assigned to *false*. And this is performed by the verification tool *Lesar*. Let us take the example of the *PushButton* interactor defined in

3.1 :

```
node PushButton (push,Release,On:bool)
  returns (Activation, Revvideo, Modif_Pres:bool);
 var
  Selected : bool;                              -- internal state flow
let
 Selected = Push and On;;                       -- internal state of selection
 Activation = false → pre Selected and Release;
 Revvideo = from_to(selected, not selected);    -- Presentation flows
 Modif_pres = edge(Revvideo) or edge(not Revvideo);
tel;
```

Let us try to verify that the behaviour of this interactor has a reactive behaviour. Informally, by reactive, we mean that a change of the internal state of the interactor implies a change of its rendering.

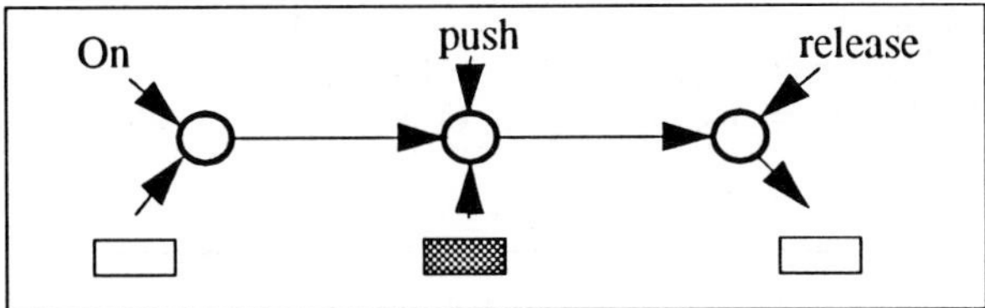

Fig. 16 A reactive behaviour of a button

On the example of Fig.16, the rendering of the button changes when the button becomes selected: the mouse is *on* it and the mouse button is *pushed*. When the mouse button is released, the interactor is deselected and its rendering changes also. This is a reactive behaviour.

Assume now that the cursor is moved onto the button, as depicted by Fig.17. The user pushes the button and moves the cursor outside the area attached to the button without noticing a modification of the presentation. The internal state changed because the button was deselected, but its rendering did not. Clearly, this is not a reactive behaviour.

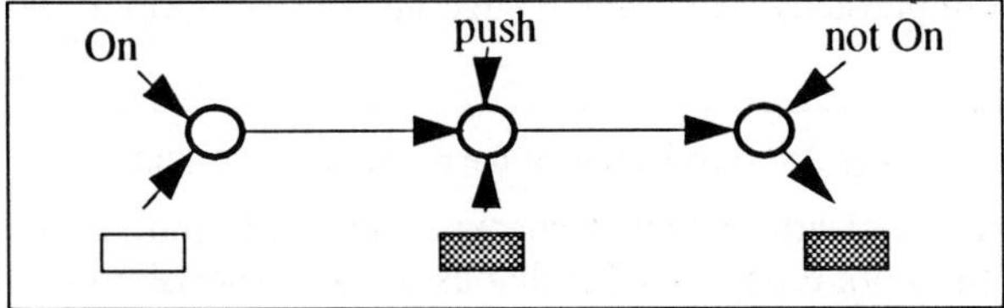

Fig. 17 A not reactive behaviour of the button

More formally, the flow *Selected* in the node *PushButton* denotes the internal state of the interactor. A change of this state is denoted by a change of the value of this flow. The Lustre expression *edge(Selected) or edge(not Selected)* denotes a flow that reflects such a change. The flow *Modif_pres* denotes a change of its rendering. So, the Lustre expression

implies(edge(Selected) or edge(not Selected),Modif_Pres)

expresses the required property. The operator *implies* is the classical logical implication and is defined by the following Lustre node:

```
node implies(a,b:bool) returns (imp:bool);
let imp = not a or b; tel
```

Then, a new verification node is built. Its name is *Verify_Reactive_PushButton*. It has the same input flows as the node *PushButton* and only an output flow *Reactive*. Its internal flows are the output flows of the *PushButton* node that is verified. The flows that were associated to the internal flows of *PushButton* are converted in output flows of this node and are included in the internal flows of the verification node.

```
node PushButton (push,Release,On:bool)
   returns (Activation, Revvideo, Modif_Pres,Selected:bool);
let
  Selected = Push and On;
  Activation = false → pre Selected and Release;
  Revvideo = from_to(selected, not selected);
  Modif_pres = edge(Revvideo) or edge(not Revvideo);
tel;
```

The relevant flows are then connected to the Lustre expression of the verified property:

```
node Verify_Reactive_Pushbutton(push,Release,On:bool)
   returns (Reactive:bool);
  var
   Activation, Revvideo, Modif_Pres : bool;
   Selected : bool;
let l
  (Activation, Revvideo, Modif_Pres,Selected) = PushButton(push,Release,On);
  Reactive = implies(edge(Selected or edge(not Selected),Modif_Pres);
tel;
```

Fig. 18 Verification of the reactivity of the button

This approach is clearly a model checking approach and can be compared, for example, with the work reported in [3] that uses CTL expressions to define properties that need to be verified on a finite state representation of dialogues. It can also be compared with the work of Paterno et alii that are reported in [19], [20] and [21]. The main difference with these previous similar works is that we use the same language Lustre both for describing the interactive system and for expressing properties. The language permits to build temporal operators and to express properties by using these operators on flows. As an example, let us define a node *Once_from_to*, taking three boolean input flows *A*, *B* and *C*. This node returns one boolean output flow *Oft* such that *Oft* is *true* if and only if *A* is *true* at least once in any time interval starting when *B* is *true* and ending when *C* is *true*.

```
node once_from_to(A,B,C:bool) returns (Oft:bool);
let Oft = implies(C, once_since(A,B)) ; tel
```

The equation defining *Oft* uses two auxiliary nodes. The node *implies* and the node *once_since*. This node has two input flows and it returns *true* if and only if, either its second input has still not been *true*, or its first input has been *true* at least once since the last *true* value of the second input :

```
node once_since(A,B:bool) returns (os:bool);
let os = if B then A else (true → (A or pre os)); tel
```

Using these operators, an expression of the Reactive property to verify on the Pushbutton interactor could be described by:

Reactive = once_from_to(Modif_Pres, Selected,not Selected) and
once_from_to(Modif_Pres, not Selected,Selected);

We use generic properties that were defined on our generic model of interactor. In fact, the expression of these properties is a Lustre expression on flows of this generic model. We built some expressions by using definitions found in [20] or that are similar to definitions given in [3] and in [23]. Verifying properties consists of instanciating their expression on Lustre nodes and of building verification nodes: from an engineering point of view, this is done by performing syntactical transformations on Lustre programs.

An other main advantage in using Lustre both to model the system and to express properties is in getting such verification nodes. These are programs that are compiled and translated in an automaton representation. This automaton uses only flows that are involved in evaluating the property to verify.

So, the model checking process does not operate over the whole representation of the system but only on the small automaton that was obtained. This allows to avoid state explosions and permits to perform verification on large models. We plan to develop also a graphical property editor that permits the user to define his own particular properties by combining temporal operators as defined in Lustre and events and states that are associated with UIL interactors.

5.2 Getting test cases

Verifying a property is done by checking that a particular Lustre expression is always *true*. The expression denotes some configurations of flows. For example, the Reactive expression has the structure *Reactive=* $a \Rightarrow b$ where *a* denotes *edge(Selected) or edge(not Selected)* and *b* denotes *Modif_pres*. The fact that *Reactive* is a property of the *PushButton* node means that the only possible configurations of the *PushButton* node are: $a \wedge b$, $\neg a \wedge b$, or $\neg a \wedge \neg b$.

A possible way to generate some tests cases that are related to this property is to ask the verification tool Lesar to check that each one of these configurations is not reachable. For example let us take the configuration $a \wedge b$. The following test node is submitted to the verification tool Lesar to verify that the configuration $\neg(a \wedge b)$ is not reachable.

```
node Test_Reactive_Pushbutton(push,Release,On:bool) returns (Test:bool);
  var
    Activation, Revvideo, Modif_Pres : bool;
    Selected : bool;
let
  (Activation, Revvideo, Modif_Pres,Selected) = PushButton(push,Release,On);
  Test = not ((edge(Selected or edge(not Selected)) and Modif_Pres);
tel;
```

Fig. 19 A node generating test cases

In fact, the result of the verification is obviously *false*. But the control automaton generated during the verification process is very useful to exhibit some sequences of signals that allow to reach the configuration a ∧ b. These sequences are possible test cases of the Reactive property.

The automaton generated for the example of Fig.19 is the following where *P* and *O* denotes a value *true* for the *Push* and *On* flows. The expressions inside squares denote event configurations that make false the ¬(a ∧ b) formula. In other words they show how the tested configuration (a ∧ b) can be reached.

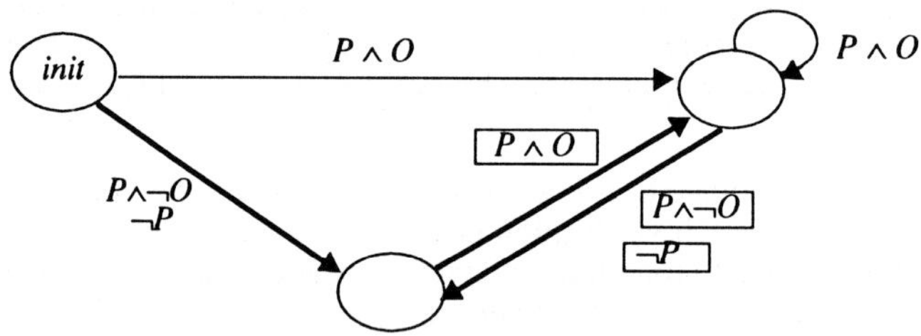

Fig. 20 Test automaton

Then, on Fig.20, at least four sequences of signals can be used to test the reactive property of the *PushButton* interactor. These sequences are built by following the bold arrows along the automaton:

((Push)* (¬Push)*)* (Push ∧ On)
((Push ∨ ¬Push) ∧ ¬ On)* (On and ¬Push)+ (Push ∧ On)
((Push)* (¬Push)*)* (Push ∧ On) ¬Push
((Push)* (¬Push)*)* (Push ∧ On) (Push and ¬On)

The first two sequences test the fact that when the button becomes selected, its rendering must change. The two last sequences permit to test the fact that when the button becomes unselected, the rendering changes also. Other sequences can be built. And the process can be applied on other configurations in order to get a set of sequences that will permit to test that a final button interactor satisfies the reactive property. This process can also be extended to more complex interactors as the *Compose* interactor for example. But the idea remains the same.

6 Conclusion

In this paper, we argued that describing the behaviour of interactive systems by using flows is a possible way for building formal models of these systems. The language Lustre permits to denote such flows and therefore to describe the behaviour of such interactive systems. This behaviour is described by a set of equations and assertions. The main advantage of the language is that properties can be expressed by using the same formalism. The verification tool Lesar performs the specified verifications. An interesting side effect of the verification process can be used for building tests.

We showed also how translating automatically an UIL description into Lustre; this translation is based on generic interactors and on their composition. In fact, it permits to include a formal verification process inside an informal development process rather based on prototyping. This formal added step is interesting in order to build a prototype that behaves exactly as intended. This step allows also to produce some test cases that are related to required properties. This will permit to evaluate whether a final product implements well these intended properties.

Our ongoing work consists in achieving the development of tools that support this approach: a translator of UIL descriptions into Lustre models; the automated verification of properties; a property editor that helps in building in expression of some particular properties; a tool for exhibiting some relevant test cases.

Acknowledgements

This work was sponsored by the French National Center for Studies in Telecommunications (CNET) whose support is gratefully acknowledged. We acknowledge also the referees for their comments on draft versions of this paper.

References

1. G.D. Abowd, A. J. Dix. Integrating status and events phenomena in formal specifications of interactive systems. *SIGSOFT'94*, December 1994
2. G.D Abowd. Formal Aspects of Human Computer Interaction. PhD thesis, University of Oxford Computer Laboratory: Programming Research Group, 1991.
3. G.D. Abowd, H.M. Wang and A.F. Monk. A formal technique for automated dialogue development, *in Proceedings of the First Symposium on Designing Interactive Systems,DIS'95*,Ann Arbor,MI, August 1995, ACM Press.
4. P. Bumbulis, P.S.C. Alencar, D.D. Cowan, C.J.P. Lucena, "Combining Formal Techniques and Prototyping in User Interfaces Construction and Verification" *in Proceedings of the Eurographics Workshop DSVS-IS '95*, Toulouse, France, June 95, Springer Computer Science, P. Palanque and R. Bastide eds,
5. P.Caspi, D.Pilaud, N.Halbwachs and J.Plaice. Lustre: a declarative language for programming synchronous systems. *In 14th ACM Symposium on Principles of Programming Languages*, January 1987
6. J. Coutaz "Interfaces Homme-Machine: un regard critique " TSI vol 10(1), 1991
7. D.J. Duke, M.D. Harrison "Event model of a human system interaction" in Software Engineering Journal, January 95
8. G. Faconti, F. Paterno "An approach to the formal specification of the components of an interaction" in Eurographics 90

9. D.J.Duke and M.D. Harrison. Abstract Interaction Objects. *Computer Graphics Forum*, 12(3):25-26, 1993
10. D.J. Duke and M.D. Harrison. Event Model of human system interaction. In *Software Engineering Journal*, Janauary 1995.
11. D.J. Duke, G.Faconti, M.D. Harrison, and F. Paterno. Unifying views of interactors. In *Proceedings of Advance Visual Interface'94 International workshop*, Bari, 1994
12. J.R. Mc Graw. "The VAL Language: description and analysis". *TOPLAS, 4(1)*, January 1982
13. M. Green. "A survey of Three Dialogue Models", *ACM TRansactions on Graphics* vol 5(3), July 1986
14. M.D. Harrison, D.J. Duke A review of Formalisms for describing Interactive Behaviour. In Proceedings of the ICSE'94 Worksho, R.N.Taylor and J.Coutaz, editors, Software Engineering and Human Computer Interaction; LNCS 896; May 1994
15. N. Halbwachs, P. Caspi, P. Raymond, D.Pilaud. The synchronous dataflow programming language LUSTRE. *Proceedings of the IEEE 79(9)*:1305-1320, September 1991.
16. N. Halbwachs, D. Pilaud, F. Ouabdesselam and A.C. Glory. Specifying, programming and verifying real time systems, using a synchronous declarative language. In *Workshop on automatic verification methods for finite states systems, LNCS 407*, Springer Verlag, June 1989.
17. D. Heller, P.Ferguson and D.Brennan. *Motif Programming Manual*, 2nd edition, February 1994.
18. G.Kahn. "The semantics of a simple language for parallel programming". *In IFIP 74*, North Holland, 1974
19. F.Paterno and G.Faconti. On the use of Lotos to Describe Graphical Interaction. In A. Monk, D.Diaper, and M.D. Harrison, editors, People and Computers VII: HCI'92 Conference, pages 155-174. BCS HCI Specialist Group, Cambridge University Press, 1992.
20. F.Paterno, Definition of Properties of User Interfaces Using Action-Based Temporal Logic, *in Proceedings of the 5'th International Conference on Software Engineering*, pp 314-319, San Francisco, June 1993
21. F.Paterno and M.Mezzanotte, Formal Verification of undesired behaviours in the CERD case study, *in Proceedings EHCI'95 Conference*, Wyoming, August 1995.
22. D.Pilaud and N.Halbwachs. "From a synchronous declarative language to a temporal logic dealing with multiform time". *Formal Techniques in Real-Time amd Fault tolerant Systems, LNCS 331*, Springer Verlag, September 1988
23. B. Sufrin and J. He, Specification, Analysis and refinement of interactive processes, in M. Harrison and H. Thimbley editors, Formal Methods in Human Computer Interaction, Cmabridge Univeristy Press, 1990.

Prototyping Device Interfaces with DSN/2

Gerd Szwillus, Klaus Kespohl

Universität - GH Paderborn, Fachbereich Mathematik/Informatik (17)
Fürstenallee 11, D-33102 Paderborn, Germany
Tel. +49 5251 60-6624, szwillus@uni-paderborn.de

Abstract: The specification language DSN/2 and the corresponding tool KAP are presented in this paper. They are designed for prototyping user interfaces of electronical technical devices, such as answering machines, CD players, or VCRs. The language DSN/2 is used to specify the control model of the user interface, including sound effects and timing conditions. The tool KAP adds a visual component to this abstract specification and allows editing, debugging, and end user testing with the defined model. Using this tool set, the designer of a user interface is enabled to create a realistic looking, sounding, and "feeling" model of the system under construction without the need of programming. In practical experience, the system was found suitable for performing user tests on the software model of a CD player. The results were verified against user testing on the real device.

Keywords

Prototyping, formal specification, design techniques, development tools, usability testing, evaluation

Introduction

Electronic devices known as "end-user programmable devices" today are more widespread than ever. Electronic clocks, telephones, answering machines, calculators, VCRs or CD players, come with sophisticated functionality brought to the human user through computer-based device interfaces. We all know that it can be very hard to control these devices, and that a considerable percentage of their owners is not able to exploit all functions, giving up at the sight of numerous buttons and a voluminous user's manual.

In the light of these problems, one could assume that sound and practical techniques for evaluating this type of interface exist. Within the HCI community these interfaces always had their share, especially the typical case of VCR programming [9]. Prototypes used for these studies, however, are mostly created with systems such as HyperCard™ or SuperCard™ [8], which were created for more general purposes. Using these systems, the functionality of the device under consideration can not be modelled with an explicit and sufficiently abstract specification; rather is it described through an ad-hoc implementation in an interpreted programming language, with the code scattered over a collection of event handlers.

We propose a technique for specifying the (internal) behaviour of a device with the abstract specification language DSN/2, and the (external) user interface appearance with an appropriate "decoration" of such a specification. DSN/2 and its tool KAP can be used to prototype device interfaces without programming effort involved. A model of the underlying application's functionality is created, design variants for the user interface are added and the model is executed with an animator tool.

Specifying with DSN/2

The application model is based on the concept of states and transitions between them; it is written in a notation called DSN/2, which is strongly based on the *Dialogue Specification Notation (DSN)* [4], which in turn has its roots in the *Propositional Production System (PPS)* [11]. We chose the name DSN/2, because we think that our modifications and additions to the original version of DSN are significant giving a new quality to DSN specifications and turning it into a language powerful enough for application to real-world problems - still it "looks and feels" DSN.

Basically, the DSN/2 specification describes the control model of the system's functionality. This covers a large part of the user's perception of the system; hence it is very desirable that a HCI expert - and not a programmer - is enabled to specify this model. DSN/2 and the possibility to test DSN/2 specifications on this abstract level can easily be mastered and requires no prior knowledge of programming concepts and techniques.

Prototyping with KAP

Apart from animating the text of DSN/2 specifications, the tool KAP enables the user interface designer to link a graphical outfit to the application model. The input events and states defined there can be mapped onto various simulations of typical input and output elements of devices, such as push buttons, sliders, keys, lights, or small displays. Decorating or guiding elements, such as key inscriptions, grouping of lights, or photographs of real devices can be added to let the simulated prototype resemble the "real thing". Currently, KAP maps all input events to mouse operations, rendering the model vulnerable to the problems of mouse usage, but we plan to use touch-screen technology to eliminate this artifact.

Prototyping device interfaces is much simpler than prototyping user interfaces of computer software as they are less dynamic. The panel of a device is fixed, as well as the number and positions of buttons and other input/output devices. DSN/2 in its current form, together with the KAP tool, meets the needs for this type of static device interfaces. Underlying state transitions as well as optical appearance and sound feedback are dealt with, enabling the designer to create user interface variants and to evaluate them through user testing, or heuristic evaluation, on the basis of a running prototype, created within hours.

In the following, we give a short introduction to DSN/2; then we describe the KAP animator tool and how a running prototype is created. After that, we report some of our experiences of user testing with the prototype, and give an outlook on what we think feasible with the approach.

1. Behaviour Modelling with DSN/2

The externally experienced behaviour of a device can roughly be described as a sequence of "situations", with a situation being given in terms of the visible appearance, the sounds produced, and the provided input possibilities[1]. The sequence of the situations is governed by rules based on the actions performed by the user and the device. Expressed in this terminology, the most severe problems human users experience stem from controlling the situation sequences, rather than from understanding the single functionalities of a device. For example, in using a CD player, one might be quite sure about what it means to "play track #7", "skip the running track", or even "play the tracks in an individually selected order" (i.e. "programming" the CD player). But it remains obscure to a large group of users which sequence of actions leads to the desired result, for example which keys the user has to push in what order to "add track #5 to the current program".

The following two effects describe frustrating, common experiences, typical for this type of "lost-in-state-space"-problem:

- A button is pushed and "nothing" happens;
- a button is pushed and something different from what the user expected happens.

In the first case, the user wonders why nothing happens, as the button exists and hence must have a meaning. The system, however, is not in a state to react to this button push - the user has a misconception about the system's state. In the second case, the button push does have an effect, but not the one the user anticipated. Again, the user's knowlegde about the system state is incorrect. In computer science terms the user has to "understand" a potentially large, finite automaton with attached functionality. The user of a device must build an internal automaton model matching the behaviour of the device.

---

[1]) Of course a device can activate other human senses: it can start to produce an ugly smell, to shake violently, or get - literally - hot. For obvious reasons, we abstracted from these effects.

Even worse, the user in fact has to understand a collection of automata simultaneously; typically a device is built up from modular subsystems containing separate state spaces and transitions. In a CD player, for instance, there exists one state machine controlling the hearable running of tracks and another one for the currently stored program. As interface space is sparse, there are usually mechanisms for switching between these (and other) subsystems (e.g. the "PROGRAM" key) and interpreting the same keys differently dependent on the currently active subsystem (e.g. the number keys). Mathematically, the global state space of such a device is a crossproduct of subspaces, with local state machines controlling only some of the components, containing transitions within and between these subsystems.

Of course, it is the task of a good design to visualize states to the user and construct the interface as "fault-tolerant" as possible, but this is not an easy task. We seek to support the designer in supplying a testbed to enable testing and evaluation of the design in this respect. The definition technique for describing the control model with DSN/2 is inherited from its predecessors PPS [11] and DSN [4, 5].

1.1 The Control Model

1.1.1 State Space

The state space specified within a DSN/2 specification is defined as a crossproduct of subspaces; the elements of the subspaces are explicitly enumerated (as so-called flags, labelled with #) in field specifications. A DSN/2 specification might, for instance, contain the following fields describing the system states

```
SYSTEM STATES
          Mode          (#Off   #On)
          Action        (#A   #B   #C)
          Function      (#X   #Y   #None),
```

resulting in a global state space containing all 18 triples made up from the three fields, i.e.

```
(#Off, #A, #X)    (#Off, #A, #Y) (#Off, #A, #None)
(#Off, #B, #X)    (#Off, #B, #Y) (#Off, #B, #None)
(#Off, #C, #X)    (#Off, #C, #Y) (#Off, #C, #None)
(#On, #A, #X)     (#On, #A, #Y)  (#On, #A, #None)
(#On, #B, #X)     (#On, #B, #Y)  (#On, #B, #None)
(#On, #C, #X)     (#On, #C, #Y)  (#On, #C, #None).
```

With this interpretation of fields, at any time exactly one flag of every field is valid - the flags are mutually exclusive. The initial configuration of the machine in terms of valid flags, one from each field, when the animation starts, is defined in a start vector, e.g.

```
START  (#Off   #A   #None).
```

The control model is triggered by events. There are two classes of events, user input events and system events. They define possible actions the user can perform or the system can suffer. For example:

USER INPUT EVENTS
OnOffSwitch (iPushOnOff)
ActionSelect (iA iB iC)
FunctionXButton (iXOn iXOff)
FunctionYButton (iYOn iYOff)
SYSTEM FEEDBACK EVENTS
Failure (sXFails sYFails).

1.1.2 State Transitions

To model the behaviour of the system, we specify transitions between the system states defined above. This is done with transformation rules of the form

PreConditions Event --> PostConditions.

Such a rule stands for the behaviour "Given that the PreConditions hold and the event Event takes place, modify the state to ensure the PostConditions to hold." Other than in simple transition networks, a single DSN/2 transformation rule can represent a potentially large set of state transitions. The basic idea is that only the flags relevant for a rule are written down; all elements not mentioned are assumed to be unimportant for this transition. In our example, the OnOffSwitch is implemented as a toggle between the system being switched on or off. Hence, the rules

1. #Off iPushOnOff --> #On
2. #On iPushOnOff --> #Off,

capture exactly the toggle behaviour and nothing else. Rule 1. defines 9 state transitions containing all possible combinations from the subspaces Function and Action:

(#Off,#A,#X) — iPushOnOff —› (#On,#A,#X)
...
(#Off,#C,#None) — iPushOnOff —› (#On,#C,#None)

Similarly, we define the selection of Action #A through the ActionSelect button as

3b. #B iA --> #A
3c. #C iA --> #A.

As only one flag of a field can be valid at a time, in the case of rule 3b., for instance, #B is switched off, when #A becomes valid.

A set of DSN/2 rules, together with specifications of user input events, system feedback events, and an initiating start vector describe the kernel of a system's control model. This corresponds to the original DSN specification technique, presented in an early version [5] of [4], which in turn is based on PPS [11].

In a user interface class we tried to apply DSN to real-world problems (in this case an answering machine) and met big problems. Although the underlying ideas of PPS/DSN are clear and appropriate for the task, the language DSN has severe problems with expressiveness. Step by step we enhanced the language and ended up with DSN/2 in its current version, which we tried out, finding that it is feasable and easy to model realistic device models. In the following we will describe our enhancements of DSN, leading to DSN/2.

1.2 From DSN to DSN/2

1.2.1 Field and Rule Expansion

When writing transformation rules one often encounters regular structures, basically because a set of flags has to be treated similarly. Hence, we included some list based specification features to allow a compact notation. The basic building block is the parameter definition, which associates a name of a list, say ActionType, to its elements, say A B C, as in

```
PARAMETER
        ActionType     ( A   B   C ).
```

Now we can use this list name to create expanded field and rule definitions. For example,

```
SYSTEM STATES
        Action          ( #Action(ActionType) )
USER INPUT EVENTS
        ActionSelect    ( iSelect(ActionType) )
```

defines the field Action with the flags #Action/A, #Action/B, and #Action/C, and the user input event field ActionSelect with the events iSelect/A, iSelect/B, and iSelect/C.

Using the parameter lists in rules leads to rule expansion. For instance, the DSN/2 rule

```
  #On   iSelect(ActionType)  -->  #Action(ActionType)
```

is expanded to the three simple rules

```
  #On   iSelect/A  -->  #Action/A
  #On   iSelect/B  -->  #Action/B
  #On   iSelect/C  -->  #Action/C.
```

The parameter field name can be used like a loop variable in a programming language to create state fields, user events, system events, rules, and other parameter fields. By using more than one parameter field in a DSN/2 rule, large sets of combinations of rules can be created with a single line. To enable flexible definitions of these loops, we included constructs to exclude single cases (EXCEPT and BUT keywords), and to take advantage of the order of elements in the parameter list (NEXT and PREV keywords). Although the techniques of field and rule expansion are straightforward and simple to use, they enhance the readability and reduce the size of specifications drastically. In practice, we found high factors for the reduction of number of rules.

1.2.2 Flexible Preconditions

Flags are always combined by logical AND to form preconditions in DSN rules. In practice, we found that it is useful to generalize this and allow the use of OR (disjunction) and NOT (negation) to form preconditions. The single DSN/2 rule,

```
NOT( #Action(ActionType) )  iSelect(ActionType)  -->
        #Action(ActionType),
```

for example, replaces the six rules (two of them were shown as rules 3b. and 3c. above) necessary for switching with iSelect events between the different actions. Again, this technique decreases the number of rules. Apart from the reduction effect, the DSN/2 rules express more explicitly than their DSN counterparts the intended meaning. Editing a specification with rules logically belonging together but being written down separately is cumbersome and error-prone. This becomes obvious, when one thinks of adding another action: In DSN/2 only the parameter field would be changed; without this feature, the number of rules grows quadratically in the number of actions (12 rules for 4 actions, 20 rules for 5 actions etc.).

1.2.3 Epsilon Rules

One of the assumptions underlying the behaviour of a DSN specification is that nothing happens unless a user event or a system event is triggered. Hence, every rule must contain an event. We introduced event-free or, as we call them, epsilon rules to overcome the structural problems resulting from this strict rule in DSN. Assume, for instance, that in our hypothetical example machine, we want to switch the machine completely off, when either function #Y is performed with action #C selected or function #X is combined with action #A. With epsilon rules allowed, we can write

```
#On  #Y  #C -->  #Off
#On  #X  #A -->  #Off
```

to achieve the effect. The functionality described elsewhere in the specification is left untouched, and the automatic switch-off conditions are given explicitly.

Without the epsilon rule construct, the designer has to take into account every possible rule potentially leading to one of the two constellations and explicitly add a switch-off rule mixed with the "normal" operation rule. This leads to a confusing mix of rules, which easily results in unreadable specifications. This style of interleaving design tasks is one the most serious drawbacks of DSN, as this type of dependency frequently shows up in realistic systems.

1.2.4 Timer Events

The only point where the system's functionality shines through in original DSN is the definition of system events. It is undefined, however, when they occur; the only facts controlled by the specification are the situations during which a system event can happen, and what the resulting state will be. This is appropriate for undeterministic, mostly erroneous behaviour, the typical example being the paper-out-signal in a DSN model of a photocopier. Realistic systems, however, show time-delayed, deterministic reactions, which can not be modelled that way. For instance, in the case of a CD player, when the user pushes the button for closing the drawer with a CD inside, it takes a few seconds for the machine to actually close the drawer, start spinning the CD, find and interpret the directory, and display the number of tracks and overall duration.

To model such a behaviour we introduced a new class of events, the timer events. Again, timer events are specified in fields and the timer duration in milliseconds added in square brackets. For example,

```
TIMER EVENTS
        ActionDuration  ( tA[2000]   tB [5000] ).
```

This specifies two timers, tA and tB. The first timer generates the event tA 2000 milliseconds after it was started, which happens as soon as the precondition of a rule containing tA as event is fulfilled. To specify, for instance, that whenever action #A is started, it is automatically finished after 2 seconds with the machine switching to #B, we write the rule

```
        #A   tA   -->   #B.
```

Timers enter races with other timers and user or system events occurring while they are "counting down". Only the winning rule fires; the losers have no effect at all.

Using simple timers allows to define delayed deterministic behaviour of a machine. Frequently, however, the situation is even more complex, with the outcome and the timing of a machine action being unpredictable; we included random timers to describe this type of nondeterministic behaviour with the optional possibility to define the probability of the different cases.

1.2.5 Practical Considerations

The enhancements of DSN/2 over DSN were developed with the practical applicability in mind to enable us to perform realistic user testing on DSN/2 models animated with an appropriate tool. We soon found out that original DSN was only applicable on a very abstract level or for toy examples. The working CD player model we tested (see below) is described in 64 DSN/2 rules; our tool KAP translated this internally into 599 DSN rules. To write such a specification with original DSN by hand would be a major effort, if at all possible, as the designer would be forced to scatter design decisions over several rules and repeat regular structures of fields and rules. Using DSN/2, we specified, among other examples, an electronic alarm clock, a pocket calculator, traffic lights, and a tic-tac-toe-game, with the CD player being the most complex and interesting model.

Constructing model-based user interfaces with languages like DSN/2 or others [16] faces the problem of being specified on an appropriate abstraction level. The model must be abstract enough to not force the designer to define too detailed information early in the design process, but still say something. Hence, the language must isolate a significant design issue and must allow easy and compact definition of its effects. We think that state space definition, state transition, and the handling of incoming events are a relevant design issue.

2. Animating the model with KAP

The specification language DSN/2 is intended for the interface designer to define the control model of a user interface. Developing such a specification necessitates the possibility to "see how it works" to verify the correctness of its behaviour. Another possibility is the implementation of analysis tools, but we preferred to pursue simulation, because we aimed at user testing of animated DSN/2 models in first place with simple analysis features of the specification contained in the tool. The tool supporting the work with DSN/2 is referred to as *KAP (Kespohl's Automaton Prototyper)*, which is working under MS Windows™ and is presented in more detail in a companion demonstration session.

2.1 Default Animation

The language DSN/2 itself does not include information about the visual appearance or acoustic feedback of the user interface. Hence, the simplest animation possible is defined in terms of flags, fields, and rules. KAP provides two default animations on this level, a textual animation and a graphical default animation. Both exist without any additional effort for every DSN/2 model loaded into KAP and both are not intended for user testing, but only as tools for the user interface designer, working on the DSN/2 model.

2.1.1 Textual Animation

Within the textual animation the situation of the DSN/2 model at any time is presented by displaying the raised flags, the rules fireable in this situation, and the

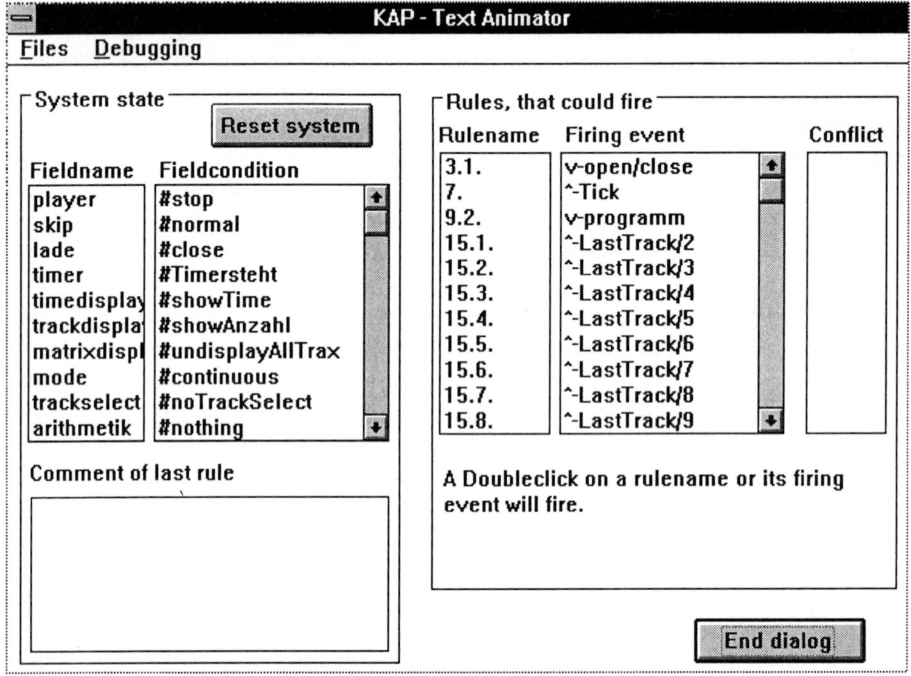

Figure 1 Textual Animation

corresponding firing events. The designer can click on rules or events and watch the changes in the system's fields. Figure 1 gives an impression of what textual animation looks like. The *Dialogue Design Tool (DDT),* as created by the authors of DSN [6] worked on the same level with DSN specifications and had a comparable functionality.

2.1.2 Graphical Default Animation

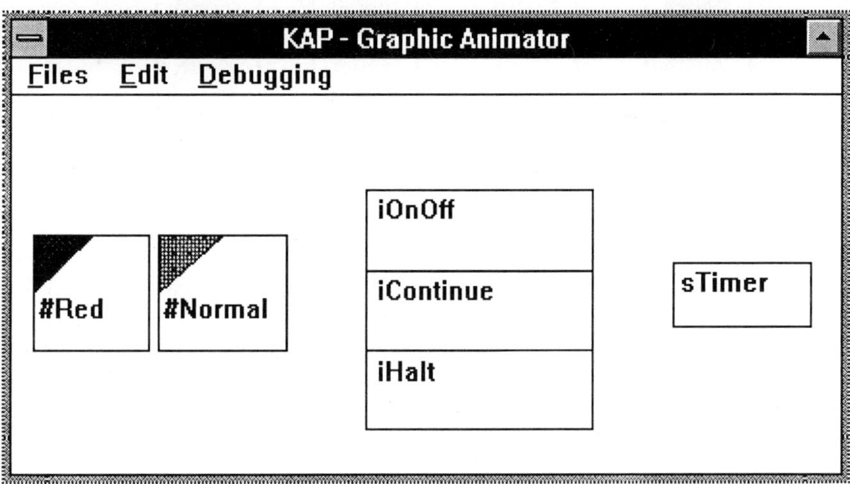

Figure 2 Graphical Default Animation

Basically, the graphical default animation contains the same information as the textual animation, with the exception of rule names. However, all elements are displayed graphically. All events possible at any time are shown as rectangles; to trigger them, one clicks on them. All fields are also displayed as rectangles, with the currently valid flag and a colour code inside. Figure 2 gives an impression from the traffic light example.

Triggering events is done by clicking on event rectangles; the resulting colour and flag changes in the field rectangles can be observed. In general, this picture is very different from a visually realistic model of the device under consideration. The appearance of the graphical animation, however, can easily be modified.

2.2 Specific Graphical Animation

The basic mechanism for creating a specific graphical animation is to replace the standard rectangle representation of states and input fields by individual graphics. Consider, for instance, the example shown in figure 2, which represents a very simple traffic light example. The state of the lights is given as rectangle, showing the names of the flags „#Red“, „#Yellow“, „#Green“, and „#AllOff“. Instead of using this, we could produce individual representations using a standard drawing tool such as Paintbrush to create four drawings visualizing these four states. They might be called „RED.BMP“, „YELLOW.BMP“, „GREEN.BMP“, and „BLACK.BMP“, and might look like this:

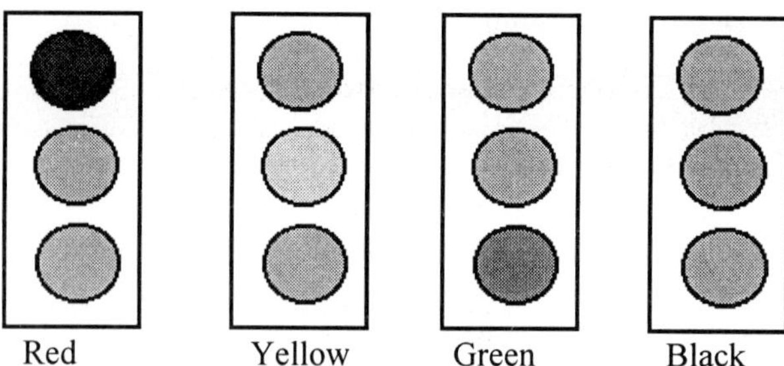

Figure 3 Visualization Drawings for the Traffic Light State

Selecting the state rectangle (the one showing the word „#Red" in figure 2) with the right mouse button brings up a menu allowing to modify the visual appearance of this object. We select the option „Add Drawing", which enables us to assign the drawing „RED.BMP" to the flag #Red, „YELLOW.BMP" to the flag #Yellow, and so forth. After all four graphics have been assigned, the system does no more show the standard rectangle with the word „#Red" etc., but the drawing „RED.BMP" instead. This applies to all four possible light states, such that now, when we operate the traffic light, the red, yellow and green lights seem to go on and off as expected. The same mechanism of overlaying standard elements with individual graphics is also applied to the input elements. Apart from just adding the pictures, we also turn the standard display off - with the effect that the user seemingly operates with the overlayed drawing, while in reality he or she pushes the underlying invisible buttons. The effect is very realistic if we use scanned photographs of real devices.

To support this mechanism further, the rectangles for states and events can interactively be moved, resized, and hidden, so that we can in fact produce a realistic looking user interface. One can visualize state transitions through lights turning on and off, changing their colours, or displays showing different symbols. The behaviour of the input event buttons can be changed to model small key matrices, toggle buttons, repeat keys, sliders, or similar devices, just by selecting the appropriate input device from the pop-up menu activated with the right mouse button. By adding bitmaps to these input elements, we could depict all input elements we met on the modelled devices quite realisticly. The modified graphical animation can be stored and loaded, hence it can systematically be developed from the default animation. Figure 4 shows a graphical animation we created of the simulated CD player. It is defined according to a real CD player and resembles it strongly.

As sound carries important feedback clues for the user, we included the possibility to

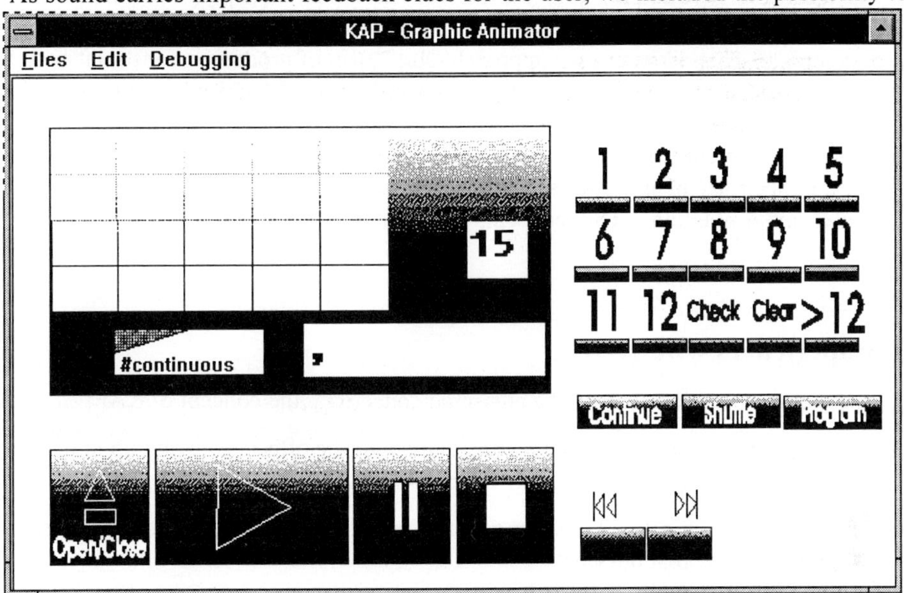

Figure 4 Graphical Animation of the CD player

link sounds to the raising of flags. This can be done interactively in the KAP tool, or it can be defined in the DSN/2 specification by inclusion of sound annotations to flags. Additionally, the designer can add external functions to be called when flags are raised, to add functionality to the system. In the case of the CD player we did so, to direct the internal CD drive of the computer; the mechanism, however, is generally available.

Taken all these features together, a designer can create a very realistic prototype of the user interface of a device. It looks like the real thing, it produces the same sounds, its state behaviour can be observed and tested, and if functionality is added (as was easy with the CD player, the alarm clock, and the calculator), it does in fact produce the same effects. Adding the animation requires only experience in using MS Windows™-like programs, hence it can be mastered by non-programmers.

2.3 Related Work

We think that KAP has significant advantages over environments such as HyperCard™ or SuperCard™ [8] typically used for prototyping purposes, as they necessitate programming expertise. The same argument applies to prototyping done with environments like VisualBasic™. An interface designer working with KAP must be familiar with the concepts of state and rule-based state transition. These, however, are central design issues anyway, so only the notation has to be learned.

Tools such as GARNET [10] and its successors or Whizz'Ed [7] urge the designer to familiarize with constraints and data-flow specifications, or having to retreat to LISP programming. This, like our own approach OBJECTION of producing constraint-based application-specific widgets [14], is too much influenced from a programming style of user interface development for being applicable by non-programming interface designers. From a methodological point of view, we see some commonality with Carr's approach [2] of unifying state transitions, constraints, and visual appearance within a single notation. This concept, however, is not supported by a simulation tool.

3. Experiences with KAP

Two important questions have to be answered concerning the concept of KAP:

- Is it possible to simulate all or at least enough of the effects showing up in device interfaces, resulting in sufficiently realistic models ?
- Does user testing of the DSN/2 model make sense, or are the results too different from testing the real device ?

In the following we deal with both questions, speaking from our experience with the KAP tool, as existing so far. With the term "KAP model" we refer to a DSN/2 specification together with a specific graphical animation created and used within KAP.

3.1 Expressive Power of the KAP Model

In the beginning of the development of KAP we simply started to model devices and invented new features (input or output), when what we had was inadequate or something was missing. The more examples we constructed, the less additional elements we needed. It is our feeling now that the set of possiblities is sufficient for most cases. This can, of course, not be proven but just experienced when trying the (n+1)-st device. In some cases, however, we truely met the limits of a computer simulation; we want to give two examples.

First, there is the case of keys on a real device being too small, and hence hard to find, or being too hard to press. Finding small items on a computer display is different from finding them on the front of a device. Even more severe is the problem of keys being hard to press for different reasons, such as mechanical resistance through construction. Mapping every key press onto the mouse button push defines this difficulty away, or generates a new one. Although we have a perfect match between computer operation and real operation (both are button presses), the evaluation results will be different for those reasons. In these cases the designer must well be aware that the model does not express problems of the lowest level (and that all keys should always be constructed for easy use...).

Second, there is the case of input one does not normally do to a computer, such as inserting a currency bill. This can only roughly be simulated as an action, as it must be mapped onto discrete user events available on computers. Physical problems with entering the note, such as "upside down" or "crumpled", can only be modelled abstractly with user event fields such as

```
InsertBill ( iCorrect  iWrongWay  iCrumpled ).
```

We plan to include drag-and-drop-technology for - at least - simulating the movement of something (the bill) from a source (the purse of the user) into the machine (the slot). So, we will translate real movements into mouse-based move operations. Still, it remains an abstraction and an interface designer must be aware of this artificial situation when interpreting user tests.

3.2 User Testing of the KAP Model

To find out about the usability of KAP as a test bed for device interfaces, we performed usability tests on the KAP model of a CD player and on the real device itself, and then compared the results. We do not claim that we made a very large study, which would necessitate the use of several different devices, and a larger number of subjects. We did neither have the time nor the ressources to do so. But we tested a handful of people on the software model and the real device, compared the outcome, identified some problems, but mostly found very promising results.

We think that the reason for this positive result is that we restrict ourselves to work on the control model with step-wise addition of visual and acoustic elements, until we end up with a realisticly animated control model. Hence, it is clear what we call the functionality we model: state transitions with time effects added, in some cases external functions. This circumvents the problem that prototypes frequently remain too much on the (optical) surface, with a not quite clear layer of functionality below [1].

In [15] the misconception between a system S and a vehicle V representing it (such as the desktop metaphor representing a work space, a file system etc.) is discussed at length. Speaking in terms of this paper, there are properties that the system does not have (denoted as S-), but the vehicle has (denoted as V+) and vice versa, their claim being that (S-V+)-properties are the most confusing to the user. But, of course, (S+V-)-properties are disturbing as well.

With respect to the latter, we found none in the CD player case. Apart from low-level effects of buttons being hard to find, which was easier on the computer screen, the simulation did not hide any problem encountered with the real system. This gives rise to the hope that the problems of a real device can in fact be found in the corresponding KAP model.

The (S-V+)-type of property showed up in two usability problems. First, we found one big problem with subjects who were unfamiliar with using the mouse. They had serious difficulties in positioning the mouse correctly and described the movement of the mouse as stucky and disturbing. For instance, had they problems to continue with the test when the mouse inadvertently had left the mouse pad. Or, they lifted the mouse and were surprised that nothing happened, while moving the mouse through the air. Moreover, they held the mouse in a wrong angle towards the monitor, so that movement up was not up and right was not right. We hope to remove this problem at least partly when we switch to touch-screen technology in the near future, with the low-level modelling problems mentioned earlier remaining.

The second problem we found is that people clicked into display areas, which were no buttons at all - an effect that never happened with the real CD player. Obviously, our representation of the display was not clearly enough recognizable as a pure output area. However, we encountered this only with subjects being experienced computer users (where the distinction between input and output areas is frequently blurred). Also, our graphical feedback of button pushes (currently just "computer-infected" inverting of the sensitive area) was criticized.

Overall, we expected more problems of this type than we encountered. We ended up with the feeling that most subjects could easily lift themselves on this almost concrete but slightly abstracted level we presented them, and then work on the common problems of the KAP model and the real device in terms of the control model.

3.3 Some Insights about the CD Player

Although the CD player we tested is not relevant for this paper, we want to give an impression about the usability problems we found to demonstrate the results feasible with our system. We prepared a standardized task sheet with 12 simple tasks to be performed with the CD player, ranging from simple steps, such as "play track #1", "skip to the next track" or "stop the player", to the programming of the CD player in "store the tracks #2, #5, #10, #6, #16 and #8 in this order in a program and play the first seconds of each track in the program". People were given the usual instructions for a thinking aloud protocol [17] (their utterances were recorded and later transcribed) and were told that and why they were given no advice from us. After the test, subjects were interviewed and answered a questionnaire. Classified into groups, we found the following five problems of the CD player:

- Some (English) key inscriptions were completely obscure to non-English-speaking and non-computer experienced users (e.g. CHECK);
- the same group of subjects was unaware of what "programming" means (and could not think of a reasonable use of the PROGRAM key).

- When programming a sequence the display does not show the selected tracks in the order they are played. Subjects were unsecure whether the system stored the tracks in their intended order.
- Some subjects were not aware of the fact that the initial display after entering a CD or after stopping the player shows the number of tracks, the interface giving no visible clue.
- The technique of entering numbers bigger than the number of keys on the device with the help of a ">12"-key is confusing.

Conclusion

We plan to work more on DSN/2 to enhance it towards an object-oriented specification language. It would be very useful to encapsulate the behaviour of "small" devices into objects, and stick them together to form a higher-level system. This would enable the construction of object libraries containing input, output, or functional elements. In doing so, we think about making DSN/2 more dynamic, hence, allowing object creation and destruction. With these features, and more sophisticated input-output elements, we think it feasible to prototype software user interfaces as well. Experiments in this direction have started.

We mentioned already the switch to touchscreen technology, as the mouse seems an input device which needs too much training for computer novices (and it is these subjects that we want most). Additionally, KAP could be enhanced towards an automatic evaluation environment. From the DSN/2 specification we could define tasks and correct "paths" through the model to solve these tasks. By adding instruction screens for the subjects and a (very!) simple user interface, we could put test users in front of such a system and perform semi-automatic user tests.

As a DSN/2 specification is the kernel of a model-based user interface, we will work on analysis tools to find design problems. This has been done partly for PPS [12] and other systems [3], [13]. The problem with these approaches is that they might fail on realisticly sized systems, or only find very trivial or extreme errors in the design. We will include some of these tests in future versions of the KAP tool, but it is still unclear to what extent.

References

1. Attwood, M E; Burns, B; Girgensohn, A; Lee, A; Turner, T; Zimmermann, B: *Prototyping Considered Dangerous,* Proceedings of INTERACT'95, pp 179-184, June 1995.

2. Carr, D A: *Specification of Interface Interaction Objects,* Proceedings of CHI'94, pp 372-378, April 1994.

3. Coutaz, J; Nigay, L; Salber, D; Blandford, A; May, J; Young, R M: *Four Easy Pieces for Assessing the Usability of Multimodal Interaction: The CARE Properties,* Proceedings of INTERACT'95, pp 115-120, June 1995.

4. Curry, M B; Monk, A F: *Dialogue modelling of graphical user interfaces with a production system,* Behaviour & Information Technology, Vol. 14, No. 1, pp 41-55, 1995.

5. Curry, M B; Monk, A F; Maidment, B A: *Task-Based Interface Specification,* Technical Report, Department of Psychology, University of York, and Data Logic, Harrow, Middlessex, UK, 1991.

6. Curry, M B; Monk, A F: *The dialogue design tool: better programming by design,* presented at the NATO Advanced Research Workshop: Cognitive Models and Intelligent Environments for Learning Programming, Santa Margherita, Italy, 1991.

7. Esteban, O; Chatty, S; Palanque, P: *Whizz'Ed: A Visual Environment for Building Highly Interactive Software,* Proceedings of INTERACT'95, pp 121-126, June 1995.

8. Gookin, D: *The Complete SuperCard Handbook,* COMPUTE! Books, Radnor, Pennsylvania, 1989.

9. Gray, W: *VCR-As-Paradigm: A Study and Taxonomy of Errors in an Interactive Task,* Proceedings of INTERACT'95, pp 265-270, June 1995.

10. Myers, B A; Guise, D A; Dannenberg, R B; Vander Zanden, B; Kosbie, D S; Pervin, E; Mickish, A; Marchal, P: *GARNET: comprehensive support for graphical, highly-interactive user interfaces,* IEEE COMPUTER magazine, pp 71-85, November 1990.

11. Olsen, D: *Propositional Production Systems for Dialog Description,* Proceedings of CHI'90, pp 57-63, 1990.

12. Olsen, D; Monk, A F; Curry, M B: *Algorithms for Automatic Dialogue Analysis Using Propositional Production Systems,* To appear in Human Computer Interaction, 1995.

13. Palanque, P; Bastide, R: *Verification of an Interactive Software by Analysis of its Formal Specification,* Proceedings of INTERACT'95, pp 190-196, June 1995.

14. Pöpping, M; Szwillus, G: *Constraint-Based Definition of Application-Specific Graphics,* Proceedings of INTERACT'95, pp 85-90, June 1995.

15. Smyth, M; Anderson, B; Knott, R; Alty, J L: *Reflections on the Design of Interface Metaphors,* Proceedings of INTERACT'95, pp 339-345, June 1995.

16. Sukaviriya, N; Kovacevic, S: *Model-Based User Interfaces - What are They and Why Should We Care ?,* Proceedings UIST'94, Panel, pp 133-135, November 1994.

17. Tognazzini, B: *Tog on Interface,* Addison-Weshley, 1991.

Toward more understandable user interface specifications

David A. Carr

Department of Computer Science and Electrical Engineering
Centre for Distance-spanning Technology
Luleå University
S-971 87 Luleå, Sweden
David.Carr@sm.luth.se http://www.sm.luth.se/~david/

Abstract. Many different methods have been used to specify user interfaces: algebraic specification, grammars, task description languages, transition diagrams with and without extensions, rule-based systems, and by demonstration. However, none of these methods has been widely adopted. Current user interfaces are still built by writing a program, perhaps with the aid of a UIMS. There are two principal reasons for this. First, specification languages are difficult to use. Reading a specification and understanding its exact meaning is very difficult. Writing a correct specification is even more difficult. Second, most specification languages are not executable. This means that after the user interface programmer makes the effort to write a specification, the user interface must still be coded. As a consequence, most programmers have little incentive to do a specification. A pilot study into the comprehensibility of specifications is described. The results of this study suggest that user interface specifications are difficult to interpret manually. A possible solution to this problem, specification animation, is also described.

Keywords: specification comprehensibility, specification animation, user interface specification.

Introduction

Researchers have been formally specifying user interfaces for over fifteen years. However, very few systems outside of research labs have been formally specified. Common practice for user interface development is predominately based on programming. Most user interfaces are written with code and perhaps with the aid of a User Interface Management System that is little more than a layout editor. Almost all dialog is described only in the program. Whatever specification that is done is usually a text description, and it is rarely updated. Development organizations resist formal specifications because they are difficult to understand, are usually not executable, and therefore, are viewed as unnecessary additional work.

In order for specification to be widely adopted, the perceived benefits of specifying a system must outweigh the costs. To fulfill this goal specification systems must become more understandable and must be executable. An understandable specification system will allow user interface designs to be communicated between design team members and can serve as the basis for discussion about improvements and problems.

An executable specification allows the development organization to skip the laborious and error prone processes of translating a specification into code. It should be noted that even an executable specification would require an understandable static representation. This representation would be used both for debugging and for communicating the design in situations where execution is not practical. A static representation can also help the designers visualize how different parts of the system relate to each other. Visualization is very difficult to achieve with a purely machine readable specification.

Of course, one must ask what "understandable" means. In this paper understandable is taken to mean that the specification reader can correctly determine the behavior of the user interface by reading the specification.

This paper first discusses various methods that have been used to specify user interfaces. It then describes a pilot study to compare the understandability of two different specification methods. Finally, it proposes specification animation as a way to improve specification understandability.

1 Previous Research

1.1 Interface Specification Methods

Over the years a number of methods have been used to specify user interfaces. These include algebraic specifications, grammar-based languages, transition diagrams (including extensions like Petri-nets and statecharts), rule-based systems, and specification by demonstration.

Algebraic specification of window systems was introduced by Guttag and Horning [8]. They proposed the design of a windowing system based on axiomatic specification of abstract data types. This method permits formally proving properties of the user interface. However, algebraic specifications have serious drawbacks. They are very difficult to read and require considerable time and training to understand. They are even more difficult to write. Specifically, the declarative nature of an algebraic specification provides little support for the sequential nature of user interface dialogs. This makes it difficult to construct or follow sequences of user actions, and therefore, makes algebraic specification unsuited for communicating interface behavior.

Systems that are grammar-based include Shneiderman's[27] multiparty grammars, Payne and Green's[22] task-action grammars, and Siochi, Hartson and Hix's User Action Notation (UAN) [10, 12]. Multiparty grammars were designed to model complete human-computer dialogs for command languages. A multiparty grammar divides non-terminals into three classes: user-input, computer, and mixed. User-input and computer non-terminals represent user actions and computer responses, respectively. Mixed non-terminals represent sequences in the human-computer dialogs. While reasonable models for command languages, multiparty grammars do not model the inherently non-sequential nature of direct manipulation well. Task-action grammars and UAN take a user-task point of view to specify command sequences. The operation of the system is organized into user tasks and the interface software is supposed to recognize sequences of user actions as a task. UAN includes support for specifying parallelism and other time dependencies[7, 11], and therefore, is better suited for specifying direct manipulation. The problem with the task-description strategy is that users don't always do exactly what the task description states. Thus, deriving a system from such a specification means that the translation tool must provide meaningful default error handling. This is not a simple job.

Another approach to modeling user interfaces is the transition diagram [32]. In this approach the transitions represent user inputs, and the nodes represent states of the interface. Computer outputs are specified as either annotations to the state or the transitions. However, transition diagrams suffer from a combinatorial explosion in the number of states and transitions as system complexity increases. Jacob [16] solved part of this problem by allowing concurrent states to coexist as parallel machines or co-routines. Others have used extensions of the basic transition diagram. These extensions include statecharts and Petri-nets. Statecharts[13] were designed as a formal solution to the combinatorial problems with transition diagrams. The statechart adds the concept of a meta-state. Meta-states group together sets of states with common transitions that are inherited by all states enclosed in the meta-state. Since meta-states may enclose other meta-states, a complete inheritance hierarchy is supported. A special history state is supported to return the meta-state to its previous status on return transitions from events such as invoking help. Meta-states are divided into two types: parallel or AND-states and sequential or XOR-states. Meta-states enclosed within AND-states may execute in parallel and fulfill the function of co-routines. As originally defined, statecharts do not incorporate data flow or abstraction. Wellner[33] adapted statecharts to user interfaces by adding state entry and state exit actions to produce output from the specification. Rouff[26] used a statechart-like language called the Interface Representation Graph (IRG) as the underlying representation for his Rapid Program Prototyper. IRGs extend the statechart to represent dialog by introducing data flow, constraints, and communication with the application's back end. My own Interaction Object Graphs[2, 3] extend statecharts for widget specification by adding representations for display changes and widget attribute data. Another extension of the state diagram, the Petri-net, has also been used for user interface specification, for example Palanque and Bastide's Interactive Cooperative Objects[24]. The transition diagram and its extensions have two clear advantages as a formalism. First, they are easily translated into an executable form. Second, they have a graphical representation that can be easily animated to enhance user understanding of the system's execution.

Rule-based specification has been used in UIDE[5], PPS[21], and Sassafras[14]. All of these systems work in a manner similar to UIDE which uses pre- and post-conditions to control user interface dialog. User input causes one or more preconditions to become true. As preconditions become true, associated actions are executed. These actions include post-conditions which may cause further changes in the interface state. Rule-based systems distribute the interactions between system components as interactions between different rules. This property makes them difficult to understand once they reach non-trivial size. It also makes them difficult to modify without introducing inconsistencies and unintended side-effects.

A final technique used for specifying human-computer dialog is to do so by demonstration. Using this technique the designer places widgets on the screen demonstrating state changes, and the system makes inferences about the intended design. "Druid" [30] and "Peridot" [20] are two systems which build user interfaces by demonstration. Druid provides for interactive layout and dialog definition. However, its dialog model does not include constraints between interface objects or the modification of interface objects. Peridot is a similar system which includes constraints and modification by redrawing. Neither of these systems has a clear visual representation for the user interface, although one could probably be developed. In fact, a state-based representation that was initially developed by demonstration and then further refined by an editor might be a very useful tool.

1.2 Animation

Animation has been used as an aid to understanding for data structures, programs, and algorithms. Systems which were built to illustrate changes in data structures include Incense[19] and VIPS[28]. The PECAN system[25] used multiple, graphical views of a program including both data and an animated flowchart. Visual programming systems such as Pict/D animated the program diagrams by highlighting the currently executing icon or flow path[6]. Pict/D programmers drew a flowchart of the program which was executed directly. Other systems such as VIPEX[9] are designed on a data flow paradigm. VIPEX nodes were connected via links at input and output pads. The programmer could selectively view data details at these pads while nodes would change color based on their status.

Other systems have concentrated on animating algorithms. With these systems the programmer constructs a program and a custom graphical representation of its abstract behavior. These systems include Balsa[1] and TANGO[31]. The Lens system[18] extends these systems by combining algorithm automation into a source-level debugger.

2 A Pilot Study on Understanding Specifications of Widget Designs

One of the most important uses of a specification is to communicate a design to others. However, how well a specification communicates that design is very difficult to measure. Yet, one would like to evaluate specification methods and determine which were better at communicating a design.

The specification methods with visual representations that are described above fall into two categories, diagrammatic and textual. The diagrammatic methods: transition diagrams, statecharts, Petri-nets, IRGs, and IOGs, are all based on some form of state machine. The textual methods are: rule-based specification, task-action grammars, multiparty grammars, and UAN. It would be useful to compare these two categories for understandability.

First, one must define "understandability". Since communication of the design to others is the goal, understandability should be defined in terms of the functions required to communicate the design. One measure of understanding is whether the reader can implement the widget. However, not all readers of the specification would be programmers. Other reasons to read the specification include: evaluating the design, documenting the widget, and verifying implementation correctness. A more useful measure of understandability would be performance in answering the following questions:

1. Given a sequence of user inputs: What is the status of important widget attributes?

2. Given a sequence of user inputs: What is the appearance of the widget on the display?
3. What sequence of user inputs is required to accomplish a given user task?
4. Given two similar specifications: What is the difference (if any) between them?

Answering the first two questions would be required for implementation, evaluation, documentation, and verification. The third question is important for evaluation and documentation. The last question would be important to someone evaluating successive designs of the same widget or trying to understand a modification to an existing widget.

The question remains whether there are any differences between text-based specifications and diagrammatic specification when communication is considered. A study comparing PASCAL syntax diagrams with BNF descriptions suggests that diagrams are superior for identifying syntax errors and for identifying the next legal language construct[4,17]. This study indicates that diagram-based specifications should be superior in answering the first two questions. Intuition would say that a specification method such as UAN should be superior in answering the third question. It is designed to give this type of information. The fourth question would probably be answered more quickly from a diagram, as long as the changes were more than the addition of an arc or two. However, it is quite possible to just rearrange the diagram and give the impression of change when none exists.

The rest of this section contains a brief description of a text specification method, UAN, a diagram specification method, IOGs, and a discussion of a pilot experiment comparing them. The experiment tries to measure differences in the number of correct answers and the time used on questions similar to the four above.

2.1 User Action Notation (UAN)

UAN is a text-based specification language for user interfaces that has its roots in user task description. A UAN specification is represented as a table divided into three columns: user actions, interface feedback, and interface state.

The first column is reserved for user actions. In this column the dialog designer writes a user action sequence in an event description language. The three most important symbols in this language are *Mv*, *M^*, and *~[Name]*. *Mv* and *M^* represent pressing and releasing the first mouse button, respectively. *~[Name]* represents moving the mouse into the context of the object *Name*. User action sequences may be prefaced with a precondition which is enclosed in parentheses. A precondition must be true before the user action sequence will have any effect. For example, the user action, "*(Sw=OFF): Mv*", means that the interface state variable *Sw* must have the value *OFF* before any interface changes beginning with the user pressing the mouse button will apply.

The second column in a UAN table describes the interface feedback. This is done with a text description of the changes such as "*display(ActiveHeat)*". This states that the image named *ActiveHeat* will be shown on the display. There are other feedback operators described in the references[10, 11, 12].

The third column in a UAN table is labeled "interface state". This column specifies changes in state variables that are used in the preconditions described above. The variables represent state information about the user interface or the application. An example of this is "*Sw = OFF*" which assigns the value *OFF* to the variable *Sw*. UAN state variables may be of any type.

UAN specifications are read by starting in the top-left and proceeding row-by-row from left-to-right. Complex actions are specified by composing simple tasks. For example, the *Operate* task in Figure 1. Figures 1 and 2 specify a heating control switch that is common in the U. S. This switch was used as part of the pilot study described below. In addition to the UAN, pictures representing the displays were provided.

Task: Operate = FirstDisplay (HeatOff | OffCool | CoolOff | OffHeat)*

Task: HeatOff		
User Action	Interface Feedback	Interface State
(Sw = HEAT): ~[Heat]Mv	display(ActiveHeat)	
((~[Off2Heat]	display(Heat-Off)	
~[Off]	display(ActiveOff)	Sw := OFF
M^)	display(Off)	
| (~[Off2Heat]	display(Heat-Off)	
M^) | M^)	display(Heating)	

Task: OffCool		
User Action	Interface Feedback	Interface State
(Sw = OFF): ~[Off]Mv	display(ActiveOff)	
((~[Off2Cool]	display(Off-Cool)	
~[Cool]	display(ActiveCool)	Sw := COOL
M^)	display(Cooling)	
| (~[Off2Cool]	display(Off-Cool)	
M^) | M^)	display(Off)	

Task: CoolOff		
User Action	Interface Feedback	Interface State
(Sw = COOL): ~[Cool]Mv	display(ActiveCool)	
((~[Off2Cool]	display(Off-Cool)	
~[Off]	display(ActiveOff)	Sw := OFF
M^)	display(Off)	
| (~[Off2Cool]	display(Off-Cool)	
M^) | M^)	display(Cooling)	

Fig. 1 UAN specification of a common thermostat.

Task: OffHeat		
User Action	Interface Feedback	Interface State
(Sw = OFF): ~[Off]Mv	display(ActiveOff)	
((~[Off2Heat]	display(Heat-Off)	
~[Heat]	display(ActiveHeat)	Sw := HEAT
M^)	display(Heating)	
| (~[Off2Heat]	display(Heat-Off)	
M^) | M^)	display(Off)	

Task: FirstDisplay		
User Action	Interface Feedback	Interface State
(Sw = OFF):	display(Off)	
(Sw = HEAT):	display(Heating)	
(Sw = COOL):	display(Cooling)	

Fig. 2 UAN specification of a common thermostat continued.

2.2 Interaction Object Graphs

Interaction Object Graphs (IOGs) are designed to add widget specification to Interface Representation Graphs. They combine the data flow and constraint specifications of IRGs with the statechart, transition-diagram, execution model. This expands the statechart to show data relationships as well as control flow. IOGs add a display state and a representation for widget attribute data in order to permit specification of low-level interaction objects which cannot be specified by Interface Representation Graphs. Below is a brief description of the IOG state diagram, and a transition description language used to specify transition conditions.

The IOG state diagram traces its lineage from UAN, statecharts, and IRGs. Statecharts added four new state types to the traditional state diagram. These states are used in IOGs. They are: the XOR meta-state, the AND meta-state, and two types of history state.

The meta-states can contain both normal states and other meta-states. Transitions from meta-states are inherited by all contained states. This helps reduce the problem of arc explosion. The XOR meta-state contains a sequential transition network. Exactly one state inside of an XOR meta-state is active when the XOR state is active. On the other hand, an AND meta-state contains more than one transition network. Each of these networks executes in parallel.

A history state can only be contained in an XOR meta-state. Whenever a transition transfers control from a meta-state, the history state remembers which state was active immediately before the transition. If a later transition returns control to the history state, the meta-state is returned to the remembered state. History states help control state explosion. To see this, consider a specification of a help system which is independent of the user interface. An ordinary transition network would require replicating the help-system specification once for every state in the user interface.

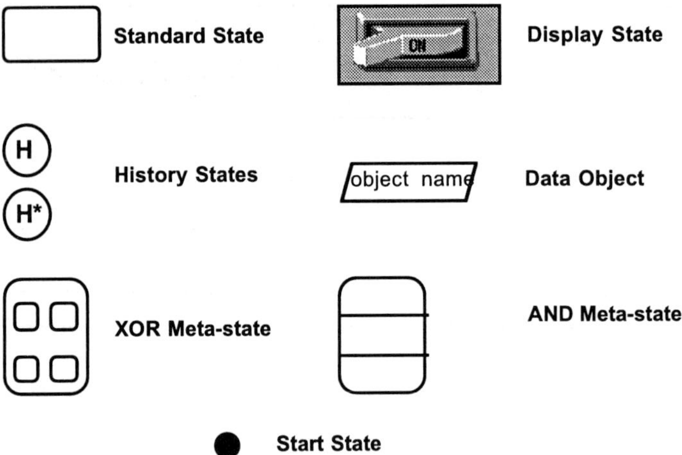

Fig. 3 IOG node symbols.

Otherwise, there would be no way to return to the user interface state that was active before help was requested. A statechart history state could receive the return transition from the help system, and only one copy would be required. There are two types of history states. They differ in how they treat a return when the last active state was a meta-state. The *H* state restarts meta-states at their start state and provides one level of history. On the other hand, the *H** state restarts meta-states at their history state, when they have one, thereby allowing multilevel history. Figure 3 shows the representation of the new states.

IOGs add two additional node types to the statechart, data objects and display states. Data objects represent the storage of a data item, and control is never passed to them. They can only be destinations for the data arcs discussed below. Display states are control states that have a change in the display associated with them. In IOG diagrams a picture of the display change is used whenever possible instead of a program-like statement such as "draw(ActiveON)". Data objects are represented as parallelograms. (Figure 3.)

IOGs also add two special arc types, the event arc and the data arc. Events allow the designer to define "messages" which may be lacking in the underlying specification model. For example when specifying the trash can in the Macintosh interface, one needs to know when a file is being dragged over it as opposed to when the pointer is being dragged over it. One way to do this would be for the file icon to generate a "dragging started" event and a "dropped" event. The trash can would then be highlighted whenever the pointer was over it between a "dragging started" event and a "dropped" event. An event is represented by a special transition passing through an *E* in a diamond. (Figure 4).

Data flow is represented in a manner similar to events – an arc passing through a *D* in a circle (Figure 4). A data flow arc may have any state as a source and can only terminate at a data object or have an unspecified termination. In addition, at least one end must be attached to a data object. Data flow arcs with data objects as a source, whose destination arrow is unspecified, and whose destination is outside of the

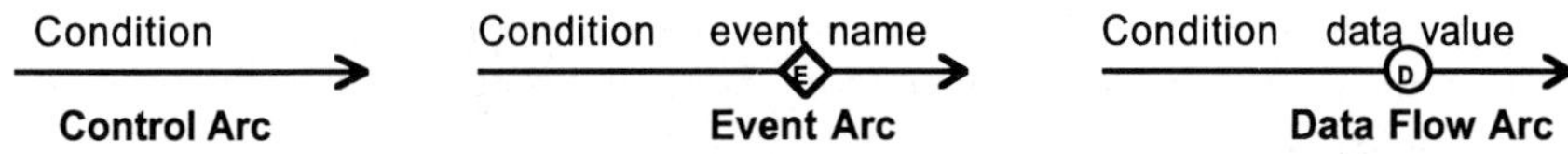

Fig. 4 IOG arc symbols

containing interaction object, indicate externally-readable data (Figure 5). This data may be used by the application or attached to other user interface components as a more complete specification is constructed. Data flow arcs with data objects as destinations represent updating the data object. If the arc's source is a control state, it represents a change in the value when the arc conditional is satisfied. In this case, the data flow arc is labeled with the new object value. An arc without a source represents externally-writable data.

Constraints are useful in specifying one attribute of the user interface in terms of others. With constraints it is simple to restrict an icon to be contained within a window or to map the values of a slider to a specific range. IOG data arcs support a form of one-way constraints by expressing the data value as an equation in terms of other attributes. Together with the condition on the data arc, these equations provide a means to constrain one attribute in terms of another with a Boolean guard. For example, specifying that in a given state (call it *S*), the image of a widget will follow the mouse cursor can be done with a constraint. This is accomplished by drawing a data arc from *S* to the data object representing the widget's location. Now, specify that the new value for the location is: "old location + the change in the mouse position". The condition for the arc would be: "the mouse position changes". This results in the widget following the mouse while the widget control is in state *S*.

In order to describe the transitions between states, an abstract model of the user interface and a description language for that model are required. IOGs abstract the interface into the following objects: Booleans, numbers, strings, points, regions, icons, view ports, windows, and user inputs. A brief description of these objects follows.

Booleans, numbers, and strings (BNS) are the usual abstractions with the usual operations. It should be noted that numbers contain both the real and integer data types. In addition, any of these may be converted into an icon representation by the operator *icon(BNS, point, font, fontsize)* or *icon(BNS, region, font)*. Both operators convert the Boolean, number, or string *BNS* to a text representation, and then convert the text representation into a picture. If specified with a point, the resulting icon is as

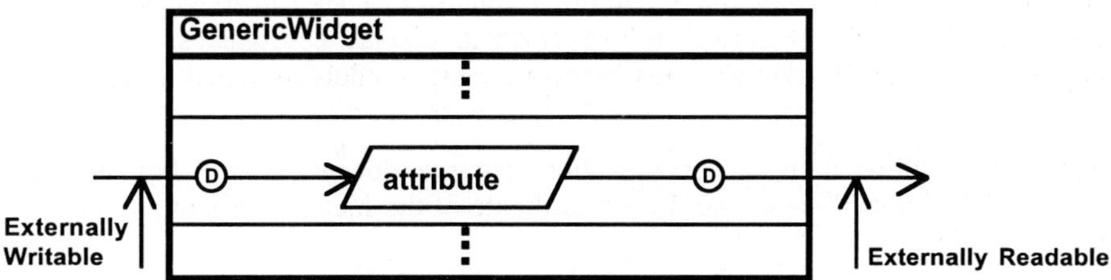

Fig. 5 A readable and writable widget attribute in an IOG.

big as it needs to be to hold the picture of the text representation. If a region is specified, the icon is the size of the region. Fonts and font sizes may be omitted. In that case defaults will be used.

Points are an ordered pair of numbers (x,y). Points have the algebraic operators which are normally associated with them. A point may be assigned a value by writing *p=(x,y)*. In addition, *p.x* and *p.y* represent the x and y coordinates from the point *p*.

A region is a set of display points defined relative to an origin called the *location.* The location of the region is always the point (minx,miny) where minx and miny are the smallest x and y coordinates in the region. Regions have a *size* operator which returns the height and width of the smallest rectangle which covers the region. Regions also have an *in* operator which tests if a point is in the region. This is written *Region.in(pt)* and returns a Boolean value. Although regions are not restricted to be rectangular, rectangles are most commonly used. Note, a region cannot be visible on the display. There is no drawing operation associated with a region.

Icons are regions with pictures. That is, some points in the region have a color number attached to them and are shown on the display. Icons add the operations *draw* and *erase.* In addition, if the origin of the icon is changed, there is an implicit *erase-draw* operation sequence. Unless otherwise specified, the region associated with an icon is a rectangle. So, *icon("text", pt, default, default)* would produce an icon with the upper-left corner at *pt*. The picture would be the word "text" in the default font and size. There would be a region associated with the icon. This region would be a rectangle with its upper-left corner at *pt* and of sufficient area to cover the text.

A view port is a region with an associated mapping function for some underlying application data. The mapping would be in two parts, conversion to a world-coordinate-system graphics representation and projection onto the display. For example, text would first be converted from ASCII to a font representation and a location on a page. The page would then be projected onto the display. The mapping is controlled by a projection function (*proj*), a translation point (*translate*), and a scale-change point (*scale*). If *convert* is the conversion function for some object in some view port, then the function *translate + proj(scale, (convert(object)))* would be the view port mapping. Parts of objects projected to points not in the region are not displayed, and objects in view ports are addressed relative to the view port location.

Windows group the above objects together. They add a level attribute which determines window stacking relative to other windows. They can be viewed as view ports containing only objects already mapped to display coordinates. A window assigned a lower-level value obscures an overlapping window assigned a higher-level value.

Objects are addressed in the specification using a dot notation. For example, "win.icon1.location.x" would be the x coordinate of the location of icon, "icon1", in window, "win".

User inputs are mapped to IOG events, numbers, points, and Boolean variables. Keyboard input is represented by quoted strings when the text is important ("quit↵" when the word "quit" is typed and followed by a carriage return) or key events,

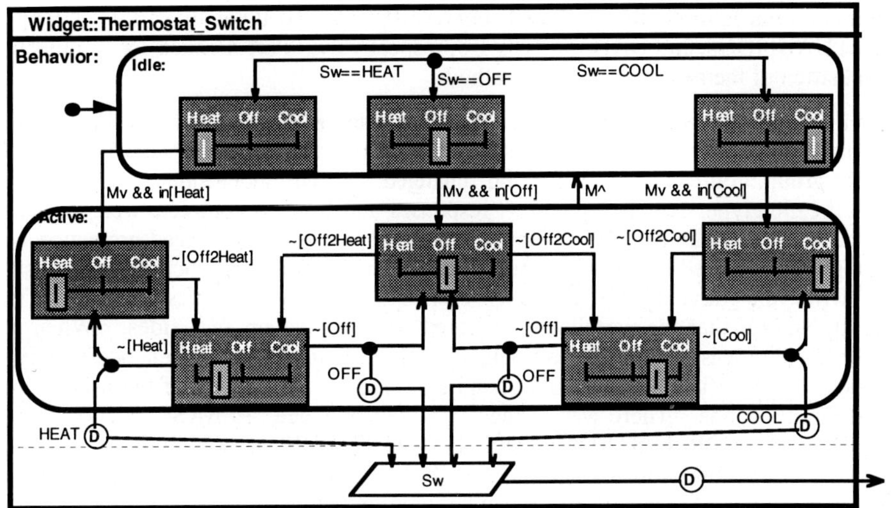

Fig. 6 IOG specification of a common thermostat.

similar to those in UAN, when the event is important (*LShiftv* for left shift key pressed). The mouse is mapped into a point for location (*M@*), a point for relative change (*MΔ*), a Boolean indicating it moved (*ΔM*), button change events (*Mv, M^, M2v, ...*), and button status variables (*Mdn, Mup, M2dn, ...*). Since the value of the mouse location is tested frequently, *in[Region]* is written as a shorthand for *Region.in(M@)*. The special notations *~[Region]* and *[Region]~* describe the events of the mouse entering and leaving the *Region*. These symbols can be combined in expressions. The operators from the 'C' language are used for these expressions. (Most commonly, "&&" for logical AND, "||" for logical OR, and "!" for logical NOT.)

Figure 6 shows the IOG specification for the same heating-control switch as was shown for UAN in the previous section.

2.3 Pilot Experiment Description

The experiment utilized a between-groups design. There were two groups of participants. All participants were graduate students from a graduate class in programming languages and software engineering. A large segment of this class is devoted to software specification methods. Participants were given a reading assignment of a conference paper[29] on UAN and a conference paper[2] on IOGs as homework prior to the day of the experiment. All participants simultaneously listened to a forty-five minute lecture on UAN and IOGs. Other than this lecture, the experimenter had no contact with the class. During the lecture a secure-switch specification (Figure 7) was presented for both methods, and the participants were given a tracing task as practice for both UAN and IOGs. The participants could ask questions at any time. After the lecture, participants had the option of declining to participate in the remainder of the experiment. Those that elected to remain were randomly divided into two groups, A and B. The experiment was divided into four parts. Group A used IOGs in Part 1, UAN in Part 2, and IOGs in Part 3. Group B used UAN, IOGs, and UAN for Parts 1-3, respectively. Part 4 asked for a subjective evaluation of the two specification methods.

Each group received a questionnaire which presented a specification and then a series of questions about that specification. The questions and widgets were identical for each group, only the specification changed. The independent variable was specification type. Number of correct responses and time to complete were measured as dependent variables. Time values were displayed on an overhead projection of a computer-generated screen with elapsed minutes and seconds. Participants were asked to write the time before and after each question in the questionnaire. The questionnaires that the participants received were copied on two sides. When the questionnaire was opened like a book, the left-hand side always contained a copy of the current specification. The right-hand side had the description of the user actions and the questions. There was a page with the message "PLEASE, STOP HERE AND WAIT FOR INSTRUCTIONS TO CONTINUE" after both Parts 1 and 2. (Due to space limitations only the specifications for Part 2 are presented here as Figures 2 and 6. All specifications can be found in [3].)

Part 1 asked participants to trace through a specification and to determine the system's response. They were given an initial state and a sequence of user actions and asked questions about the final state of the interface. The widget specified in Part 1 was a file icon similar to that used in the Macintosh desktop. However, the widget specification contained an error that gave the file icon a different dragging behavior than a Macintosh file icon. Participants were warned that an error existed, and that they should trace specification behavior, not expected behavior. Participants were given ten minutes to study the specification and answer seven questions. The participants were presented with a sequence of user actions followed by one or more questions. There were three user-action sequences with one, one, and five questions. Because the experiment was given in a seventy-five minute class period, time for Part 1 was limited to 10 minutes.

Part 2 also asked participants to trace through a specification and to determine the system's response. The specification in Part 2 was of a three-position, heating/cooling switch similar to the Heat-Off-Cool switch found on many residential thermostats. As in Part 1, there were three user-action sequences but with two, three, and two questions. Time for Part 2 was also limited to ten minutes.

Part 3 specified a modified secure switch with added feedback when the user moved outside of all switch regions just before releasing the mouse. Two questions were asked in Part 3. The first asked, "What has changed?" The second question asked the users to just write down the sequence of user actions required to turn the switch from "off" to "on". The first question tests if the visual nature of IOGs helps participants remember previous specifications and spot modifications of those specifications. The second question might be easier using UAN as participants must simply write the sequence from the user task description while with IOGs they must trace a series of arcs.

Part 4 asked each participant two questions of the form: "Rate the XXX specification method for understanding and ease of use." Where "XXX" was IOG and UAN, and the rating was on a scale of one-to-nine with nine as best. Participants had unlimited time for Parts 3 and 4.

2.4 Results

Twelve graduate students elected to participate in the experiment. They were randomly assigned into group A or group B with six in each group.

Tab. 1 Individual and total number correct for Part 1. (1 = Correct).

IOG

Subject ID	A1	A2	A3	A4	A5	A6	Total
Question #1	1	0	1	1	1	1	5
Question #2	1	0	1	0	0	0	2
Question #3.1	1	0	1	1	0	1	4
Question #3.2	1	1	1	0	0	1	4
Question #3.3	1	0	1	0	0	1	3
Question #3.4	1	0	1	0	0	1	3
Question #3.5	0	0	0	0	0	0	0
Total Correct	6	1	6	2	1	5	21

UAN

Subject ID	B1	B2	B3	B4	B5	B6	Total
Question #1	1	1	1	1	1	0	5
Question #2	1	0	0	0	0	0	1
Question #3.1	0	1	1	1	1	0	4
Question #3.2	0	1	1	1	1	0	4
Question #3.3	0	1	0	0	0	0	1
Question #3.4	0	0	0	0	0	0	0
Question #3.5	0	1	0	0	0	0	1
Total Correct	2	5	3	3	3	0	16

In Part 1, the IOG group (A) got a higher number of correct answers (21 vs. 16). (See Table 1). However, the result was not statistically significant using a t-test (t = .504). In addition to the t-test, Fisher's exact tests were computed to see if any individual questions had significant differences for the number correct between the IOG and UAN groups, and none were found at the α=.05 level. What is interesting to note is that the error rates were very high, 50% for IOGs and 62% for UAN. When questions that were not answered are not counted, the error rates still remain high at 36% for both groups combined.

For Part 2, the IOG group (B) again got more correct than the UAN group (36 vs. 33). (See Table 2). However, the result was not statistically significant using a t-test. As was done for Part 1, Fisher's exact tests were computed to see if any individual questions had significant differences for the number correct between the IOG and UAN groups. In Part 2, there were no questions with significant differences between specification methods. Part 2 was an easier question, and the participants had more practice, but the error rates were still high at 16% for IOGs and 22% for UAN. When unanswered questions are ignored, the error rate still remains at 18% for both groups combined.

Tab. 2 Individual and total number correct for Part 2. (1 = Correct).

IOG

Subject ID	B1	B2	B3	B4	B5	B6	Total
Question #1.1	1	1	1	1	1	0	5
Question #1.2	0	1	1	1	1	1	5
Question #2.1	1	1	1	1	1	1	6
Question #2.2	1	1	1	1	1	1	6
Question #2.3	1	1	1	1	1	1	6
Question #3.1	1	1	1	1	1	1	6
Question #3.2	1	1	0	0	0	0	2
Total Correct	6	7	6	6	6	5	36

UAN

Subject ID	A1	A2	A3	A4	A5	A6	Total
Question #1.1	1	1	1	1	1	1	6
Question #1.2	0	0	1	1	1	1	4
Question #2.1	1	1	1	1	0	1	5
Question #2.2	1	1	1	1	1	1	6
Question #2.3	1	1	1	1	0	1	5
Question #3.1	0	1	1	1	0	1	4
Question #3.2	0	1	1	0	0	1	3
Total Correct	4	6	7	6	3	7	33

Part 3 asked two questions designed to test different aspects of the difference between diagrams and linear text specifications. The first question presented a modified version of the secure switch and asked the participants what had changed. The participants were shown both UAN and IOG specifications of the secure switch during the lecture. It was expected that visual memory of the IOG specification would result in more IOG participants being able to describe the changes made to the IOG specification. Although the four IOG users were correct as compared with only one UAN user, Fisher's exact test did not indicate that this difference was significant (Table 3).

The second question asked participants to write the sequence of user actions required to turn the switch from off to on. Since UAN was designed to specify user tasks, it was expected that UAN would be better than IOGs for this question. However, there was no significant difference between methods for this question. (See Table 3.)

Tab. 3 Fisher exact test for Part 3.

Recognize Changes

	Right	Wrong
IOG	4	2
UAN	1	5
Fisher's exact test P(UAN = IOG)	0.12	

List User-Action Sequence

	Right	Wrong
IOG	5	1
UAN	4	2
Fisher's exact test P(UAN = IOG)	0.50	

The final part of the questionnaire asked participants to rate UAN and IOGs for understandability and ease of use. Question one was, "Rate the IOG specification method for understanding and ease of use." Question two simply substituted UAN for IOG. A scale of one-to-nine was used. Instead of analyzing the raw scores, a difference was computed by subtracting the UAN rating from the IOG rating (Table 4). This difference score gives a relative, subjective rating of the participants' opinion of the differences between UAN and IOGs.

The mean difference was computed for both groups, and a t-test was performed to test the hypothesis that both groups had the same mean. The mean differences were not the same for each group (t-test, P(IOG=UAN) = .015). Therefore, a t-test was made for each group to see if the difference was significant (Table 5). For group A, no significant difference was found. However, a significant difference was found (p = .029) for group B.

Tab. 4 Subjective ratings.

Subject ID	A1	A2	A3	A4	A5	A6	B1	B2	B3	B4	B5	B6
IOG Rating	3	7	7	6	3	7	8	7	4	7	7	9
UAN Rating	2	6	3	6	5	6	5	3	1	5	2	4
Difference	1	1	4	0	-2	1	3	4	3	2	5	5

Tab. 5 Subjective ratings statistics.

	Group A	Group B
Number of Subjects	6	6
Mean Difference	0.83	3.67
Standard Deviation	1.95	1.21
t-score	0.43	3.03
P(IOG=UAN)	.686	.029

It is possible that there was some undetected difference between groups. However, the participants were randomly assigned from a unusually homogeneous population. All were graduate students in the same first-level graduate course, had been given the same reading assignments, and had exactly the same training lecture. Therefore, the most likely reason for this is that Group B had UAN for both the "file icon" questions and the "what's different" questions. These would have been more difficult with UAN, and this probably lead the group B participants to rank IOGs higher. In particular, the "what's different" question would have pointed out the difficulty in spotting modifications in UAN's tabular format. They also used UAN first and would have had poorer scores with UAN because of inexperience.

2.5 Conclusions

The pilot experiment gives some support to the theory that diagrams are better for understanding specifications. The software-engineering students did better with IOGs than with UAN. Although no statistically significant results were found for correctness, participants with IOGs got the correct interpretation more often. Those participants who used more UAN specifications preferred IOGs.

The pilot experiment might have had more significant results if there had been more participants. With only twelve participants, statistical significance is hard to obtain for any experiment involving humans. In addition, a longer questionnaire and more time would have been helpful. More questions would have given participants more experience. Removing time pressure might have lead to lower error rates and shown significant differences in time to complete. However, more time and more participants were not available.

Perhaps the best way to compare specification methods would be to have competing teams of equal skill developing the same widgets using different specification methods. This would be a natural next step as part of a more extensive experiment. Another useful test would be field trials with professional programmers working on commercial products.

The experiment did point out the difficulty in tracing specifications by hand. For questions that they answered, participants had error rates of 36% and 18%, for parts 1 and 2, respectively. Clearly, computer support for understanding specifications would be beneficial.

3 Animating Specifications

The pilot experiment in the previous section illustrates the difficulty in tracing a specification. Error rates were between 18% and 36%. In addition, my own personal experience indicates that it is very difficult to explain a specification to others. Experience with specifications also shows that it is rather difficult to specify "good" interactions. That is, a specification that looks good on paper turns out to be rather difficult to use.

These problems motivated a search for a way to make specifications easier to understand and test. Since IOGs are executable and have a graphical representation, it seemed natural to try animating an IOG specification diagram while simultaneously executing the widget. With an animated specification the readers can test the specification and improve their understanding of its meaning. In addition, the widget designer could directly observe specification behavior while designing and debugging the specification.

If we consider this animation design with respect to the four usability questions below:

1. Given a sequence of user inputs: What is the status of important widget attributes?
2. Given a sequence of user inputs: What is the appearance of the widget on the display?
3. What sequence of user inputs is required to accomplish a given user task?
4. Given two similar specifications: What is the difference (if any) between them?

We can see that the first two questions can be answered directly by running the animation on the user-input sequence. Animation is not as much help answering the third question. It can be used to verify a sequence will accomplish the desired result, but it cannot be used to guarantee that it is the shortest. If many questions of this kind need to be answered, then a path search on IOG transitions would be better. For the final question, animation allows the reader to test the differences between two specifications by applying the same sequences of input to each, but again the reader cannot obtain a direct answer.

3.1 Design of IOG Animation

The first problem when animating IOGs is to decide what to display. Clearly, one would want to display the widget which has been specified. One also needs a IOG diagram. IOGs have two main components, arcs and nodes. One possible animation would be to highlight each node which is active and highlight each arc as it activates. For example, each active state could be outlined in red. As the IOG interpreter executes, arcs attached to each active state are tested. When one of these arcs has a true precondition, that arc could then be changed to red. After all arcs have been tested, the IOG interpreter updates the list of active states. Each newly active state could be outlined while the outline was removed from the deactivated states. At this time, the arcs would be returned to their normal appearance.

However, there are some efficiency and aesthetic problems with the above approach. The bottleneck in any user interface system is updating the display. This is especially true of the IBM PC clone used to display the animation. As more arcs are activated, the operation of the widget lags farther behind the user. This results in very poor operation of the widget. In addition, there are problems with data nodes. They are not really ever active. They could be flashed as they update, but this affects the system response when objects are being dragged. Also, there is the question of outlining meta-states. For the secure switch, this would result in three meta-states being permanently outlined: the entire widget diagram (an AND meta-state), the area of the diagram not containing the data objects, (a XOR meta-state), and either "Static States" or "Operating States". This seems excessive. Rather than impair the performance of the widget itself, a more limited animation was used. Only active, atomic states were outlined for the test system. This seems to give a good idea of how the widget operates without cluttering the display with outlines and without slowing widget execution.

IOGs are an executable specification method and a library of C++ classes modeling the IOG abstractions had been implemented. A program that used these classes to demonstrate a few widgets was augmented to produce the animated versions of these widgets.

3.2 IOG Animation Example

This section gives a brief description of how the animation will appear. The widget shown is a secure toggle switch based on a design by Plaisant[23]. The switch was designed for operation on a touch screen control panel and is secure in the sense that an accidental tap on the switch cannot cause a state change. In order to cause the switch to change the user must press on the side representing the current state of the switch, drag the switch through a middle position into the side representing the new state, and finally, lift-off from the touch screen while pointing at a region corresponding to the new state. Lifting-off from the touch screen in any other region of the screen cancels the operation and returns the switch to its original position.

Figure 7 depicts the animation as the user turns the secure switch from off to on. Before the user begins to manipulate the switch, the widget and the diagram appear on the display. Figure 7a shows this state with the mouse pointing in the *Off* region.

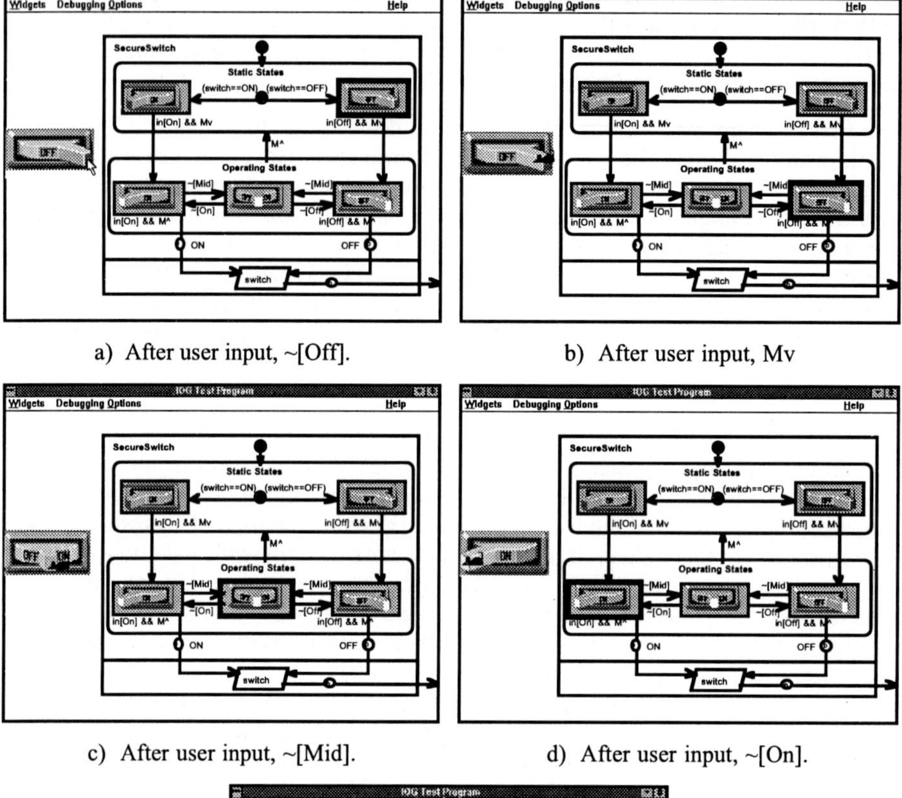

a) After user input, ~[Off].

b) After user input, Mv

c) After user input, ~[Mid].

d) After user input, ~[On].

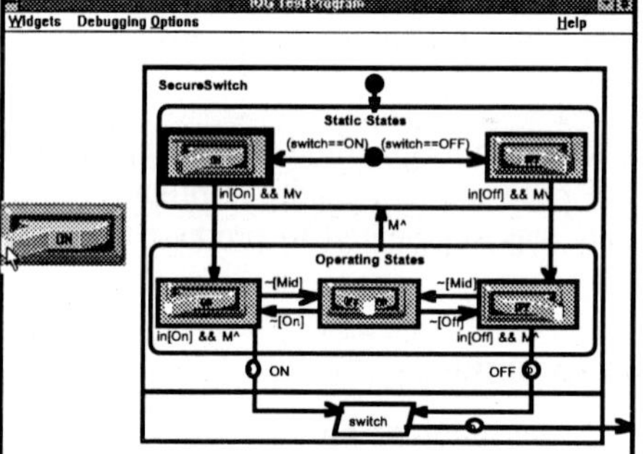

e) After user input, M^.

Fig. 7 Animation of the secure switch.

The upper-right display state is outlined in red on the display. (This appears as a wide rectangle in Figure 7a.) When the user presses the mouse button while pointing in the *Off* region, the animated diagram changes. The lower-right display state is now outlined (Figure 7b). (Note, the mouse cursor has changed to be the outline of a mouse. This change is used when demonstrating for large audiences to inform them that the mouse button has been pressed. Changing the cursor makes an invisible user action visible.) Next, as the user moves the mouse into the *Mid* region with the mouse button pressed, the IOG diagram changes so that the lower-middle display state is outlined (Figure 7c). As the user continues moving the mouse into the *On* region without releasing the mouse button, the IOG diagram changes with the lower-left display state outlined (Figure 7d). Finally, when the user releases the mouse button, the upper-left state is outlined (Figure 7e).

While no study has been done to evaluate the effectiveness of animation in improving specification understanding, animation has made it much easier to explain the IOG specification method. Animation viewers grasp the essentials of IOGs quicker. Also, they begin to find usability problems with the widgets specified and can propose solutions. (For example, the *On* and *Off* regions are not explicitly displayed for the user, and there is no state explicitly indicating that the switch will revert to its original value. This can be fixed by adding two more display states showing that the switch will revert.)

Conclusion

Current specification methods are difficult to understand. In addition, many are not executable. If specification is to become an accepted practice in the software industry, both problems must be overcome. Preliminary work suggests that a possible solution to these problems would be an executable specification with a diagrammatic representation that could be animated as the specification was executed. However, more research is required. An experiment to determine if animation actually helps is necessary. If this proves to be the case, much work is required to scale the animation and specification method to larger interfaces. The class library itself needs to incorporate the animation more directly, and better tools for browsing and editing the specification are required. In any case I believe that the animation technique could be applied to any specification based on a transition diagram or one of its extensions.

Acknowledgments

I wish to thank Carl Rollo for his work in implementing the Microsoft Windows' interface and implementing the animated specification. I would also like to thank him for proofreading drafts of this paper. Special thanks also go to Sylvia Sheppard and Christopher Rouff of NASA Goddard Space Flight Center for their encouragement and support. Much of this work was carried out while I was at the Human-Computer Interaction Laboratory of the University of Maryland, and I would like to thank my colleagues there for their suggestions for improvements in the development of IOGs.

References

1. Brown, M. H. and R. Sedgewick: Techniques for algorithm animation". *IEEE Software*, 2(1), Jan. 1985, 28-39.

2. Carr, David: Specification of interface interaction objects. *Proceedings of ACM CHI'94 Conference on Human Factors in Computing Systems.* 372-378.

3. Carr, David, *A Compact Graphical Representation of User Interface Interaction Objects*. University of Maryland, Department of Computer Science, Ph.D., 1995.

4. Fitter, M. and T. R. G. Green: When do diagrams make good computer languages?. *International Journal of Man-Machine Studies,*11(2), 1979, 235-261.

5. Gieskens, Daniel and James Foley: Controlling user interface objects through pre- and postconditions. *Proceedings of the ACM CHI'92 Conference on Human Factors in Computing Systems*, 189-194.

6. Glinert, Ephraim P. and Steven L. Tanimoto: PICT: An interactive graphical programming environment. In: *Visual Programming Environments: Paradigms and Systems*. E. Glinert, ed. IEEE Computer Society Press, Los Alamitos, CA, 1990, 265-283.

7. Gray, P. D., D. England, and S. McGowan: XUAN: enhancing UAN to capture temporal relationships among actions. In G. Cockton, S. W. Draper and G.R.S. Weir, ed. *People and Computers IX*, 1994, Cambridge University Press, Cambridge, UK., 301-312

8. Guttag, John and J. J. Horning: Formal specification as a design tool. *Proceedings of the Seventh ACM Symposium on Principles of Programming Language.* 1980, 251-261.

9. Haarslev, Volker and Ralf Möller: VIPEX: visual programming of experimental systems. In: *Visual Languages and Visual Programming*. Shi-Kuo Chang, ed. Plenum Press, 1990, New York, NY, 185-212.

10. Hartson, H. Rex, Antonio C. Siochi, and Deborah Hix: The UAN: a user-oriented representation for direct manipulation interface designs. *ACM Transactions on Information Systems*, 8(3), July, 1990, 181-203.

11. Hartson, H. Rex and Phillip D. Gray: Temporal aspects of tasks in the user action notation. *Human-Computer Interaction*, 1992, vol. 7, Lawrence Erlbaum Associates, 1-45.

12. Hartson, H. Rex and Kevin A. Mayo: A framework for precise reusable task abstractions. *Interactive Systems: Design, Specification, and Verification.* F. Paterno, ed. Springer-Verlag, 1994, 279-297.

13. Harel, David: On visual formalisms. *Communications of the ACM.* 31(5), May, 1988, 514-530.

14. Hill, Ralph D.: Supporting concurrency, communication, and synchronization in human-computer interaction - the Sassafras UIMS. *ACM Transactions on Graphics,* 5(3), July, 1986, 179-210.

15. Hudson, Scott E.: Graphical specification of flexible user interface displays. *Proceedings of the ACM SIGGRAPH Symposium on User Interface Software and Technology,* 1989, 105-114.

16. Jacob, Robert J. K.: A specification language for direct-manipulation user interfaces. *ACM Transactions on Graphics*, 5(4), October, 1986, 283-317.

17. Jensen, Kathleen and Nicolas Wirth: *PASCAL User Manual and Report.* Springer-Verlag, New York, 1975.

18. Murkherjea, Sougata and John T. Stasko: Toward visual debugging: integrating algorithm animation capabilities within a source-level debugger. *ACM Transactions on Computer-Human Interaction*, 1(3), Sept. 1994, 215-244.

19. Myers, Brad A.: A system for displaying data structures. *Computer Graphics.* 17(3), July, 1983, 115-125.

20. Myers, Brad A.: *Creating User Interfaces by Demonstration.* Academic Press, Boston, 1988.

21. Olsen, Dan R. Jr. and Kirk Allan: Creating interactive techniques by symbolically solving geometric constraints. *Proceedings of the ACM SIGGRAPH Symposium on User Interface Software and Technology,* 1990, 102-107.

22. Payne, Stephen J and T. R. G. Green: The structure of command languages: an experiment on task-action grammar. *International Journal of Man-Machine Studies,* 30(2), Feb., 1989, 213-234.

23. Plaisant, Catherine, and D. Wallace: Touchscreen toggle design. *Proceedings of the ACM CHI'92 Conference on Human Factors in Computing Systems.* 667-668 (Video).

24. Palanque, Phillipe, and Rémi Bastide: Petri net based design of user-driven interfaces using the interactive cooperative objects formalism. *Interactive Systems: Design, Specification, and Verification.* F. Paterno', ed. Springer-Verlag, 1994, 383-400.

25. Reiss, Steven P.: PECAN: program development systems that support multiple views. In: *Visual Programming Environments: Paradigms and Systems*, E. Glinert, ed. IEEE Computer Society Press, Los Alamitos, CA, 1990, 324-333.

26. Rouff, C. and E. Horowitz: A system for specifying and rapidly prototyping user interfaces. *Taking Software Design Seriously*, J. Karat, ed., Academic Press, 1991, 257-272.

27. Shneiderman, Ben: Multiparty grammars and related features for defining interactive eystems. *IEEE Transactions on Systems, Man, and Cybernetics,* 12(2), 1982, 148-154.

28. Shimomura, T. and S. Isoda: Linked-list visualization systems. *IEEE Software*, 8(3), May, 1990, 44-51.

29. Siochi, Antonio C., and H. Rex Hartson: Task oriented representation of asynchronous rser interfaces. *Proceedings of the ACM CHI'89 Conference on Human Factors in Computer Systems*, 325-330.

30. Singh, Gurminder, Chun Hong Kok, and Teng Ye Ngan: Druid: a system for demonstrational rapid user interface development. *Proceedings of the ACM SIGGRAPH Symposium on User Interface Software and Technology,* 1990, 167-177.

31. Stasko, John T.: "TANGO: a framework and system for algorithm animation. *Computer*, 23(9), Sept., 1990, 27-39.

32. Wasserman, Anthony I.: Extending state transition diagrams for the specification of human-computer interaction. *IEEE Transactions on Software Engineering,* SE-11(8), August, 1985, 699-713.

33. Wellner, Pierre D.: Statemaster: a UIMS based on statecharts for prototyping and target implementation notation for specification. *Proceedings of the ACM CHI'89 Conference on Human Factors in Computing Systems,* 177-182.

Towards an integrated proposal for Interactive Systems design based on TLIM and ICO

Philippe Palanque[1,3], Fabio Paterno[2], Rémi Bastide[3], Menica Mezzanotte[2]

[1] CENA
7 avenue Edouard Belin
31055 Toulouse, France
Tel: +33 62 25 95 91
palanque@cena.dgac.fr

[2] CNUCE - CNR,
36, via Santa Maria,
56128 PISA, Italy
Tel: +39 50 59 32 89
F.Paterno@cnuce.cnr.it

[3] LIS - IHM
Université Toulouse I
31042 Toulouse, France
Tel: +33 61 63 35 88
bastide@cict.fr

Abstract The importance of applying formal methods in the design and development process of Interactive Systems is increasingly recognised. However it is still an open issue the identification of systematic methods able to support designers and developers in specifying and demonstrating properties of user interfaces. TLIM and ICO are two formal methods which have been used for this purpose with interesting results. They address similar concepts but also have different features which allow us to consider useful their integrated use to obtain synergistic and complementary results. In this paper we show their application to some examples in order to discuss similarities and differences and we outline a proposal for their integrated use.

1. Introduction

In the domain of interactive systems, there is a lack of structured methods which can drive the work of designers and developers especially for applications which require sophisticated interaction techniques. We think that a valid answer to these problems can be found only by expressing both tasks and system models in a formal way. This is precisely the aim of both TLIM and MICO methods that have already been introduced in [31, 28].

The purpose of this paper is twofold:

- to compare the two methods (TLIM and MICO) for the design, specification, verification, development and evaluation of Interactive Systems both based on the use of formal techniques;
- to identify elements useful for a proposal which takes features from the two previous methods and indicates a new method, with related tools and techniques, for improving the design and development of user interfaces.

We believe that the goal to obtain a single notation powerful enough to express all the relevant aspects in user interface design will not find a successful result because such a notation should be very complicated to develop and to use. This complication in most cases would be useless as usually designers would be interested only in a specific subset of its functionality as they often need to focus only on some aspects which depend on their current goals. Thus we think that a more successful approach is to identify a set of interesting notations and to indicate clearly when one is better than the others and how concepts from a notation can be mapped onto concepts of another notation.

Two aspects are important in the specification of Interactive Systems: parallelism and temporal ordering among actions. Thus we will consider two notations based on these concepts: Petri Nets [29] and LOTOS [14] and the goal of this work is to evaluate two methods based on them, to check whether they are both useful, or complementary, and can be used with synergistic effects.

The global goal of this research work is to be able to both give precise descriptions of the relevant aspects in the design of Interactive Systems and provide the possibility to reason about properties which are important in the design, development and evaluation of user interfaces.

2. Related work

UAN [12] is an example of successful notation which has been used to describe both tasks and external behaviour of corresponding user interfaces. We aim to describe also the software component controlling the external user interface in order to provide support to both developers and designers.

Sutcliffe and Farady [36] have developed an interesting approach which starts from the result of task analysis and provides useful suggestions for the design of multimedia user interfaces taking into account aspects such as the type of information to be presented, the communicative goal and the resources available. However they do not provide precise specifications of the design rules and the resulting specification.

Gray and Johnson [11] have discussed the requirements for the next generation of user interface specification languages. They argue that many existing approaches cannot easily be used to capture the temporal properties that characterise interaction with distributed systems. They indicate Branching Time Temporal Logic, XUAN and Petri Nets as examples of notations which avoid this limitation and recognise the need to develop a hybrid notation from the best features of several existing formalisms.

The idea of an integrated use of LOTOS and Petri Nets has already been recognised useful. Sisto and Valenzano [38] have defined how to map Petri nets with inhibitor arcs into basic LOTOS specification preserving strong bisimulation equivalence. However, both basic LOTOS and Petri nets with inhibitor arcs have an expressive power too low for an easy modelling of interactive systems. Another approach for translating Full LOTOS specifications into Petri nets can be found in [15].

Instead of comparing formalisms in general terms, this paper compares two methods addressing the specific requirements of interactive systems design. These methods are based on higher level models which are Full LOTOS and Interactive Cooperative Objects (a formalism based on Petri nets and objects) respectively.

3. Basic concepts for formal design of interactive systems

Many studies (such as [6] p. 404) have highlighted that the design of Interactive Systems depends on three models:

- **user model**, a specification to describe the user behaviour and characteristics, mainly from a cognitive point of view;

- **tasks model**, the abstract description of how the user can achieve some desired state modifications which are associated with his/her goals;
- **system model**, the identification of the basic architectural components which model the implementation.

In all these three models the key elements are:

- **requirements**, which can be considered as specific properties that have to be satisfied; for example temporal requirements that express temporal behaviour and constraints where time is considered from a quantitative point of view;
- **notation**, the formalism used to describe the different models. These notations can be either formal or informal but as far as system model is concerned, we consider notations developed in the software engineering field ;
- **tools**, whether there exit automatic support for the building of both the specification and the implementation and for further processing such as verification, simulation and execution of the specification;

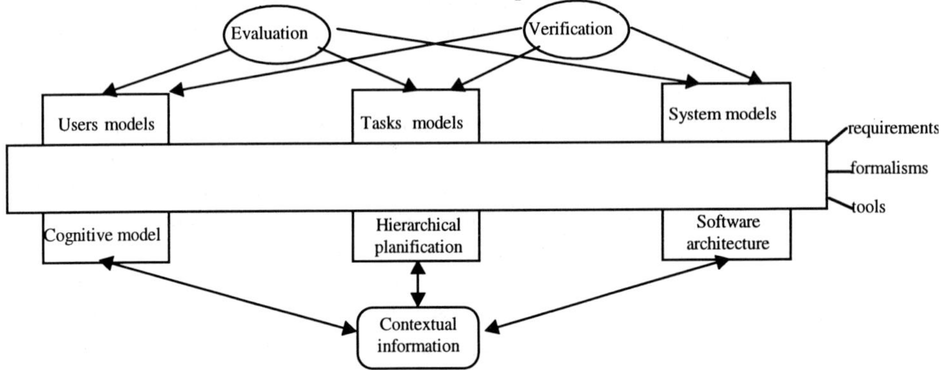

Fig. 1 Basic concepts for formal design of Interactive Systems.

Once the design of a specific Interactive System has been obtained by integrating the three models various results are possible:

- **verification**, verifying whether specific properties or requirements are supported by a specification;
- **contextual information**, it is possible to exploit a model in order to have automatic generation of information or help which take into account the current context;
- **evaluation**, the possibility to evaluate specific aspects (such as complexity, time performance, task support, etc.) using the specification.

Figure 1 shows the different models involved in the design of interactive systems. Each model (user, task and system) relates to different underlying model written in the bottom of the square boxes. The rounded box shows one immediate benefits of using embedded models in interactive systems design. The ellipses show processes that can be performed on the models. Verification can be performed if the models are built using a formal approach offering analysis techniques. Performance evaluation can only be computed if the models are built using formal approaches offering both quantitative aspects modelling and performance evaluation mechanisms.

The three layers (requirements, formalisms and tools) are shown in an horizontal bar stating that they are meaningful for the three different models.

4. The design process

In this paragraph we give a short outline of the methods which we consider and we discuss both of them with respect to the issues indicated in the previous section.
In the TLIM method the tasks are identified at the beginning together with their constraints which are described by LOTOS operators. Tasks are described in a hierarchical way using LOTOS operators to indicate their temporal relationships. For each task it is possible to provide tables with objects and actions. Next we have to identify how tasks are associated with software components. The transformation is driven by the tasks relationships and we obtain an interactor-based description of a system which comply the task constraints. This architectural description can be graphically described and it can be translated into a corresponding LOTOS expression. The LOTOS specification can be used as the model to check automatically properties expressed by Action-based temporal Logic. The result of the verification is useful to evaluate the design performed. Depending on the result of the evaluation both the task specification and the formal specification of the software architecture can be modified. Once a satisfying design has been obtained the prototyping of the implementation is almost a mechanical exercise because we developed an interactor-based toolkit which provides software objects with the same architecture as the objects used to structure the specification.
In the MICO method we can start the design process using either a preliminary task model or a preliminary system one. This preliminary task (respectively system) model is then used in order to build the system (respectively task) model. Each time a model is built, verification is performed on it in order to ensure its soundness. Then the task and system models are evaluated in order to check the conformance of the system model with respect to the task model. This is done by automatically building a single model by merging both tasks and system models. The analysis of this model will determine whether or not the models are compatible (see [24] for a complete description of the process). If the models are compatible (all the tasks are supported by the system model) some performance evaluation (such as time to achieve a task) is computed. This cycle is performed until the result of the performance evaluation is good enough. This only describes how the different stages of the MICO refer to the design process (see for a complete description [23]). The precise description of how the system models are built according to the three classical layers (Presentation, Dialogue and Functional Core) can be found in [28].
One relevant difference is that in TLIM the task model is used as an abstract specification which is used to model the system specification, while in MICO they are considered at the same abstraction level and the designer can start to work with one of them without any difference.

4.1 The Formalism

In TLIM there is a mix of formal notation and informal graphical representations. The LOTOS notation is used to perform specification of the three main elements (user, task and system). It is oriented to express temporal ordering of actions that a concurrent system can perform. Often an informal graphical representation is used for describing the software architecture in a more immediate way with respect to the LOTOS specification (we have boxes corresponding to each single LOTOS process and arrows associated with the synchronisation gates). Requirements and properties are expressed using Action-based temporal logic [7] which is a branching time temporal logic which allows designers to reason about actions that the entity considered can perform.

The MICO method heavily relies on one formalism: the Interactive Cooperative Objects. This formalism is based on both Petri nets and the object oriented approach. This formalism is used for the description of tasks, users and system models. Petri nets concepts are used in order to describe the dynamics in the models while object-oriented ones are used for structuring concerns. The object oriented approach provides a set of other notations such as inheritance trees, use relationship diagrams that are used in order to describe the software architecture and to structure the models. At the moment we have no notations in order to describe properties or requirements. We plan to use temporal logic to describe them and to use work that has already been done in the Petri net community in order to prove that the Petri nets models verify the properties expressed in temporal logic.

4.2 Task models

In TLIM the task specification is used to describe the set of possible user tasks and their temporal constraints. LOTOS operators (such as interleaving, enabling, disabling, synchronisation) are used to describe these constraints. In the task specification only constraints related to the task level are expressed without considering possible constraints introduced by the user interface implementation. In the specification there is a hierarchical decomposition of the possible tasks. It is possible to give further information about each task indicating what the related semantic objects and actions are.

In MICO task models are used in order to describe user's action on a system. Those actions are structured according to goals. A goal is a high level abstraction of task. For a given goal there are several possible task models describing the possibilities to achieve this goal. In a task model we describe the sequences of actions that the user has to perform on the system in order to achieve his/her goal. Task models are described using the ICO formalism. This formalism based on Petri nets and objects allows to describe temporal relationship between actions and the synchronisation of several flow of control usually called multi threading which is frequently encountered in interactive systems. Time extensions of Petri nets are added to the task description. This is used during the performance evaluation phase when it comes to find out the complexity of a task.

4.3 Software architecture

In TLIM software architectures are modelled following the interactor model. It is an abstract model to describe software objects which have to interact with users. It is characterised as an entity able to support a bidirectional information flow from the

user side to the application side and vice versa. Its communication with the outside is structured by six types of channels which are classified depending on they are used for information communication or control events (triggers) or if they are used to receive (or to send) information from (to) the user or the application. On the implementation side this model has been implemented in an object-oriented programming language in such a way to maintain the properties of the abstract model.
In MICO we address software architecture in two different ways.
The first one is to describe what a basic component of the system is: the behaviour, the data structure and the services. For services and data structure we use the object oriented constructs while for the behaviour we use Petri nets.
The second way is to describe how the components cooperate. This is done using a formal client server protocol expressed in Petri nets. This protocol is used for the cooperation between objects in the system model but also between system model and the other models (user and task). The resulting software architecture can be directly implemented using any classical object-oriented environment.

4.4 Verification

In the MICO method we verify properties on the quality of the design of the models. We address properties such as absence of deadlocks, reinitialisability, computation of reachable states. This verification is done on system, tasks and user specifications. According to the software architecture, this verification can be done either on a precise component in order to verify its soundness (what we call unitary verification) or on a set of components cooperating together (what we call cooperation verification). Unitary verification is performed using results available from the Petri net theory while cooperation verification is done using results on the formal client server protocol. Algorithms for the verification of properties are available and it is thus straightforward to add a verification module in a design environment. A full study about both unitary and cooperation verification can be found in [24]. Another verification can be performed using performance analysis techniques. Indeed, undesirable behaviours of the system can be found by looking at the performance analysis results.
In TLIM verification is performed by general tools for model checking: the LOTOS specification of the Interactive System is automatically translated into a corresponding finite state transition system. Next ACTL properties can be automatically checked against this model and the tools can indicate whether or not they are verified and in the negative case give information about why they are not verified. ACTL has been used to prove:

- usability properties (such as continuous feedback, visibility, ...);
- task-related properties (such as task reachability, possibility to perform a task at any state, task equivalence);
- safety properties (whether some bad behaviour can happen).

One limitation in this approach is that there are LOTOS expressions which correspond to transition systems with an infinite number of states thus the model checking approach needs to be replaced with new techniques which are being developed and which, at this time, are able to provide answers only in a subset of cases. A discussion about when applying these types of techniques for verifying user interface properties can be found in [19]

4.5 Time

Temporal aspects are fully incorporated in the MICO methodology. Indeed, the use of Petri net for describing the behaviour of objects in models allows us to describe qualitative temporal relationships between actions. These temporal relationships can be before, after, meanwhile as well as concurrent behaviour. Moreover, it is possible to use temporal extensions of Petri nets in order to describe quantitative temporal relationships both at a static and a stochastic level. Static temporal aspects allow descriptions such as this will happen in 10 seconds or this will long 1 minute. Stochastic temporal aspect allow us to model the stochastic behaviour of the user when interaction is concerned. For example it is possible to use probabilistic laws in order to describe the time that will be elapsed before the user presses a button after a message appears. The mathematical foundations of Petri nets can be used in order to analyse the behaviour of the models with those quantitative temporal descriptions in order to compute performance evaluation both on individual components and on the set of components cooperating together (see [27] for more information on performance evaluation). The quantitative time aspects are useful for describing multimodal interaction and a concrete example is given in [1].

In TLIM quantitative time related problems have been addressed in limited way. Most of the LOTOS specifications have been performed using standard LOTOS which does not include time. However, some work has been developed using a time-oriented version of LOTOS which now is being subjected to the standardisation process. This work [17] has shown that it is possible to specify formally timed-oriented interactors which are useful especially to describe multimodal interactions. The main problem with time-oriented LOTOS is the current poor automatic support which limits its application. Stochastic-oriented extensions of LOTOS have been developed as well but they have not been applied to the Interactive System field.

4.6 Tools

In TLIM general purpose tools for formal specifications are used and new specific tools have been developed. The general purpose tools more often used are those which check correctness of the specification, which allow designers to simulate its behaviour, and those related to automatic verification of properties by model checking. Furthermore some automatic tools have been specifically developed to support graphical editing of the tree of tasks, to translate this tree into a corresponding LOTOS expression, to support transformation from the task specification to the architectural specification, to edit graphical representation of the architectural specification. Finally a toolkit, following the interactor model, has been implemented where the available interaction classes are mainly classified depending on semantic, task-oriented aspects. Now a new version of this tool is being implemented by using the Java programming language.

At present time there is no environment available for supporting the design of interactive systems according to the MICO methodology. However, this development is on the way and part of it is already available. The kernel of the environment (called PetShop for Petri net Workshop) has been developed so that it is now possible to execute several object communicating together (see [4] for the description of the PetShop). The link with the user interface management system is not achieved yet so there is no relation between the objects and user's actions. However, algorithms for going from the specification to the implementation have been designed and can be found in [5]. Besides, the tools for the verification of models are available as they are used as is. This is the same for performance evaluation which is fully supported by available Petri nets environments.

4.7 The User Model

In TLIM the first applications of user models have been accomplished by describing the user as a LOTOS expression able to perform both internal and external actions. This type of structure of the specification is meaningful for both the user model and the LOTOS model. The resulting specification can give a complete description of the set of traces of actions which can be generated by user and system interactions. A more structured specification of the user has been developed [18]: it has been described as a set of LOTOS processes which share the same behaviour and which are associated with specific cognitive subsystems following the ICS model [2].

It is possible to derive guidelines from user models, such as ICS, to drive the mapping from abstract user tasks to software objects which have to support them in a multimedia environment.

When we want to obtain a complete description of an Interactive System we can just compose the corresponding LOTOS expressions indicating synchronisation for those user actions which activate a specific behaviour on the system side. This is obtained by exploiting the strong support to compositionality provided by LOTOS: the construction of complex systems as the combination of simpler components obtained by just applying the composition operators for processes composition.

In the ICO approach the users models we build are timed Petri nets. At the moment we use a very simplistic approach to user modelling, it is the one proposed in the human processor proposed by Card, Moran and Newel. This allows us to have quantitative values about what the user is able to (both according to the physical and psychological point of view). This model is then used in order to evaluate whether or not the task and system models are adequate with respect to the user characteristics [27]. Another, way we followed for modelling user’s cognitive behaviour is to use a common Petri net for system, device and user [16]. The next step in user modelling is to take into account higher level cognitive models such as ICS [2] in order to evaluate tasks and system models wrt more domain related information.

4.8 Contextual help

As MICO is a model-based method and as the models are available and executed at run time it is possible to use the models in order to provide contextual information about (for example) why an action is or is not available (see [21] for a precise description). Besides, we have improved this first level of help by cloning each model (task or system ones) and decorating them with more specific information. The use of

the task model as an input for help allows the users to have information about how to terminate a task thus how achieve a goal. On the other side the use of system model allows the users to have information about how the system is working and why the user is encountering difficulties.
In TLIM there is a mapping between tasks and the interactors in the formal specification used to perform them (which can be a 1 to 1 or a 1 to n mapping). Finally there is a 1 to 1 mapping between interactors in the formal specification and in the implementation. These relationships are useful at run-time in order to provide task oriented information about what the user can do, and how can do it. A specific algorithm to navigate in the task tree in order to identify how to enable a task has been defined [32]. At this time the help messages are generated by composing pieces of pre-written text which are composed by connectors such as *and* or *or* which are chosen depending on the relationships among the corresponding tasks.

4.9 Evaluation

As a consequence of the TLIM method a prototypal version of an evaluation tool (TASM) has been developed [33]. In this case the basic idea is to take input from the task-driven specification of the system and the file logs which store the events generated by the user in a specific session. The tool is able to map the physical events which occur during the user session in actions of the LOTOS specification so that next it can evaluate them with respect to the tasks performances (whose completion are associated with other actions in the LOTOS specification) and give some evaluation about user preferences or difficulties in task performance which can be used to improve the user interface design.
While designing interactive systems with the MICO, evaluation is performed after each stage in the design life cycle. This evaluation is performed at two different levels. The first one delivers result about the quality of the design and heavily relates to the verification of properties. The second one delivers results about the compatibility between the system model and the task model by doing performance evaluation on the models. This evaluation is used in order to detect when the design process has to be stopped because the actual design is satisfactory.

5. Integration of the design methods

In this section we will first position the two methods with respect to the classical V design life cycle in order to express the complementarities, the overlapping or where some work is still necessary in order to have a well defined method covering the various stages of the design life-cycle. Among the models used in software engineering we consider the V cycle model because it gives a clear indication of the phases in the development and in the evaluation processes and of their relationships.

5.1 Position of the methods with respect to the V cycle

In the TLIM method the statement of requirements can be expressed by a set of ACTL properties. The design description is a graphical representation of the interactor needed indicating their compositions. The module design is the LOTOS specification from which it is possible to derive an object-oriented implementation. The integrating testing is the evaluation which can be carried out by using the logs of the user events and the information gathered from the architectural specification. The verification of ACTL properties can be considered the acceptance testing phase.

The MICO method does not provide specific information and processes for the requirements analysis phase. This is usually done by following object-oriented approaches such as [34] or [37]. The architectural design phase structures the objects in the three main classes (passive classes, cooperative classes and interactive cooperative classes). This phase details for each of these classes the services they offer, the data they handle and the "specification" of the object control structure. This specification is modelled using Petri nets which only describes the availability of services according to the possible states of the classes. This phase can be named the what phase as it only deals with what the objects offer to their environment and not how they offer it. The detailed design phase deals with the how part of the design i.e. how objects provide services and how they implement it. This is called "implementation" part of the object control structure. This distinction is not very important while dealing with simple examples such as the ones presented in this paper but as far as real applications are concerned, this allows an easier design and understanding of the system.

The coding phase can be automatically performed as Petri net specification can be directly executed using interpreted [4] or compiled ways [28]. As shown in Figure 3 and Figure 6 the unit testing is done by the formal verification of the individual Petri net models. The "integration" and "integration testing" phases correspond to the formal verification of the Petri nets cooperating together according to the client server protocol. This protocol has been formally specified using Petri nets and an example of the verification of cooperating objects can be found in [26].

5.2 Complementarities and overlapping

As the methods address the same aspects of software engineering i.e. the formal design of interactive systems it is not surprising that they overlap widely on most of the phases of the design life cycle.

However, they can be seen as complementary on several aspect:

- only the TLIM method provides specific treatment of the requirements phase by modelling requirements using the temporal logic dialect ACTL. Thus integration could be to use the same mechanisms in the MICO methodology. This would enable development teams to use the same requirements but to have different way of specifying a system meeting those requirements.
- during the V cycle different teams with different backgrounds might be involved in the design of the system and thus the two methods can provide information for the design of interactive systems for both Petri nets and LOTOS specialists.
- it is possible to use the two notations in an integrated way, for example in [38] it is proposed to include Petri Net specifications of basic building blocks into a LOTOS specification of the whole system. This may be useful to exploit the compositionality of LOTOS in order to combine basic blocks specified as nets and it may be useful in case the specification of a module is easier or more convenient to develop in Petri nets rather than LOTOS as, sometimes, Petri nets are more intuitive, especially for small specifications.

6. Example of application

In this section we consider three examples of user interactions. We describe them by both approaches and finally we discuss the usefulness to apply both approaches. We do not consider the task models related to the examples in order to simplify the discussion and as we want to focus on the comparison of the two methods in specifying user interfaces.

6.1 A simple sequence of buttons

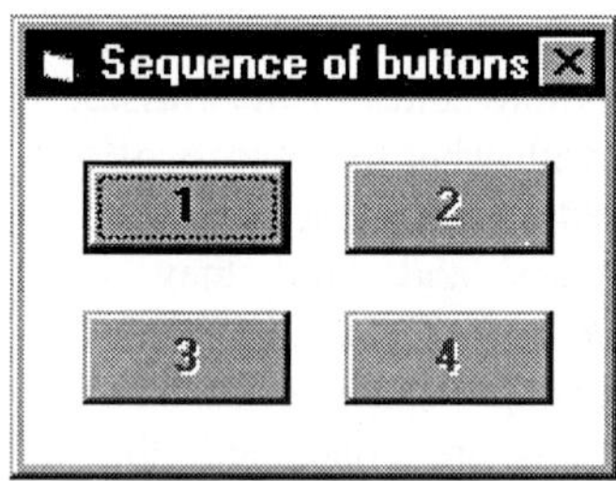

Fig. 2 Presentation part the simple sequence of buttons application

This small example (see Figure 2) aims at showing how it is possible to relate interactive system's specification using Petri nets with LOTOS. This very simple example behaves as follow:

- at the beginning, only the first button is available (the three other ones being deactivated),
- by clicking on the available button (i) deactivates it (shown as greyed out) and activates the next one (i+1),
- each time a button is triggered by the user, a page must be sent to the printer. It is supposed that the function Print(x) which print the page number x of a given document is provided. The specification will thus only show how this operation hold by the functional core can be triggered by the user, using the presentation shown in Figure 2.

6.1.1 ICO specification of the sequence of buttons.

Using the ICO formalism this simple application is modelled using two different classes :

- a non interactive class corresponding to the functional core of the application named **NFObject**
- an interactive classed related to the user named **Window-button**.

The specification of these two classes is given below.

The class NFObject features one internal operation (it cannot be requested by other classes) Print (n) which corresponds to the functional core operation. It offers four services (they can be requested by other classes) to the environment each of these being able to print a given page on the document (Print1 is able to print the page 1 of the document). The ObCS is trivial as it consists only in the mutual exclusion of the four services. As this object is not aimed at being triggered by the user, it does not feature a Presentation part.

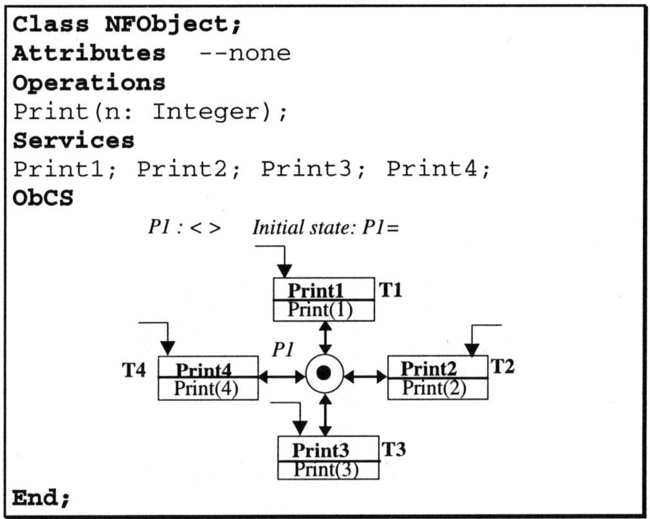

```
Class NFObject;
Attributes   --none
Operations
Print(n: Integer);
Services
Print1; Print2; Print3; Print4;
ObCS
```



```
End;
```

The class Window-buttons is a full fledged ICO as it features a Presentation part. The user can use it by interacting on four different widget (of the kind PushButton) as shown both in Figure 2 and in the Presentation part of the class definition. The ObCS of the class (see Figure 3) describe the dialogue of the application (what is the available sequencing of user's actions) and how the class communicate with the other class of the specification (this is modelled in the action part of the transition by the code `x.Print`). In the same time it relates to the presentation part by showing to the user the available actions by greying out the widgets that are not currently available.

When using Petri nets for describing interactive applications, transitions are used for describing actions (usually performed by means of interactors). At the opposite, places describe state variables and the distribution of tokens in the places describe the actual state of the model.

In Figure 3 the initial state is depicted by a single token in place P1. Thus only the transition T1 is available. This allows us to describe directly the dialogue of the simple application. Indeed, as each transition is related to an action on a widget of the application, the availability of the transition fully states the availability of the widget action. For the particular example of the buttons in the simple application, there is only one possible action on each button directly mapped to a transition in the Petri net model (this is described by the activation function of the class).

```
Class Window-buttons;
Attributes --none
Operations  --none
Services  --none
ObCS (see Figure 3 right part)
Presentation (see Figure 2)
```

```
  Widgets           name: B1, caption: "1";
            name: B2, caption: "2";
            name: B3, caption: "3";
            name: B4, caption: "4";
  Activation function (see Figure 3 left part)
End;
```

Widge t	User's actions	Trans ition
B1	Click	T1
B2	Click	T2
B3	Click	T3
B4	Click	T4

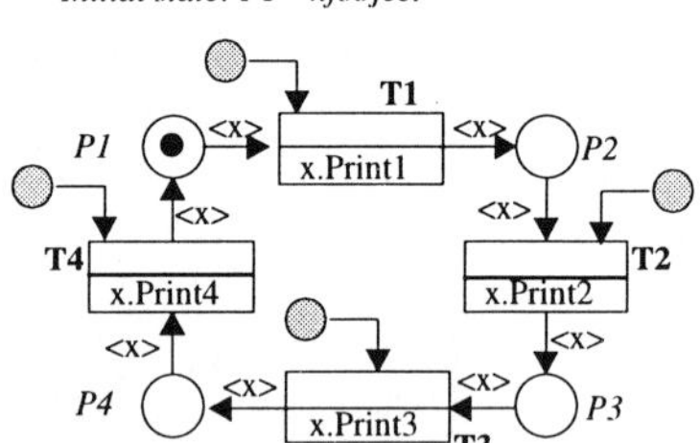

Fig. 3 Petri net describing the behaviour of the sequence of buttons (right part) and the activation function (left part)

The structure and the physical behaviour of the objects directly represented on the screen (buttons and the window) are not described here as they are well known and this description is the main point of the most user interface builders. However, most of the widgets available in the norm CUA [13] have been modelled using the ICO formalism. This can be found in [20].

6.1.2 LOTOS specification of the sequence of buttons.

The LOTOS specification follows the interactor model. Thus we have four interactors, one for each button and each interactor is described by one LOTOS process. Their behaviour is the same so we only need four instances of the same process. The difference among the instances of the process definition is in the gates that they use to communicate.

The process which we define to describe the button behaviour is:

```
process int_but[input_receive, input_trigger, input_send, output_send, output_trigger]
(enabled: Bool): noexit :=
input_receive; output_send; interactor_button [...] (enabled)
[] input_trigger; ([enabled eq true] -> output_send; input_send; interactor_button [...]
(false)
                  [] [enabled eq false] -> interactor_button [...] (enabled))
[] output_trigger; output_send; interactor_button [...] (true)
endproc
```

This LOTOS expression describes a button which can receive in the *input_receive* gate the event informing that the cursor is in the button area, next there is an ouput_send event which provides feedback of it (for example by highlighting the border of the button) and then there is a recursive call to its behaviour without

modifying the state of the button. The state of the button is represented by a Boolean variable (*enabled*). Alternatively ([] is the choice operator) it is possible that the button receives either an *input_trigger* event or an *output_trigger* event. The input trigger is associated with the button pressing in the button area, if the button was enabled (this is indicated by the boolean guard [enabled]) then there is a feedback (*output_send* event) which means that the button changes its colour to indicate that it becomes disabled, it provides to the outside by the *output_send* gate the information that it has been selected and it changes its state into false, otherwise nothing happens and it calls recursively its behaviour. Finally we can have the *output_trigger* event which in this case indicates that the button becomes reactive thus it changes its colours (*output_send* event) and its state becomes true.
The specification of the four buttons behaviour is obtained by creating four instances of the int_but process, each of them with a specific set of gates. At the beginning only one process has the Boolean variable defining its state sets to true indicating that it is reactive. In the composition of the four processes we have to indicate which gates are used for synchronisation (the |[x, y]| operator). The synchronisation occurs on the gate which is used to communicate to the outside that the button has been selected. This communication is provided to the next button too and it provides an output trigger for it as when it occurs the colour of the button changes other than becoming active. Thus we obtain the following expression:

```
specification four_buttons [ ... ] : noexit :=
int_but[cursor_in1, sel1, print1, pres_but1, print4] (true)
                |[print1, print4]|
(int_but[cursor_in2, sel2, print2, pres_but2, print1] (false)
                |[print2]|
int_but[cursor_in3, sel3, print3, pres_but3, print2] (false)
                |[print3]|
int_but[cursor_in4, sel4, print4, pres_but4, print3] (false))))
```

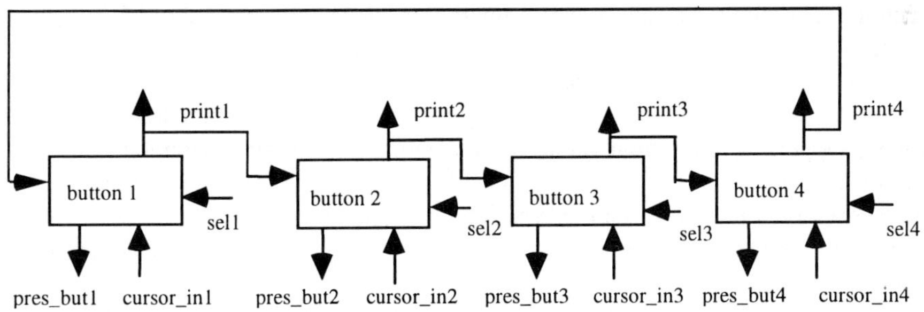

Fig. 4 Graphical interactor representation of the four button example

The LOTOS expression describing the int_but process is a Full LOTOS expression as it uses both process and data algebra. The data algebra is used to describe the state of the process, which is simply one Boolean value in this case. If we want to verify ACTL properties over this specification the Full LOTOS specification has to be translated into a Basic LOTOS specification, which means to remove the data algebra part. This translation can be done in automatic way but the result would be not completely equivalent to the initial specification because the tool remove the Boolean guards ([enable eq ...] -> operators). Thus the behaviours indicated after the two guards can occur at any time while we want that this depends on the user selections. To solve this problem we can modify the FULL LOTOS specification into an equivalent basic LOTOS specification but without using the automatic tool. The result is described by two Basic LOTOS expressions which are mutually recursive in such a way to describe the two possible behaviours of the buttons. It becomes:

```
process int_but_ab [input_receive, input_trigger, input_send,
output_send,output_trigger] : noexit :=
input_receive; output_send; int_but_ab [...]
[] input_trigger; output_send; input_send; int_but_disab[ ... ]
[] output_trigger; output_send; int_but_ab [...]
endproc

process int_but_disab [input_receive, input_trigger, input_send, output_send,
output_trigger] : noexit :=
input_receive; output_send; int_but_disab [...]
[] input_trigger; int_but_disab [...]
[] output_trigger; output_send; interactor_but_ab [...]
endproc
```

As you can see in the int_but_ab process is active, after an input trigger event occurs then the int_but_disab process is enabled. Vice versa when the int_but_disab process is active, after an output trigger event occurs then the int_but_ab process is enabled.

6.1.3 Discussion.

In the TLIM method the basic idea is to have the LOTOS specification of an interactor as a template which encapsulates the main behaviour of an interaction technique and then when one specific interactor has to be specified only the specific actions should be provided to this template. There is a set of well defined small modifications of the general interactor behaviour which can be used as refinements of the general template (for example, only input or only output interactors, interactors which disappear once an input value for the application has been generated).

The modifications of the presentation of an interactor are represented by the occurrences of the *output_send* action. The communication with the functional core (in this example the calls to the command for printing a page) is represented by the *input_send* action.

In the given example the behaviour of the widgets (only PushButtons here) is very simple and consists only in one user action with a predefined feedback (the widget changes its colour when clicked by the user). For this reason there is no modelling of the behaviour of the widget. However, this low level of interaction can be modelled using the ICO formalism as described in [20, 28]. The models make clear the software components that will exist in the implementation. The functional core is modelled by a class as well as the window. Their communication is fully and formally expressed (see the [3, 35] for a description of this protocol), even though no generic communication between widgets is described as in the interactor model of TLIM. That could be a first way of integration for the two approaches where the TLIM interactors could be used as a generic concept during the actual design of the behaviour of the widgets.

6.2 A simple temperature translator

The display of the example is shown in Figure 5. This example allows users to type in a temperature in (Celsius or in Fahrenheit) in one of the two text boxes. If the temperature is entered in the Celsius box then the button called CeToFa becomes available and if the user clicks on it the translation to Fahrenheit will be triggered and the result will be displayed in the Fahrenheit box.

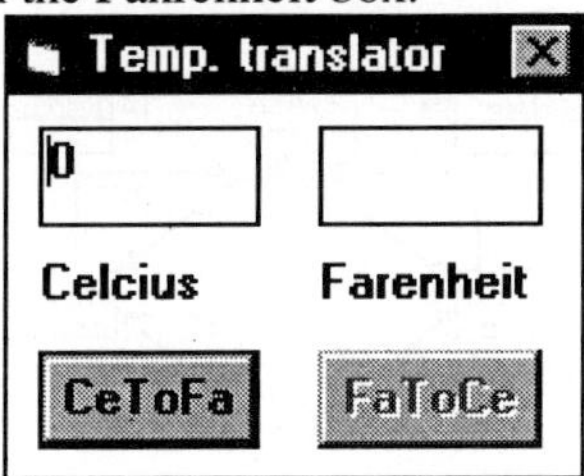

Fig. 5 The presentation part of the temperature translator application

6.2.1 Petri net specification of the temperature translator

Class Window-buttons;
Attributes --none
Operations -- internal
FaToCe (n: integer) : integer; begin FaToCe = 5/9*(n -32)end;
CeToFa (n:integer) : integer; begin CeToFa = 32 + 9/5*n end;
Services --none
ObCS (see Figure 6)
Presentation (see Figure 5)
Widgets name: B1, caption: "FaToCe"; name: B2, caption: "CeToFa"; name: Text1; name: Text2;
Activation function

Widget	User action	Transition
B1	Click	T6
B2	Click	T5
Text1	EnterKey	T2,T4
Text2	EnterKey	T1,T3

End;

The Petri net in Figure 6 describes the dialogue of the temperature translator.
At the beginning both transitions EditCelsius (T1) and EditFahrenheit (T2) can be triggered as there is a token in each of their input place. This models that the user can either enter a temperature in Celsius or in Fahrenheit. One of these transitions can be triggered when the user has pressed the EnterKey. After triggering a transition EditCelsius (T1) for example, both transitions EditCelsius (T3) and CeToFa (T5) can

be triggered. This represents that the user can either modify again (and this several times) the value in Celsius entered or directly select the translation button. By clicking on the button CeToFa, the transition T5 is fired (this is stated by the activation function of the presentation part of the class), the token form place *P2* is removed and a token is set to the place *P1*, thus putting the system back in its initial state.

P1: <>; Celcius, Franheit, P2, P3 : <Real>;
Initial State : P1 = <>; Celcius = 0; Farenheit = 32;

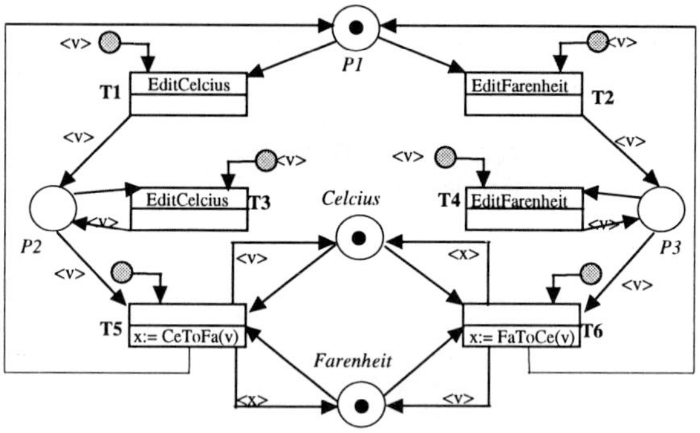

Fig. 6 Petri net modelling the dialogue of the temperature application

6.2.2 LOTOS interactor specification of the temperature translator

In this case we have two types of processes: one associated with a box for user-typed string and the other with a button. As in the previous case we need Boolean variables to control the state of the interactors in order to indicate when they are reacting to the user-generated events. The boxes for user-typed strings are interactors which receive input from the keyboard, the feedback is the presentation of the text received.
When the user provides a value in one type-in box this event is sent at the same time to other two interactors: the related button, in order to enable it for receiving user selections, and to the other type-in box in order to make it disabled to reacting at the user-generated events. Similarly the selection of a button produces an event which is an input trigger event for the related value-input box. Thus when it occurs the value-input box sends its information to the other one which receives it, translates it into the other measurement units and visualises it.
At the beginning the two type in boxes are enabled to react to the user-generated events (their control variable is set to true) while the two buttons are disabled (and the related control variables are set to false)
These composition of processes is expressed in the following way:

```
Text_in_cel [inp_cel, inp_transf1, cel, far, inp_far, pres_cel] (true)
                    |[inp_transf1, inp_cel]|
(But_CetoFa[cursor_in1, sel1, inp_transf1, inp_cel, pres_but1] (false)
                    |[cel, far, inp_cel, inp_far]|
(Text_in_far [inp_far, inp_transf2, far, cel, inp_cel, pres_far] (true)
                  |[inp_transf2, inp_far]|
But_FatoCe[cursor_in2, sel2, inp_transf2, inp_far, pres_but2] (false)))
```

The graphical representation of these composition of LOTOS processes is in the following picture.

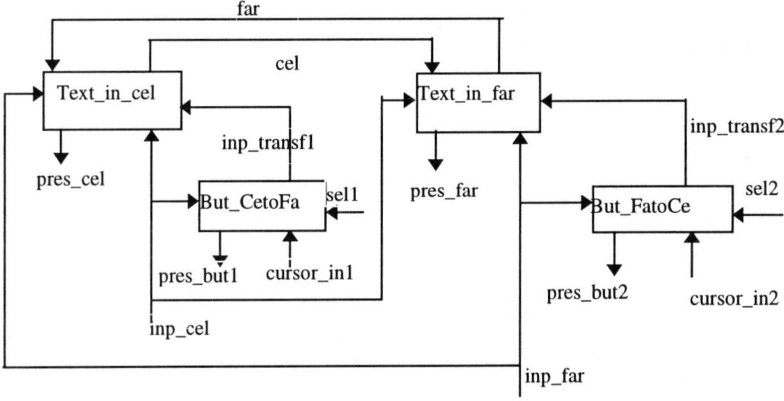

Fig. 7 Composition of interactors for the temperature translators

6.2.3 Verification of ACTL properties

We can verify properties expressed by ACTL. They allow the designer to express features which summarise the behaviour of the system considered. Here we consider simple properties as the example is a simple interaction but the approach is valid even with specification of larger dimensions.

In this case we can verify reachability properties which link user actions with modifications of the presentation such as:

AG[inp_cel] [sel1]AF[pres_far]true

This means that for all the possible futures (A operator) and for all the states (G operator) if the user performs the inp_cel action (corresponding to input Celsius value) and sel1 action (the selection of the button 1) then for all the possible futures at some time (F operator) the presentation of the Fahrenheit will be updated (pres_far action).

We can express properties which link user actions with actions associated with the user interface modifications which must happen in the next state, such as:

AG[inp_cel] (AX{pres_cel}true)

This means that whenever the user provides a Celsius value than for all the possible futures in order to get to the next state the presentation of the Celsius value has to be modified.

ACTL is a branching time temporal logics and this means that it is possible to demonstrate that at some time it is possible to perform certain actions. In this example we can check that:

[inp_cel] [pres_cel](EX{inp_cel}true & EX{sel1}true)

This means that once a Celsius value is given in input and the related feedback is provided then it is possible for the user both to provide an input value again and (& operator) to select the related button. More precisely the EX operator means that there exists a temporal evolution where in order to get to the next state the indicated action has to occur.

6.2.4 Discussion

The Petri net in Figure 6 gives an immediate indication of the activation state of each widget. The designer can understand it by just looking at the token position. The interactor-specification of Figure 7 gives an immediate indication of what the widgets used are and how they communicate. This gives us one possible way to integrate the approach: using Petri nets for indicating the activation and deactivation of the components and LOTOS interactors to describe the behaviour of each specific object.

6.3 A third example highlighting complementary analysis

This section aims at showing complementary results which can be obtained by the TLIM and the MICO methods. This is shown on a simple example that we have previously studied in [23].

6.3.1 Informal description of the example

The system is made up of two unconnected subsystems, the stop-watch and the light bulb. The light bulb is controlled by a timer. The timer is triggered by a button, and its period is 30 seconds. If the button is depressed while the light is on, the timer is reset and starts over counting down for 30 seconds. In order to use this system, the user needs to use his/her hands (to push the light and stopwatch switches) and vision in order to watch the elapsed time on the stopwatch.
Note that this information is the minimum that has to be provided to the task analyst in order to build a meaningful task model. If it is not provided, the task model designer will have to make assumptions about the behaviour of the system or he/she will only be able to discuss about general goals, at a level that will not enhance further our ability to improve the system.

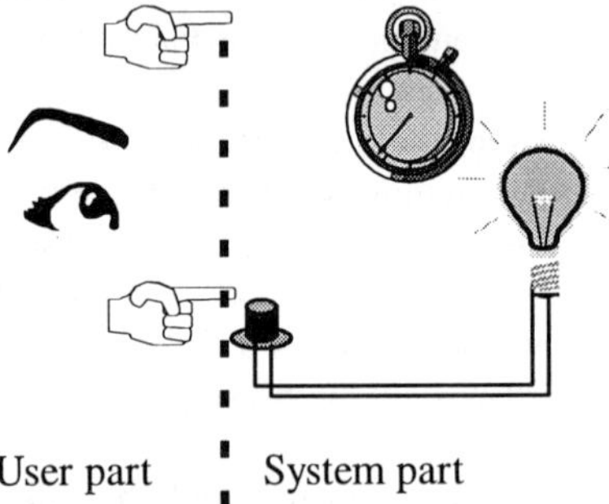

Fig. 8 The simple case study of the user keeping the light on for 3 minutes

6.3.2 Petri net modelling of the example

The model of the system is presented on Figure 9. It is made up of two components, the *TimerControlledLight* and the *Watch*. As those two components are not related, the system model is composed of two unconnected Petri nets. The Petri net corresponding to the *TimerControlledLight* is presented on the left-hand part of Figure 9. The timer controlled light offers only one affordance, a button allowing to turn it on. The model of this device also highlights other primitives of our modelling approach: In the initial state of the system the light is off which is modelled by a token (a black dot) in the place *LightOff*. From that initial state the transition PressSwitch can be triggered by the, removing the token from the input place *LightOff* and setting it in the output place *LightOn*.

The transition *PressSwitch* depicts a *user service*, meaning that the user has a way to trigger the occurrence of the transition through the associated input device. Graphically such transitions (called *user transitions*) are depicted with input and output unconnected broken arrows, which may eventually bear input or output parameters stating the data provided by the user or returned by the system. The *AutoSwitchOff* transition is an *internal transition*, not related to any user action, and performed on the system's behalf as soon as it is enabled. The arc between place *LightOn* and transition *AutoSwitchOff* is labelled with the timing inscription [30]. This means that a token staying in place *LightOn* is not available for the firing of transition *AutoSwitchOff* until it has stayed 30 time units (here, seconds) in the place. The effect of the second transition labelled *PressSwitch*, resets the time for the token in place *LightOn* by removing it and putting it back (this is graphically represented by a double arrow). This represents the resetting of the timer each time the switch is pressed. Note that two different transitions are labelled PressSwitch, which means that the button affords being pressed in two different states of the system.

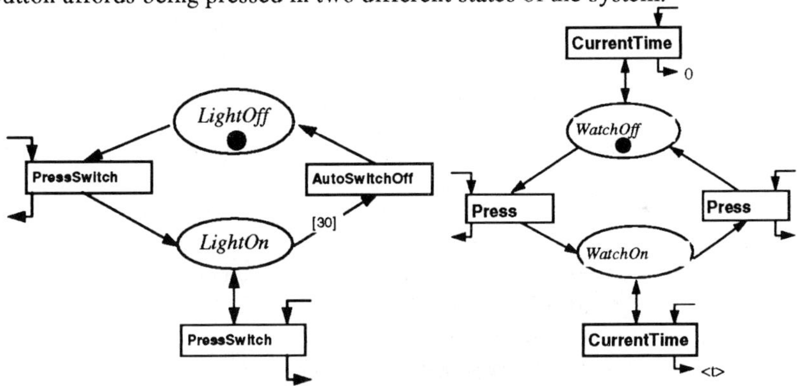

Fig. 9 **The system model of the system**

The other Petri net in Figure 9, describes the functioning of the *Watch*. The *Watch* offers two affordances: one button allowing to toggle it on and off, and the watch face itself, which allows the current time to be read. When the *Watch* is off the returned value is always zero. The watch affordances are modelled by transitions in the Petri net, labelled Press and CurrentTime. The transition CurrentTime provides of return value labelled <t>, the current time when the watch is read.

6.3.3 The task model for the Upgraded system

Given the system defined in the previous section, the task may be described informally as follows :

➊User presses the light button, and starts the stop-watch (in any order, and within a short time interval)
➋User measures the elapsing of 25 seconds,
Actions ➊ and ➋ are then repeated 6 times.
The formal task model corresponding to the above natural language description is described by the high-level Petri net in Figure 10.
Figure 10 demonstrates the complexity of the model describing the task associated with this seemingly simple goal. This complexity mainly stems from the fact that the user has to perform a polling loop to consult the stop-watch, and then evaluate the value he/she has just read to decide whether he/she has to press the LightSwitch button once again. This can be seen on the model by the cycle made up of place *ReadyToReadTime*, transition T3, place *CurrentTime* and transition T1.
From the initial state (modelled by one token in both *ReadyToStartWatch* and *ReadyToStartLight* places and seven tokens in *CountDown* place), the user can press both buttons in any order. After the firing of the corresponding transitions one token is removed form each input place and one token is set in *LightStarted* and *WatchStarted*. Then the internal transition StartCounting is triggered internally by the user meaning that from that state the user knows that the system is started and it is possible to see the elapsed time on the watch (this is represented in the model by a token set in the place *ReadyToReadTime*).

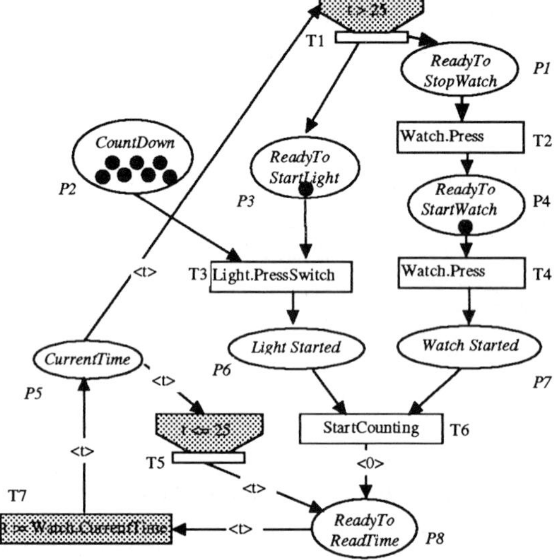

Fig. 10 The task model corresponding to the first upgrade of the system

6.3.4 LIM specification of the example

The LOTOS specification of this example is provided by associating one LOTOS expression with each component of the Interactive System considered and composing these expressions indicating the actions which require synchronization among them:

Light[...] |[light_press]| User [...] |[watch_press]| Watch[...]

The user can be specified as:

process User:= User_init[...] >> Visual[...] |[decide_to_start]| Action[...]

where User_init[...] := decide_to_start; (light_press; watch_press; exit
[] watch_press; light_press; exit)

User_init is the description of the user behaviour at the beginning of the interaction when once she/he has decided to start it then she/he presses the two buttons in any order. When terminated this initial behaviour activates (using the sequential enabling operator (>> symbol)) the normal user behaviour which is described by the Visual and Action processes.

Visual[...] := Read_watch; evaluate_time; (decide_to_start; Visual[...]
[] Visual[...])

Action[...] := decide_to_start; (light_press; watch_press; exit
[]watch_press; light_press; exit) ....

In this case we consider only two parts of the user behaviour: the Visual subsystem and the Action subsystem. They communicate because after reading the watch the user evaluates the current time and if he/she decides to start the watch he/she has to communicate this decision to the physical part performing this action. In this case the Action process has to perform three actions six times: receiving the start stimulus and pressing the two buttons.

The watch is described by Watch[...] := watch_press; Watch1[...]
where Watch1[...] := watch_press; Watch[...]
[] mod_pres; Watch1[...].

This means that it can receive the user action which starts it, after that it can either receive another pressing of the related button which stops it or it updates its presentation (mod_pres event) with the new time.
The LOTOS expression describing the light system is:

Light[...] := light_press; light_on; i; (i; light off; Light[...]
[] Light_system[...])

This means that once the button is pressed the light is turned on and there is sometime which is passed in this way, next there are two possibilities: either sometime passed again and then the light is turned off or a new button selection keeps again the light on. Note that we have used internal actions (i) to represent the elapsing of time without any external action. In this way without using operators with explicit time constraints and data we can describe the system considered.

6.3.5 Proving ACTL properties on the example

We use the LOTOS model of this simple Interactive System in order to check its properties. The possible behaviours are not very complicated so we can check properties which could be verified in other ways too.
For example we can check the possibility that at some time the light system can behaves in different ways:

AG[light_press][light_on] (<light_off>true & E[true{~light_off} U {light_on} true])

This means that whenever the sequence of the actions press_button and light_on occurs then it is possible both that the next observable action (this means that internal actions are not considered) is the light_off event and that a temporal evolution exists where no light_off event occurs until a new light_on action is verified. The second behaviour means that the user can turn on the light again before that it finishes the time related to the previous selection.
Another example of property of the user behaviour is that she/he can read the watch and then either read it again without pressing the button in the meanwhile or decide to press again the button to keep the light on. This is checked by the following ACTL expression:

AG [read_watch] (E[true{~light_press} U {read_watch}true]
& E[true{~ read_watch } U {light_press }true])

6.3.6 Comments on the third example

The specifications discussed in this section give a clear indication about complementary analysis which can be performed by the two methods. In one case, with the ICO method, it is possible to provide precise indication of quantitative time-related aspects. Vice versa with the LOTOS-based approach it possible to reason about temporal ordering of actions and demonstrate specific properties which summarise the main features of the system behaviour.

7. Conclusions

In this paper we have discussed the main concepts useful to apply formal methods in Interactive Systems design. Then we have outlined a proposal for an integrated use of two approaches based on different notations and we have applied both approaches to three small case studies. We have found that Petri Nets provide a more immediate to interpret representation of the temporal ordering among the possible actions, especially with not large specifications. Thus they are suitable to provide abstract representations of the overall behaviour of the system or descriptions of small components. The TLIM method is more oriented to provide detailed descriptions of the system behaviour: the composition operators guarantee a good modularity in the specifications performed which is very important for the development and the modification of large specifications. These specifications can be difficult to interpret at the beginning but they are useful to build models which can be analysed by automatic tools in order to check properties and to drive the development of an implementation reflecting the structure of the specification.

We believe that this integrated proposal is an interesting approach and we plan to develop it further and to apply it to an industrial case study from the air traffic control application area.

Acknowledgements

The authors would like to acknowledge the Italian CNR and the French CNRS (by the way of the GDR-PRC-CHM) which partially supported this work under the cooperation project n° 3226.

References

1. J. Accot, S. Chatty & P. Palanque. A formal description of low level interaction and its application to multimodal interactive systems, in [10], Springer Verlag 1996.

2. P. Barnard, J. May, "Interactions with Advanced Graphical Interfaces and the Deployment of Latent Human Knowledge" in [8].

3. R. Bastide. Cooperative Objects: a formalism for the design of concurrent systems. PhD dissertation, University of Toulouse 1, 1992 (in French).

4. R. Bastide & P. Palanque. A Petri net based environment for the design of event-driven interfaces. 16th international conference on Application and Theory of Petri nets, LNCS 935, Springer Verlag 1995.

5. R. Bastide & P. Palanque. Implementation techniques for Petri net based specifications of human computer dialogues. In proceedings of Computer Aided Design of User Interfaces (CADUI'96), Namur, June 1996, Presses Universitaires de Namur (Pub.).

6. S. K Card., T.P. Moran & A. Newell. The psychology of Human-Computer Interaction. Lawrence Erlbaum Associates, 1983.

7. R. DeNicola, A. Fantechi, S. Gnesi, G. Ristori, An Action Based Framework for Verifying Logical and Behavioural Properties of Concurrent Systems, Computer Networks & ISDN Systems, Vol.25, n° 7, February 1993, pp.761-778.

8. Proceedings of the first Eurographics workshop on Design, Specification and Verification of Interactive Systems, F. Paternó Ed. Springer Verlag 1995, ISBN 3-540-59480-9.

9. Proceedings of the second Eurographics workshop on Design, Specification and Verification of Interactive Systems, P. Palanque & R. Bastide Eds. Springer Verlag 1995.

10. Proceedings of the third Eurographics workshop on Design, Specification and Verification of Interactive Systems, F. Bodart & J. Vanderdonckt Eds. Springer Verlag 1996.

11. P. Gray & C. Johnson, "Requirements for the Next Generation of User Interface Specification Languages", Proceedings DSV-IS'95, Springer Verlag, pp.113-133.

12. R. Harston & P. Gray, "Temporal Aspects of Tasks in the User Action Notation", Human Computer Interaction, Vol.7, pp.1-45.

13. IBM (1989) Systems Application Architecture, Common User Access. Advanced interface design guide. Package SDK Windows - June 1989.

14. ISO (1988) Information Processing Systems - Open Systems Interconnection - LOTOS - A Formal Description Technique Based on temporal Ordering of Observational Behaviour. ISO/IS 8807, ISO Central Secretariat.

15. D. Larrabeiti, J. Quemada, S. Pav6n. From LOTOS to Petri nets through Iexpansion. Proceedings of the FORTE'96 conference Chapman & Hall, October 1996.

16. T. Moher, V. Dirda, R. Bastide & P. Palanque A bridging framework for modelling devices, users and interfaces. In [10], Springer Verlag 1996.

17. M. Mezzanotte & F. Paterno', "Including Time in the Notion of Interactor", Proceedings Workshop on Usability Aspect of Time, July '95, Glasgow and in SIGCHI bulletin vol 28, n°2, p. 57-61.

18. M. Mezzanotte & F. Paterno', "Use of Task, User and Formal Models to Support Development of Multimedia Interactive Systems", Proceedings of IFIP Working Conference Domain Knowledge for Interactive System Design, pp.213-226, May 1996, Chapman & Hall.

19. M. Mezzanotte & F. Paterno', "Verification of Human-Computer Dialogues with an Infinite Number of States", Proceedings Workshop on Formal Aspects of the Human Computer Interface, Sheffield, September'96, Springer Verlag.

20. P. Palanque. Modelling user-driven interfaces using the ICO formalism. PhD dissertation, University of Toulouse I, 1992 (in French).

21. P. Palanque, R. Bastide & L. Dourte (1993) Contextual Help for Free with the Formal Design of User Interfaces, HCI International'93, Elseiver, North Holland.

22. P. Palanque & R. Bastide. Petri net based design of user_driven interfaces using the interactive cooperative object formalism. In [8], p. 383-401.

23. P. Palanque & R. Bastide. Task Models - System Models: a Formal bridge over the Gap. In Critical Issues in User Interface System Engineering (Benyon & Palanque Eds.) Springer Verlag 1995.

24. P. Palanque & R. Bastide (1995) Verification of an Interactive Software by Analysis of its Formal Specification. Proceedings of Interact'95, Norway, Chapman et Hall.

25. P. Palanque & R. Bastide. Time modelling in Petri nets for the design of Interactive Systems. GIST workshop on Time in Interactive Systems. Glasgow, July 1995. SIGCHI bulletin vol 28 n°2, p. 43-46.

26. P. Palanque & R. Bastide. Formal specification and verification of CSCW using the Interactive Cooperative Object formalism. People and Computer X. Proceedings of the HCI'95 conference, Huddersfield, 1995, p. 213-232.

27. P. Palanque & R. Bastide. Performance evaluation as a tool for the formal design of interactive systems. IEEE Computational Engineering in Systems Applications (CESA'96) conference, Lille, July 1996, IEEE Press.

28. P. Palanque, R. Bastide, C. Sibertin, L. Dourte (1993) Design of User-Driven Interfaces using Petri nets and Objects ; CAISE'93. LNCS n° 685, Springer-Verlag.

29. J. Peterson. Petri nets and the modelling of systems. Prentice Hall, 1981.

30. F. Paterno' & M. Mezzanotte, Analysing Matis through Interactors and ACTL, Amodeus II BRA Report, sm/wp36.

31. F. Paterno' & M. Mezzanotte, "Formal Verification of Undesired Behaviours in the CERD Case Study", Proceedings EHCI'95 IFIP Working Conference, Chapman&Hall Publisher, Wyoming, August 1995.

32. S. Pangoli & F. Paterno', "Automatic Generation of Task-oriented Help", Proceedings ACM Symposium on User Interfaces Software and Technology, pp.181-187, ACM Press, Pittsburgh, November 1995.

33. F. Paterno', S. Sciacchitano & J. Lowgren, "A User Interface Evaluation Mapping Physical User Actions to Task-driven Formal Specifications", Proceedings DSV-IS'95, pp.35-53, Springer Verlag Publisher, Toulouse, June'95.

34. J. Rumbaugh, M. Blaha, W. Premerlani, F. Eddy, W. Lorensen. Object oriented modelling and design. Prentice Hall 1991.

35. C. Sibertin-blanc. A client server protocol for the composition of Petri nets. In proceedings of Application and Theory of Petri nets LNCS 691, USA, 1993.

36. A. Sutcliffe & P. Faraday, "Designing Presentations in Multimedia Interfaces", Proceedings ACM CHI'94, pp.92-98.

37. S. Shlaer, S. Mellor. Object oriented systems analysis. Modelling the world in data. Yourdon Press, 1988.

38. R. Sisto & A. Valenzano, "Mapping Petri nets with Inhibitor Arcs onto Basic LOTOS Behaviour Expressions", IEEE Transactions on Computers, Vol.44, N.12, December 1995, pp.1361-1370.

The Evaluation Of User Interface Notations

Chris Johnson

Glasgow Interactive Systems Group (GIST), Department of Computing Science,
University of Glasgow, Glasgow, G12 8QQ.
Email: johnson@dcs.gla.ac.uk
Web: http://www.dcs.gla.ac.uk/~johnson
Tel: +44 0141 330 6053

Abstract. Over the last decade a wide range of graphical, tabular and textual notations have been proposed to support the design of human-computer interfaces. These notations are intended to strip away the clutter of implementation details that frequently obscure interaction properties. Unfortunately, relatively little work has been done to evaluate the usability of these notations for 'real-world' interfaces. We have, therefore, conducted an empirical evaluation of the User Action Notation (UAN), State Transition Networks (STN) and temporal logic 'in the wild'. By this we mean that our subjects were drawn from realistic samples of users and designers. We also presented our subjects with realistic descriptions of two user interfaces. This avoids a weakness of previous investigations that have used 'toy examples'. The results of our investigation show a strong preference amongst our subjects for the use of natural language descriptions. More surprisingly, our results also suggest a link between the frequency of comprehension errors and positive attitude statements towards particular notations. In other words, our subjects made most errors with the notations that they liked the best. This suggests that while graphical notations, such as state transition networks, have a strong intuitive appeal they may also create significant problems for real-world development tasks.

1 Introduction

A vast array of notations are now available for the development of human-computer interfaces [3]. These range from purely graphical languages, such as State Transition Networks [4], through tabular notations, such as the User Action Notation [7], to textual formalisms, including grammars and logics [10]. These notations have had a relatively low impact upon the development of mass-market computer systems. They are, however, playing an increasingly important role in the design of human-computer interfaces to safety-critical applications [14]. The additional complexity and the high consequences of 'failure' are such that design notations are being exploited by government agencies, such as NASA [18] and the UK Ministry of Defence [17], as well as by corporations, including IBM and Mitsubishi [8].

The commercial application of interface design notations is complicated by the difficulty of selecting an appropriate language. Previous authors have focused upon technical comparisons, such as consistency and completeness [14]. While these issues are important, they are perhaps less significant than the usability of the notations themselves. Green emphasises this point in his work on cognitive dimensions [5]. He assesses the usability of a notation against a number of generic criteria. For instance, the 'viscosity' of a language indicates how easy it is to change a design. This work has, typically, been driven by the qualitative assessments of researchers (for example, see [21]). In contrast, we want our assessments to come direct from the designers and domain experts who must use the notations [12, 11].

2 Structure of the Paper

Section 1 has introduced the argument that is presented in this paper. Section 3 goes on to describe the methodological problems that frustrate the use of laboratory-based studies, think aloud techniques and questionnaires as means of evaluating interface design notations. It is argued that in the absence of any 'ideal' evaluation technique, it is nevertheless important that we attempt to validate our work on the specification and verification of interactive systems. Section 4. presents some of the problems that arise when recruiting users for such an evaluation. It is argued that we are facing a Catch-22 situation. It is difficult to produce statistically significant results because not enough industrial designers are exploiting formal and semi-formal notations. This makes it difficult to identify a reasonable population of users to support any evaluation. In turn, we will not develop a reasonable population of users until we can evaluate and demonstrate the benefits of interface notations. Section 5 introduces the UAN, state transition and temporal logic notations that were used in our study. Section 6 then presents the window manager example that was used to evaluate the usability of these formalisms. Section 7, in contrast, introduces the safety-critical gas turbine interface that was used in our second study of industrial interface designers. Section 8 describes the comprehension questions that were used in our two trials. It also presents the more qualitative attitude statements that were used to identify more general responses to the UAN, temporal logic and state transition specifications. Section 9 presents the results of the evaluation while Section 10 discusses the reasons for our findings. Finally, Section 11 presents the conclusions that can be drawn from this research and suggests directions for further work.

3 The Problems of Evaluation

Our aim was to assess the usability of interface notations for 'real-world' design tasks [15]. This raised a number of methodological questions. For instance, much of the previous work into the usability of programming languages has focused upon subjects drawn from University courses [20]. This is far from ideal because undergraduates may have little experience of the commercial pressures that are often cited as reasons why interface design notations are not used in the 'real-world'.

3.1 Laboratory Evaluations

The need to recruit a realistic user-population led us to focus our evaluation on industrial designers. This raised a number of further problems. In particular, it made it difficult for us to exploit the 'laboratory-based' evaluation techniques that have traditionally been used in other areas of human-computer interaction. Asking commercial designers to leave their normal work to conduct a tightly controlled experiment under laboratory conditions may not produce results that will be replicated under the pressures of everyday life [2].

Few employers are willing to release their staff for the period of time that would be required in order to complete a thorough laboratory based evaluation of interface design notations. Significant amounts of training and practice are necessary if individuals are to reach basic levels of proficiency. A further problem is that interface designers come from a range of backgrounds. Some have considerable expertise with mathematical notations. Others are 'self-taught' and have no experience with such notations. This makes it difficult to 'factor out' expertise. In order to ensure that the users have identical training we would ideally want to put them all through the same introductory course. The results of our previous investigations into the industrial application of formal methods have shown that there is little incentive for designers who are already familiar with interface notations to sit through such a course [13].

Further problems arise because the term 'usability' is too vague to produce statistically significant results. This ambiguity can be reduced. For example, we might design a series of experiments to identify the total time that it takes to learn different notations. Alternatively, we might analyse the frequency and type of errors that are made when reading design documents. Unfortunately, such approaches raise further methodological problems. In particular, those errors that might be observed under controlled laboratory conditions may have little relationship to the mistakes that people actually make when using an interface design notation 'in the wild'. Similarly, the time taken to learn an interface notation will clearly be affected by the competing tasks and deadlines that characterise the working lives of interface developers.

3.2 Think Aloud Techniques

Think aloud evaluation techniques, such as those developed by Wright and Monk [22], offer an alternative to the tightly controlled conditions of laboratory-based evaluation techniques. This approach requires that users complete a number of 'realistic' tasks under normal working conditions. They are asked to talk about the critical incidents and decisions that occur while they complete these tasks.

This approach overcomes some of the problems associated with the replication of laboratory results in people's working environments. The evaluation of a notation can take place in a realistic setting. The intention is to provide qualitative feedback. The focus is not upon the production of statistically significant results. It is, therefore, perfectly possible for an evaluation to be interrupted by the distractions that frequently occur during everyday life: the telephone may ring; colleagues may ask questions and so on. Such normal events are, typically, factored out in laboratory-based investigations where interruptions would distort the results obtained for errors or learning times.

Unfortunately, think aloud techniques do not entirely resolve the methodological challenges that are posed by interface design notations. For instance, the task of thinking aloud can affect the results that are produced by this approach. Designers are forced to reconsider their actions during the process of vocalisation. This period of introspection may actually reduce the number of errors that would otherwise have been made. In consequence, the critical incidents that are noted during the analysis might be very different from those that would occur 'in the wild' [22].

We were concerned to analyse the usability of interface design notations against two different classes of application. This decision was motivated by the argument that safety-critical applications may pose totally different challenges than mass market word processors or windowing systems. This raised further problems for the use of think-aloud evaluation techniques. It is extremely difficult to gain access to interface designers for safety-critical systems. Those designers that we could interview were developing user interfaces for oil rig workers on production platforms off the Norwegian and United States' coasts. We could not, therefore, directly observe the use of our chose notations in the manner advocated for think aloud techniques

3.3 Questionnaires And Surveys

Questionnaires avoid many of the weaknesses of laboratory-based evaluations and think aloud techniques. They do not take designers out of their normal context of work nor do they require that interface designers enter their working environment to observe their everyday tasks with an novel notation. In contrast, they are low cost and can be easily distributed with minimal overheads for the companies involved.

Unfortunately, the use of questionnaires also limits the scope of our investigation. In laboratory based approaches, evaluators can tightly control a users' working environment. In think aloud techniques, evaluators can observe users working on specified tasks. In surveys, users are free to respond to questions at any time in their daily life. They may fill in the form during a break or while waiting for a program to compile or while working on another design task. These other activities can interfere with the results that are obtained from our study.

The caveats raised above emphasise the point that there are no ideal techniques for the evaluation of interface design notations. Laboratory-based techniques filter out real-world influences. Think-aloud techniques require close access to designers who are already familiar with a notation. Surveys are unconstrained and cannot easily be controlled. With this in mind, we are exploring a range of evaluation techniques that combine elements of all of these approaches [13]. This work is still in its early stages. In anticipation of the results of this work, existing evaluation techniques must be exploited to analyse the usability of interface design notations. It is important that such investigations are attempted even if the evaluation tools are far from ideal. Without such evidence there is no means of validating current research into the specification and verification of interactive systems.

4 The Users

As mentioned, it is difficult to find industrial designers with the time and motivation to support the evaluation of interface notations. We were only able to recruit six people who were prepared to help us study the usability of these formalisms. Their expertise with interface design notations ranged from none at all up to three years experience with the application of formal and semi-formal specification languages. The group included recent graduates and experienced interface designers from an oil production company. This range of backgrounds further complicated the problem of producing reliable results. It is difficult to draw direct comparisons between users with such different levels of expertise and skill. The situation was even worse. Those designers that did have previous experience were not all skilled in the same notations. Some users were familiar with software engineering formalisms but not with user interface notations. Others claimed to have high levels of expertise in both areas. Additional problems are created because it is difficult to assess whether claimed levels of skill are actually vindicated by an individual's ability to use a particular notation to design a human-computer interface.

Given the limited sample size and the varied backgrounds of our participants, it is difficult to produce statistically significant results. Unlike many laboratory studies of small-scale user interfaces or well constrained psychological phenomena, it is simply not possible to factor out all of the potential biases and external factors that might influence our results. One solution to the problems listed above would be to recruit more users to perform the survey. It can be argued that as more and more people use interface specification notations then it will become easier to recruit users for our evaluations. Unfortunately, we are in a Catch-22 situation. Many people will not exploit formal and semi-formal notations until they are convinced of their utility. We cannot, however, gain evidence of the utility of notations until we have a large enough user population of industrial designers.

5 The Notations

UAN, STN and temporal logic were chosen for our evaluation because each represents either a tabular, a graphical or a textual approach to interface design. The following paragraphs describe the basic components of each of these notations.

5.1 User Action Notation

Hix and Hartson's [7] User Action Notation categorises user interface events according to the agent that initiated them. In this context, the agent is assumed to be either the user or the computer. Figure 1 illustrates the resulting tables. The users action of moving the mouse pointer to the button is represented in the interface by the cursor tracking to that button. The action of pressing the button down is represented in the interface by the button being highlighted. Finally, the user action of releasing the button is mirrored by the button being de-highlighted in the interface and by the corresponding action being executed in the application. The sequence of events flow from left to right and then from top to bottom. The structure of the table, therefore, not only provides important cues about the agent that is engaged in a particular activity but it also represents temporal information.

Task: Select Button

User Actions	Interface Feedback	Interface State
move to button	cursor tracking	
mouse button down	button high-lighted	
mouse button up	button un-high-lighted	execute button action

Fig. 1. Simple Example Of The User Action Notation.

We are interested in evaluating this notation because it provides a relatively simple syntax with what is claimed to be an intuitive structuring. In particular, the use of columns to organise user and system function helps the reader to keep track of their various activities. Both perform visible actions and hypothesised, invisible internal actions. These internal actions for the system include operations on the dialogue state, application actions etc. Those for the user include changes to short and long term memory, planning and other cognitive actions.

5.2 State Transition Networks

State transition networks provide a graphical notation for the development of user interfaces [19]. The state of interaction is represented by a circle. The state of the system changes when particular conditions are satisfied. In Figure 2, this would occur if the condition 'User presses mouse' were fulfilled. If the mouse were pressed then the action 'highlight button' would be executed. The state of the system would change and the button would now be highlighted. The complexity of this simple textual explanation indicates the attraction of graphical representations, such as State Transition Networks! Figure 2, also represents the hierarchical nature of state transitions networks. The box to the right of the diagram represents another state-transition network that may contain sub-sub-networks.

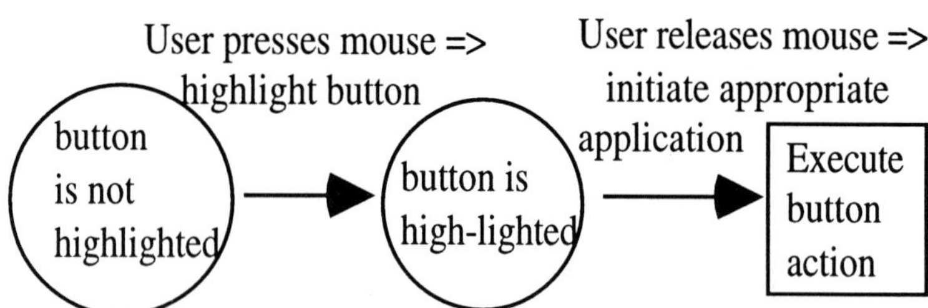

Fig. 2. Simple Example Of State Transition Networks.

We are interested in State Transition Networks, rather than other graphical notations such as Petri Nets [1] or State Charts [6], for entirely pragmatic reasons. During initial tests it was found that State Transition Diagrams could be more quickly understood from written instructions than other available formalisms. This was significant as our evaluation was to be based around a series of paper-based questionnaires. Again, our findings re-iterate the close relationship between evaluation tools and the scope of any validation exercise for interface design notations. It is difficult to validate the claims that are made for many of the more sophisticated tools and techniques because they require significant training and practice. Many designers will not meet the costs of such training and practice until the approaches have been thoroughly validated.

5.3 Temporal Logic.

Temporal logic is a purely textual notation [9]. The following example states that an action is executed if in the present interval the user moves the mouse and eventually they select a button and in the next again interval the system executes the action associated with that button. The $\bigcirc$ symbol is read as 'next', $\diamond$ is read as 'eventually':

$$\begin{aligned} &execute\_action \Leftarrow user(moves\_mouse) \wedge \\ &\quad \diamond(user(selects\_button) \wedge \bigcirc(system(executes\_action))). \end{aligned} \tag{1}$$

It may seem paradoxical to include temporal logic. We excluded Petri Nets because our users found them difficult to understand. The logic notation poses greater challenges than the networks of the graphical formalism. Our decision is justified by a concern to compare a relatively intuitive graphical notation with the more 'demanding' textual formalism. We are also interested in evaluating temporal logic because timing properties can have a profound impact upon the course of interaction [11]. For example, numerous studies have shown that variable delays in system responses lead to frustration and error [16]. It is important, therefore, that designers can represent and reason about the impact of these problems. Temporal logic provides a textual representation of the timing properties that are represented using the graphical and spatial cues of UAN and state transition networks. We are also concerned to evaluate this notation because we have limited qualitative evidence to suggest that there are a number of usability problems with the logic [3]. It can be difficult for designers to exploit such notations; a training in discrete mathematics is required in order to understand the formal underpinnings of the language. It can also can be extremely difficult to construct the complex chains of interaction that arise between a system and its user.

5.4 The Difficulties Of Comparing Notations

Our intention was to ask designers and users to provide qualitative usability assessments of UAN, state transition networks and temporal logic Such analyses have only previously been produced from the introspection and insight of academics and researchers in human-computer interaction [3]. The examples presented in the previous paragraphs illustrate some of the problems of performing such an evaluation of different interface description techniques. For example, each of the notations encodes slightly different design information. UAN explicitly represents the agents involved in interaction. This information may be included in STN and temporal logic but it is not explicitly part of the notation. Similarly, temporal logic offers a wide range of facilities for dealing with time. These can be introduced into UAN and STN but again they are not basic components of the notations [3]. Such differences make it difficult to determine whether the various descriptions of an interface are equivalent. For instance, if we chose to represent an interface to a time-critical system, such as a process control application, then we might expect temporal logic to perform well in our qualitative evaluation. If we described interaction with an agent based interface then UAN might out-perform the other notations. In order to reduce this bias, we chose to evaluate the notations against two different scenarios. The first involved the design of a graphical user interface to a window manager. The second involved human-computer interaction with a turbine on an oil-rig.

6 Trial 1: The Window Manager

A window manager was chosen because it typifies many mass-market applications. Our intention was to provide a detailed example; Figure 3 illustrates the STN from this evaluation. Our subjects were asked to use this description to answer a series of questions about the interface. There then followed a number of qualitative questions about the usability of the notation. This process was repeated for each of the remaining notations.

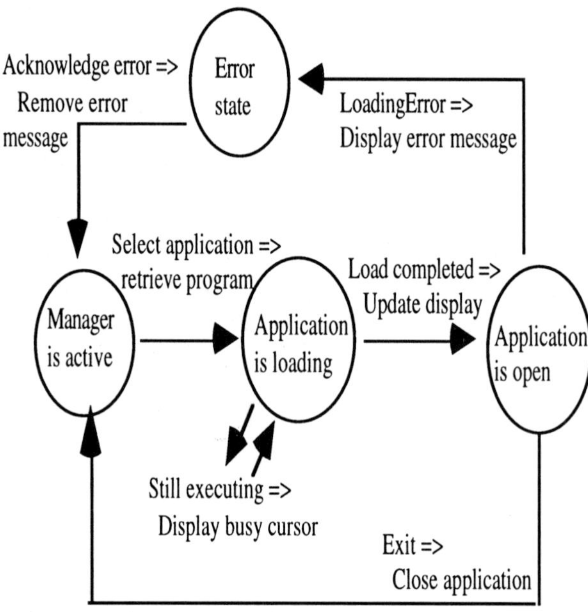

Fig. 3. STN For Opening An Application In The Window Manager.

This experimental design created a number of problems. The design of the user interfaces had to be different enough from the existing systems so that designers had to use the descriptions. Questions about qualitative preferences between notations might also have been biased if each of the notations had been introduced in the same order. Similarly, learning effects might bias answers about the interface design. Information gleaned from a description using one notation might have been used to answer questions about another description. In order to combat these effects we had to produce three sets of descriptions and three sets of behavioural questions for each of the notations. The ordering of the material was randomised.

7 Trail 2: The Turbine Control System

The human-computer interface for a turbine control system was chosen because it typifies the safety-critical applications that are increasing being developed using abstract notations such as temporal logic. As in the previous trial, we analysed the reactions of 'real' users and designers to each of the interface design notations.

In this case our subjects were drawn from a Norwegian oil production company. Figure 4 provides an example of the UAN description that was presented to our subjects. At first sight, this might seem impossibly complicated for designers and users with little expertise in interface notations. It should be remembered, however, that the subjects had a high degree of domain expertise and, as the results show, they were able to interpret descriptions at this level of detail.

Task: Respond To Status Report

User Actions	Interface Feedback	Interface State
InfoFrameStatus: event (UserSelection) -> (shutdown | warning acknowledge | reconfigure)	CASE system_shutdown; remove_warning; reconfigure_menu; END_CASE	 InfoFrameStatus: empty InfoFrameStatus: user

Fig. 4. UAN For Operator Selection In The Turbine Control System

Each of the case studies was given to a different group of subjects. In both trials we endeavored to use interface designers and users who might actually be involved in the development of such an application. In our first case study, this involved subjects from a range of backgrounds and educational experience. In the second case study, the specialist nature of the user interface created a much more homogenous user population; oil-rig engineers with experience in turbine control.

8 The Questions

We wanted to gain evidence about designer and user reaction to interface description notations. In order to do this we were concerned to ensure that the subjects made some attempt to understand the notations in each condition. A number of questions were, therefore, asked about the behaviour of each interface. Figure 5 presents the questions that were asked about the STN shown in Figure 3. The design of these comprehension questions raised a number of interesting problems. Early versions of the questionnaire provided 'Don't know' in addition to the 'Yes' and 'No' options. An initial survey revealed a strong tendency amongst the subjects to avoid commitment by selecting the 'Don't Know' option. This is an interesting phenomena for further research. The decision was taken, however, to force commitment to a binary decision. This creates the risk

Q1 Is 'Select application' the only possible condition to open an application? ☐ Yes ☐ No

Q2 Is 'Update display' the next action to be performed after the application has been loaded? ☐ Yes ☐ No

Q3 Is the application open only after the display has been updated? ☐ Yes ☐ No

Q4 Does the example describe any possible error state after the application has been loaded? ☐ Yes ☐ No

Q5 Is the application active after the error state has been ackowledged? ☐ Yes ☐ No

Fig. 5. Comprehension Questions About the Window Manager

that the subjects would rely on chance by guessing the answers. Fortunately, our results do not indicate that these tactics were not being used.

A second set of questions provided qualitative evidence about users' and designers' reactions to the interface notations. The same questions were asked in both case studies and for all of the conditions within each test. Subjects were asked to tick the boxes next to those statements that they agreed with. As

☐ The notation is easy to read

☐ The overall layout is good

☐ The interaction details would be better explained in English text.

☐ It is difficult to keep track of the sequence of interactions.

☐ The notation syntax (ie symbols used) makes the notation hard to understand.

☐ The notation is useful to describe a large number of interactions

Fig. 6. Qualitative Questions About the Notations

mentioned in the introduction, the decision to conduct our analysis 'in the wild' imposed a number of constraints upon our evaluation. The users and designers in the second trial were 'off-shore' and could only be reached by radio or fax. We, therefore, had no means of recording completion times for the survey. Subjects in all of the conditions were asked to take as much time as they liked. They were, however, asked to record how long they spent on the comprehension questions associated with each notation. Each trial was performed by six subjects.

9 Results of the Evaluation

9.1 Time Consumption

Figure 7 illustrates the average total time required to complete the comprehension questions in each of the test conditions. These figures do not include the time taken to read and understand the 'familiarisation' material. It is important to emphasise that these timings were reported by the subjects and were not monitored by the investigators. In the window manager condition, the standard

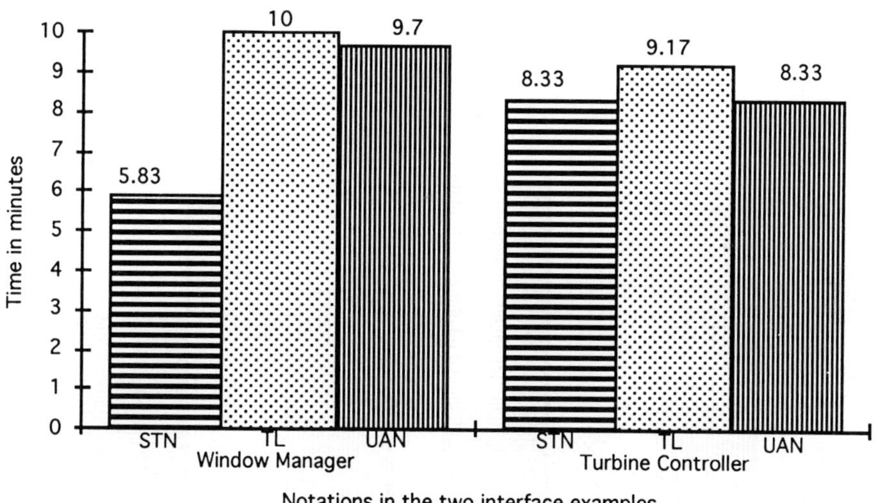

Fig. 7. Average Total Time Required To Answer The Comprehension Questions.

deviation for both the STN and the UAN conditions was 2.0. In the former case, this was caused by one of the subjects taking substantially longer than the mode of 5 minutes. In the UAN condition this was caused by one user taking substantially longer than the modal value of 10 minutes. In the temporal logic condition, the standard deviation was 0 as all of the users took longer to complete the tasks than the allotted 10 minute interval. The standard deviation for the STN and the UAN in the Turbine Controller conditions was 2.6. For temporal logic, the standard deviation was 2.0. This lower figure was due to only one user completing the task within the allotted ten minutes.

Before presenting further results is worth mentioning that the findings reported in Figure 7 help to illustrate additional problems that must be addressed during the evaluation of interface notations. The standard deviations seem high in relation to the absolute time values reported. These results are due to the strong effects that arose when many users in the trial fail to complete the task within the allotted interval. This raises serious problems. If we simplify the questions to reduce the time taken then we will be forced to evaluate our techniques against toy systems and trivial examples. If we expand the amount of time required to complete an evaluation then we increase the burdens and demands upon our scarce user population.

9.2 Score For Behavioural Questions

Figure 8 indicates the percentage of correct answers to the comprehension questions for each of the interface description languages. As mentioned in previous sections, these questions elicited yes or no responses to a series of questions about the potential interfaces. In the Window Manager trial the standard devia-

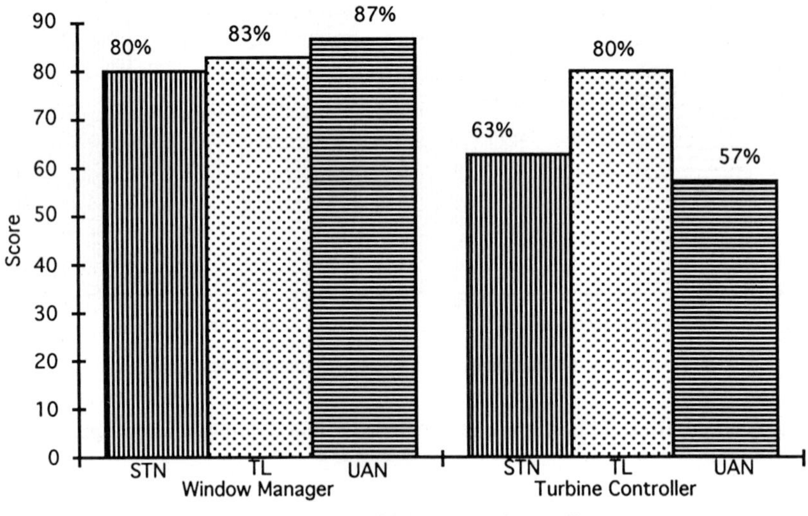

Fig. 8. Percentage Of Comprehension Questions Answered Correctly

tion for all three notations was 17.9. This homogeneity reflects in part the effects of previous skills and expertise that have been address in the earlier sections of this paper. Those with little experience of formal notations scored well below the mean. Those with considerable skills in the use of interface design notations, scored consistently above the average in each case. In the Turbine Controller trial, the standard deviation for STN was 19.7, for temporal logic it was 22 and for UAN it was 26.6. This last result is rather surprising as it reflects a genuine spread of results from 20% correct up to 100% correct. Even so, the similarity in the standard deviations again re-iterate the problems of evaluating interface notations. Greater confidence in the range of scores might have been obtained by asking a larger number of questions. This would have emphasised the differences between novices and experts, it might also have emphasised the differences between each of the notations. Unfortunately, the results from Figure 7 already raise serious questions about the amount of time that is required to conduct any evaluation of interface notations. A related issue is the question of fatigue that might begin to have a serious effect if designers were asked to apply particular notations to answer a large number of comprehension questions.

9.3 Overall Attitude Statements

Figure 9 records the summation of the attitude statements for each of the notations in the two interfaces. The scores were obtained by adding one to a total for each of the positive attitude statements that were agreed with. One was subtracted if a positive statement was denied. This total was then divided by the number of subjects in each test. In this way the most positive score would have been six, zero would have indicated a neutral attitude and minus six represents the strongest negative reaction to a notation. The standard deviation for the qualitative assessment in the Window Manager case was 1.6 for STN, 3.0 for temporal logic and 2.7 for UAN. In the Turbine Controller case, the standard deviation was 3.7 for STN, 1.5 for temporal logic and 3.7 for UAN. As before, these results can only be interpreted with reference to the methodological barriers that frustrate the evaluation of user interface notations. For example, the high variance for temporal logic in the first case study might be compared with the relatively low variance in the turbine controller. This might suggest that users were undecided about the limitations of the notation for designing a Window Manager but were convinced that it was of little benefit for a Turbine Controller. Alternatively, these results may simply reflect the a-typical views of one or two individuals in the sample that we selected. In order to be convincing these results must be replicated. This leads us back to the previous argument that conventional evaluation techniques cannot be used to provide this additional evidence. The low numbers of commercial practitioners and their varying levels of expertise create a pressing need to find alternative means of validating interface specification techniques.

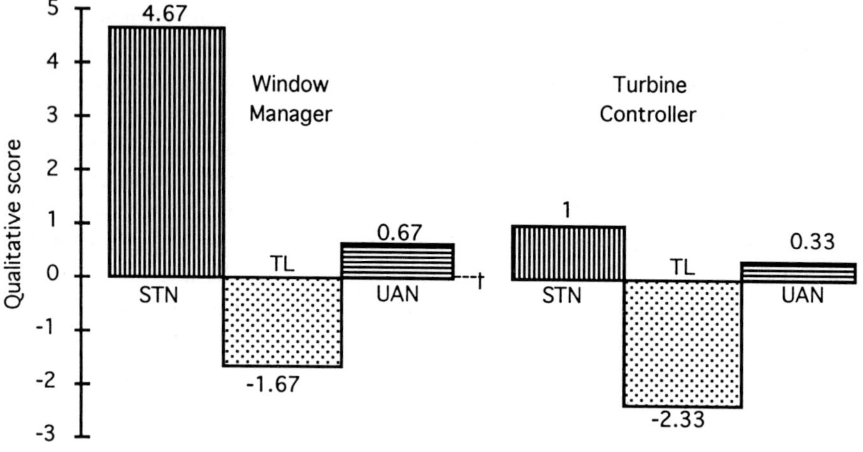

Fig. 9. Overall Summation Of Scores For Attitude Statements.

9.4 Detailed Attitude Analysis

Figure 10 presents the summation of scores for the individual attitude statements that were asked about each notation in both of the test interfaces. The absence of a bar in the figure indicates a neutral score of zero. The (1) indicates the window manager trial, the (2) refers to the turbine controller.

10 Discussion

The results shown in Figure 10 indicate a preference amongst our subjects for the use of natural language rather than UAN, temporal logic or STN. The only exception to this is a slight preference for STN over natural language in the window manager trial. This positive reaction might have been due to the highly opportunistic design for interaction with this interface [19]. The preference for natural language in the other conditions might have occurred because our examples were too easy. Interface notations are often argued to offer the greatest benefits for complex design tasks [14]. This analysis is supported by the high average score for all of the comprehension questions. It is, however, difficult to believe given the complexity of the UAN and STN shown in figures 4 and 3. Alternatively, this negative reaction may indicate that designers and users 'in the wild' remain unconvinced about the benefits of interface description techniques. This is perhaps unsurprising for the window manager trial. The use of abstract notations has made relatively little progress in this area. It is more surprising

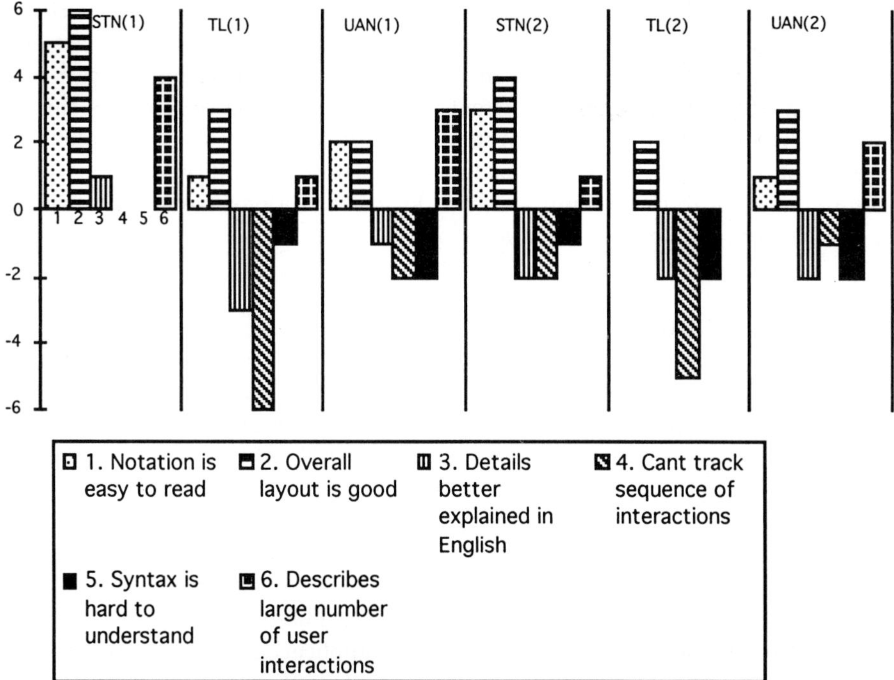

Fig. 10. Individual Summation Of Scores For Attitude Statements.

in the case of the turbine controller where issues of safety and complexity have already led to the use of formal notations by the engineers in our evaluation.

We see a strong negative reaction towards temporal logic in both case studies. This is confirmed by an analysis of variance for the average attitude scores. The relatively low variance for the negative reaction towards temporal logic in the turbine trial indicates a strong consensus on this point:

	STN	TL	UAN
Window Manager	13.3	43.3	37.3
Turbine Controller	70	11.3	67.3

The negative reaction to temporal logic is perhaps explained by the average time taken to answer the comprehension questions. Both UAN and temporal logic took significantly longer than STN in the window manager example. This, in turn, may explain the high positive reaction to STN in the qualitative assessments mentioned above. Our investigation, therefore, confirms previous observations about the high intuitive appeal of graphical notations. More interestingly, our evaluation shows that in spite of this strong preference our subjects actually made significantly more mistakes in the turbine comprehension exercise with STN than with temporal logic. Current work is attempting to replicate this result. If correct, this analysis has profound consequences for the development of human-computer interfaces. The usability problems and additional time that is required in order to understand temporal logic descriptions may be out-weighed by the benefits of increased accuracy.

11 Conclusions

This paper has argued that more attention must be paid to the usability of interface notations if they are to be applied in the 'real-world'. In order to support this argument, we conducted an evaluation of UAN, STN and temporal logic. Our intention was to ask designers and users to provide qualitative usability assessments of interface design notations. Such analyses have previously been performed by academics and researchers in human-computer interaction.

The user interfaces to a window manager and a turbine control system were described using UAN, STN and temporal logic. These descriptions were then presented to the designers and users of these applications. They were asked to answer a series of comprehension questions to test their understanding of the design. They were also asked some qualitative questions about the use of the notation. The results showed that all subjects achieved high levels of comprehension for both trials and all notations. In spite of this, our subjects expressed a strong preference for natural language descriptions.

The most surprising finding from our study was that the highest number of comprehension errors were observed for notations with the most positive attitude statements. Temporal logic gained significantly better results for the turbine interface than either UAN or STN. This was in spite of the fact that it also provoked the most negative reaction.

Much further research remains to be done. In particular, the work described in this paper has exposed the inadequacies of existing evaluation techniques for the validation of interface design notations. Laboratory studies tend to ignore the everyday pressures of designer's working environments. Think aloud techniques can be difficult to apply with novice designers who are still learning novel notations. Questionnaires provide only limited evidence about the long term effectiveness of a design technique. This is a critical problem. Unless we can validate our techniques then it will be difficult to persuade designers and companies to invest in the specification and verification of interactive systems [13].

Acknowledgements

Thanks are due to Jarle Gjosaether who helped to design and run the experiments that are described in this paper. Thanks are due to Phil Gray who helped with the application of UAN. Gilbert Cockton helped with our use of the STN notation. Steve McGowan, Paddy O'Donnell and Steve Draper also helped with the comparative analysis. This work is supported by UK EPSRC grant number GR/K55042 and by JCI Grant number SPG-9201233.

References

1. R. Bastide and P. Palanque. Petri net objects for the design, validation and prototyping of user-driven interfaces. In D. Diaper, D. Gilmore, G. Cockton, and B. Shackel, editors, *Human-Computer Interaction—INTERACT'90*, pages 625–631. Elsevier Science Publications, North Holland, Netherlands, 1990.
2. J. Campion. Interfacing the laboratory with the real world. In J. Long and A. Whitefield, editors, *Cognitive Ergonomics And H.C.I.*, pages 35–65. Cambridge University Press, Cambridge, United Kingdom, 1989.
3. P.D. Gray and C.W. Johnson. A critical analysis of interface specification notations. In *The Design, Specification and Verification of Interactive Systems*, pages 113–133. Springer Verlag, Berlin, Germany, 1995.
4. M. Green. Design notations and user interface management systems. In G.E. Pfaff, editor, *User Interface Management Systems*, pages 89–107. Springer-Verlag, Berlin, FDR, 1985.
5. T.R. Green. Cognitive dimensions of notations. In A. Sutcliffe and L. Macaulay, editors, *People And Computers IV*, pages 443–460. Cambridge University Press, Cambridge, United Kingdom, 1989.
6. D. Harel. On visual formalisms. *Communications Of The ACM*, 31(5):514–530, 1988.
7. D. Hix and H.R. Hartson. *Developing User Interfaces.* John Wiley and Sons, London, 1993.
8. A. Jack. It's hard to explain but Z is much clearer than English. *The Financial Times*, page 22, 12 April 1992.
9. C.W. Johnson. A formal approach to the presentation of CSCW systems. In J.L. Alty, D. Diaper, and S. Guest, editors, *People And Computers VIII*, pages 335–352. Cambridge University Press, Cambridge, United Kingdom, 1993.
10. C.W. Johnson. A probabilistic logic for the development of safety-critical interactive systems. *International Journal Of Man-Machine Studies*, 39(2):333–351, 1993.
11. C.W. Johnson. The challenge of time. In P. Palanque and R. Bastide, editors, *The Second Eurographics Workshop On The Design, Specification And Verification of Interactive Systems*, pages 345–357. Springer Verlag, Berlin, Germany, 1995.
12. C.W. Johnson. The economics of interface design. In K. Nordby, P.H. Helmersen, D. Gilmore, and S.A. Arnesen, editors, *Human Computer Interaction - Interact '95*, pages 19–25, London, United Kingdom, 1995. Chapman And Hall.
13. C.W. Johnson. Literate specification. *IEE/BCS Software Engineering Journal*, 1996. Accepted and to appear early in 1996.
14. C.W. Johnson and M.D. Harrison. Software engineering for human computer interaction. *SIGCHI Bulletin*, 26(2):46–48, 1994.
15. C.W. Johnson, J.C. McCarthy, and P.C. Wright. Using a formal language to support natural language in accident reports. *Ergonomics*, 38(6):1265–1283, 1995.
16. W. Kuhmann. Stress inducing properties of system response times. *Ergonomics*, 32(3):271 - 280, 1989.

17. Ministry Of Defence. *Requirements for the Procurement of Safety Critical Software*, MOD DEF-STAN 00-55, London, United Kingdom, 1991.
18. National Aeronautic and Space Administration. *Advanced Orbiting Systems - Architectural Specification For The CCSDS Secretariat*, Washington DC, United States of America, 1989.
19. D.R. Olsen. Presentational syntactic and semantic components of interactive dialogue specification. In G.E. Pfaff, editor, *User Interface Management Systems*, pages 125–133. Springer-Verlag, Berlin, FDR, 1985.
20. J. Scholtz and S. Wiedenbeck. Learning to program in another language. In D. Diaper, D. Gilmore, G. Cockton, and B. Shackel, editors, *Interact'90*. Elsevier Science, North Holland, 1990.
21. S. Buckingham Shum. Cognitive dimensions of design rationale. In D. Diaper and N.Hammond, editors, *People And Computers VI: Proceedings Of HCI'91*. Cambridge University Press, Cambridge, United Kingdom, 1991.
22. P.C. Wright and A.F. Monk. A cost-effective evaluation method for use by designers. *International Journal of Man-Machine Studies*, 35(6):891–912, 1991.

This article was processed using the LaTeX macro package with LLNCS style

Supporting Error-Driven Design

Chris Johnson and Phil Gray

Glasgow Interactive Systems Group (GIST), Department of Computing Science,
University of Glasgow, Glasgow, G12 8QQ.
Email: {johnson,pdg}@dcs.gla.ac.uk
Web: http://www.dcs.gla.ac.uk/~{johnson,pdg}
Tel: +44 0141 330 6053

Abstract. This paper argues that two limitations restrict the utility of interface specification languages. Firstly, they provide no means of capturing the cognitive conditions that lead to operator 'error'. This makes it difficult to distinguish between the normal behaviour of an expert and the mistakes that often lead to problems for novices. The second weakness is that interface notations cannot easily be used to represent and reason about asynchronous failures. This prevents designers from identifying solutions to failures that could occur at many different points during interaction. These are significant limitations because they reflect a pre-occupation with normative behaviour. Unless we have some means of analysing system failure and operator error then we will continue to have interfaces that are designed to support perfect users in perfect environments.

1 Introduction

Over the last decade a vast range of notations have been used to represent and reason about human-computer interaction. This work has extended the application of notations that were originally intended to support software engineering. These include textual notations, such as logics and process algebras, as well as graphical formalisms, such as Petri Nets [3]. In contrast, a number of other formalisms, such as the tabular User Action Notation (UAN) [11] and Task Analysis for Knowledge Description (TAKD) [5], have been specifically developed to support the design of interactive systems. All of these notations strip away the low level implementation details that frequently obscure critical properties of interactive systems [7].

Unfortunately, few formalisms provide interface designers with means of reasoning about the consequences of system failures and operator error [9]. For example, it can be extremely difficult to represent the ways in which planning errors continue to frustrate a user's interaction with a system over long periods of time. Similarly, it can be difficult to represent the uncertainty that arises when remote machines fail to satisfy an operator's request. This, in turn, makes it difficult to analyse interaction with stochastic applications, such as process control systems, as well as distributed information retrieval environments, such as the World Wide Web.

In this paper, we review the support that interface notations provide for reasoning about such sub-optimal or abnormal behaviour by both systems and their users. In particular, it is argued that temporal logics and UAN provide limited means of analysing the consequences of such breakdown. These formalisms have been chosen because they illustrate the two main trends in the development of interface specification notations. Temporal logic has been developed to support software engineering. From this its application has been extended to support interface design [15]. UAN, in contrast, was created to specifically support interface development [6]. Later sections build on this analysis to identify means of enhancing the next generation of interface design notations.

Many further questions remain to be answered. For example, the limitations of existing notations might be addressed by developing extensions to UAN or temporal logic. These extensions can capture asynchronous failures or the probable causes of operator error. Such approaches raise additional research questions. For instance, the ad hoc development of increasingly complex notations would exacerbate the usability problems that frustrate the commercial exploitation of formal description techniques. The development of tractable interface design notations that have a clear semantics is a research area in its own right. Our intention is not to present a complete solution to the problems that we describe. Our intention is, however, to address the current bias towards design techniques for error-free interfaces. Unless we have some means of analysing system failure and operator error then we will continue to design applications that only support perfect users in perfect environments.

1.1 Error-Driven Design

In many engineering disciplines, it is standard practice to focus resources upon critical or abnormal situations that are considered to pose the greatest threat to system integrity. Design progresses by first considering pathological situations and then by considering the detailed engineering of normative behaviour. Good design is facilitated by capturing and reasoning about the causes and consequences of failure. This analysis points designers towards solutions and support mechanisms. For instance, in structural engineering it is common to analyse the performance of materials under infrequent or pathological conditions that might arise only once in every two or three hundred years. Petroski [20] argues that bridge engineering has made its major advances in design as a result of its failures rather than its successes.. Systems engineers often exploit notations, such as Fault Trees and Failure Modes Analysis, that have been explicitly designed to reason about application failure. In computer systems, software and hardware engineers devote considerable resources to the identification and testing of conditions in which critical components may fail. A number of languages, such as Ada and C, have even been extended with exception handling facilities that provide programmers with suitable abstractions for responding to infrequent system failure and erroneous input from application processes. Unfortunately, this support has not been extended to the development of human computer interfaces.

This paper argues that interactive systems can be improved if designers pay greater attention to the abnormal or sub-optimal behaviours that can arise during interaction. Interactions which deviate from the expected and/or which do not further the accomplishment of the task goal are not always undesirable. Carroll [4] has argued that learning is enhanced by encountering and dealing with problematic situations. However, whether problems are to be encouraged for the sake of learning or simply cannot be ruled out, it is still important for the designer to try to provide support for recovery, damage limitation and repair.

1.2 The Limitations of Interface Design Notations

We are particularly concerned to develop notations that will help designers to represent and reason about the potential for 'breakdown' in human-computer interaction. This paper defines breakdown to occur whenever machine failure or operator error presents users with unexpected or unfamiliar situations. Our usage of the term is derived from that of Winograd and Flores [24]. Such situations are critical in the sense that they often help to determine the usability of an interactive system. If interfaces do not provide appropriate levels of support then users may not be able to rectify errors or overcome exceptional system failures. As Norman [18] argues, these situations demand appropriate levels of feedback because they expose users to the failure of automation

Many interface specification notations are poorly equipped to represent and reason about instances of such breakdown. In contrast, they have primarily been used to represent error-free behaviour. They describe the intentions and actions of an ideal user and almost entirely ignore the slips and lapses of everyday life. Such criticisms are far from novel; previous authors have reviewed the strengths and weaknesses of notations such as GOMS For example, John and Kieras [12] argue:

> "CMN, and almost all subsequent GOMS work, presents analyses and predictions based on the assumption that the user does not make errors. Since errors in computer usage are quite frequent, it would seem that GOMS family models have little to say about actual human performance. But we would argue, along the lines of ... analogy with EPA mileage estimates, that GOMS models of error-free performance do in fact provide useful design information. For example, a poorly designed systems that is difficult to learn and to use even under a no-errors assumption is almost certainly still a poor design if the user does make errors. So, optimising learning time and execution time under the no-error assumptions should result in a system that is a good design overall, given that errors do not always occur, and assuming that some reasonable error recovery is possible...At this time, GOMS models also fail to address error prediction or prevention."

It is, however, surprising that most interface specification languages are so poorly equipped to represent and reason about operator errors. This argument is illustrated by the fact that none of the interface specification papers in recent proceedings of CHI, HCI or Eurographics Workshops show sustained examples of error handling.

1.3 The Electronic Mail Case Study

To help focus the subsequent discussion, we shall restrict our examples to one application area: electronic mail. Interaction with electronic mail systems offers a fruitful source of raw material for interaction breakdowns because:

- the user population is large and varied in terms of experience and expertise. Thus, the problems encountered run the gamut from the classic and frustrating "dead-ends" of novice use to the more subtle difficulties of expert users;
- the use of email is embedded in a wide range of higher-level tasks (chatting with friends, collaborative editing of papers, broadcasting) each of which imposes its own requirements and user expectations;
- email systems exhibit many different interaction styles and techniques. Consequently, just about every type of interaction breakdown can be identified in email use.

We shall be examining several email problems in some depth and mentioning many others. To exemplify the range of breakdowns in which we are interested, consider the following two (we shall revisit them again later during our analysis).

Rapid Scanning of Mail in mail. The standard ASCII-terminal-based UNIX mail program uses the UNIX utility more for paging long text files. The interface to this mail system is illustrated in Figure 1. We are interested in this application because a recent survey has revealed that it remains one of the most popular cross-platform mail utilities for UNIX users [17]. The mail and more systems interact in a particularly unfortunate way. Namely, both use the 'q' key for quitting. Consider a user who is browsing through recent messages, looking for a particular message that cannot be identified just by the header information. One after the other, a series of long messages are scanned and then dismissed without looking beyond the first screenful, necessitating the use of 'q'. Then, along comes a message, again not the target message, but this time less than a full screenful in size. The user, now set in a routine of pressing 'q' to get the next message, presses the 'q' key again, but this time quits the mail system entirely. In many applications this has the additional side-effect of forcing an update to the user's mailbox so that all of the previous messages are removed from the mailbox and are placed in a pre-determined file. This carries a significant penalty in terms of the time that is required to write large files to disk.

Lack of response after sending a message. The tasks performed with email systems often span more than a single session. Consider the case of a user who sends a message which requires a reply. Later, perhaps in a subsequent use of

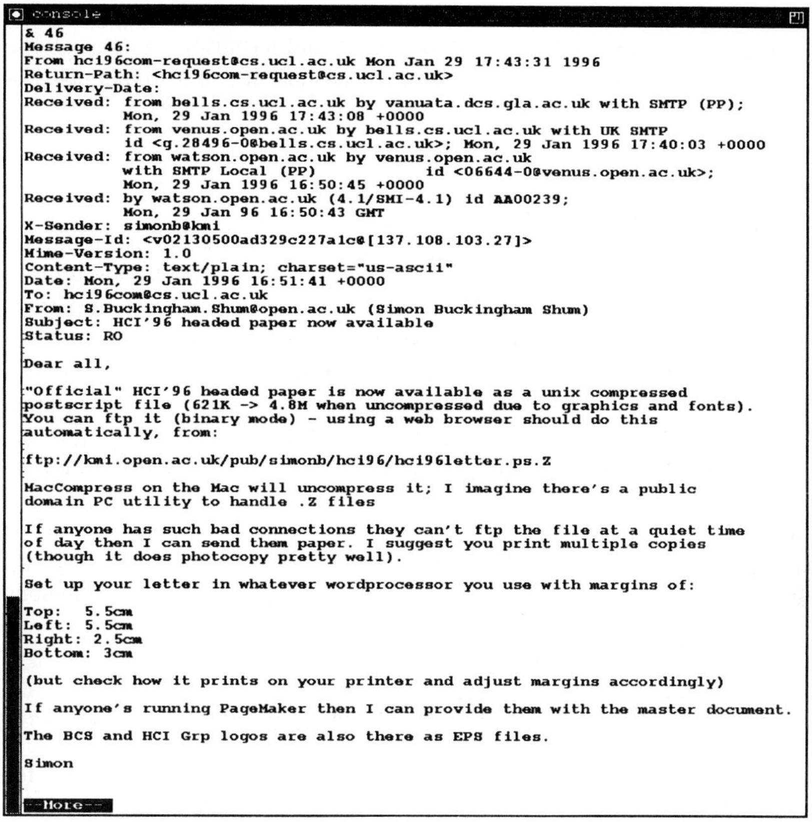

```
& 46
Message 46:
From hci96com-request@cs.ucl.ac.uk Mon Jan 29 17:43:31 1996
Return-Path: <hci96com-request@cs.ucl.ac.uk>
Delivery-Date:
Received: from bells.cs.ucl.ac.uk by vanuata.dcs.gla.ac.uk with SMTP (PP);
          Mon, 29 Jan 1996 17:43:08 +0000
Received: from venus.open.ac.uk by bells.cs.ucl.ac.uk with UK SMTP
          id <g.28496-0@bells.cs.ucl.ac.uk>; Mon, 29 Jan 1996 17:40:03 +0000
Received: from watson.open.ac.uk by venus.open.ac.uk
          with SMTP Local (PP)          id <06644-0@venus.open.ac.uk>;
          Mon, 29 Jan 1996 16:50:45 +0000
Received: by watson.open.ac.uk (4.1/SMI-4.1) id AA00239;
          Mon, 29 Jan 96 16:50:43 GMT
X-Sender: simonb@kmi
Message-Id: <v02130500ad329c227a1c@[137.108.103.27]>
Mime-Version: 1.0
Content-Type: text/plain; charset="us-ascii"
Date: Mon, 29 Jan 1996 16:51:41 +0000
To: hci96com@cs.ucl.ac.uk
From: S.Buckingham.Shum@open.ac.uk (Simon Buckingham Shum)
Subject: HCI'96 headed paper now available
Status: RO

Dear all,

"Official" HCI'96 headed paper is now available as a unix compressed
postscript file (621K -> 4.8M when uncompressed due to graphics and fonts).
You can ftp it (binary mode) - using a web browser should do this
automatically, from:

ftp://kmi.open.ac.uk/pub/simonb/hci96/hci96letter.ps.Z

MacCompress on the Mac will uncompress it; I imagine there's a public
domain PC utility to handle .Z files

If anyone has such bad connections they can't ftp the file at a quiet time
of day then I can send them paper. I suggest you print multiple copies
(though it does photocopy pretty well).

Set up your letter in whatever wordprocessor you use with margins of:

Top:   5.5cm
Left:  5.5cm
Right: 2.5cm
Bottom: 3cm

(but check how it prints on your printer and adjust margins accordingly)

If anyone's running PageMaker then I can provide them with the master document.

The BCS and HCI Grp logos are also there as EPS files.

Simon

--More--
```

Fig. 1. The UNIX mail system

the system, no reply is forthcoming. She sends a follow-up message, but still gets no answer. Discussion with the system "postmaster" reveals that the local mail system has not been working properly.

This example illustrates the fact that breakdowns need not be errors in any simple sense. The user did not do anything wrong, but had a belief about the functioning of the system that turned out to be false. Often it is not the original action which is the proper focus of attention for the system designer, but rather the consequent repair activity that takes place when a breakdown occurs. Suppose our user had talked to the systems manager and were told that the local system was functioning correctly. Only much later, after further investigation, she finds that the alias she was using was incorrect and the messages were going to the wrong person, who happened to be on vacation (with no automatic reply system in operation). Clearly, in this case, the level of disruption to the user was exacerbated by an inefficient repair strategy; she probably should have checked the email address early in her investigation.

What both these examples, and indeed all breakdowns, have in common is the fact that there is damage which must be repaired, even if that is only to restart the task as in our first example. Over-emphasis on normal interaction, and the lack of analytic tools for examining abnormal situations, increases the likelihood that users will be forced into repair situations. It can also increase the costs associated with repair activities.

1.4 Outline Of The Paper

Section 1 has introduced the argument that is presented by this paper. In particular, it has been suggested that interface specification notations provide limited means of representing breakdown in the interaction between a user and their system. Section 2 builds on this argument and uses temporal logic to illustrate the problems that arise when reasoning about abnormal traces of interaction. Section 3 shows that many of the weaknesses of temporal logic also arise when using UAN to analyse breakdown during human computer interaction. Section 4 builds on our analysis and temporal logic and UAN. It is argued that it is possible to produce a taxonomy of breakdown that might be used to build a set of requirements for future generations of specification notations. Section 5 shows how some of these requirements can be satisfied by innovative extensions that we have developed for existing design languages. Section 6 presents the conclusions that can be drawn from this work and suggests areas for future research.

2 Analysing Breakdown Using Temporal Logic

This following sections review the support that existing notations provide for designers who must reduce the potential for breakdown within a user interface. The discussion centres around the User Action Notation and Temporal Logic. This choice is appropriate because these formalisms each represent two different approaches to user interface design. UAN was specifically developed to support the design of human computer interfaces [6]. In contrast, temporal logic was first enhanced to support the software engineering of complex applications but has since been extended to support interface development [15].

2.1 Using Temporal Logic To Analyse Normative Behaviour

Logic is an attractive medium for the specification of human computer interfaces because it provides designers with a range of high level abstractions. These can be used to focus upon critical properties of an interactive system, without necessarily forcing designers to represent the low level implementation details that must be considered during an eventual implementation. For example, the following proposition uses a variation of the Horn clause notation to specify that the user exits the mail application if the mail system displays the list of message headers and the user types 'q':

$$\begin{aligned}&\mathit{quit\_mail} \Leftarrow\\&\quad \mathit{display}(\mathit{mail}, \mathit{Message\_headers}) \wedge \mathit{user\_input}(\mathit{mail}, 'q').\end{aligned} \tag{1}$$

Unfortunately, such first-order clauses lack any notion of time. The previous proposition would be true if the user typed 'q' well before the list of messages was presented to indicate that the user was in fact operating the mail tool. Temporal logic, in contrast to first-order logic, introduces a number of operators that are specifically intended to represent the timing requirements that often determine the usability of interactive systems. For instance, the $\bigcirc$ operator is read as 'next', $\Diamond$ is read as 'eventually' and $\Box$ is read as 'always'. With the introduction of these operators it is possible to represent the sequencing information that was missing in the previous clause. The user eventually inputs 'q' after the message headers have been presented:

$$\begin{aligned}&quit\_mail \Leftarrow\\&\quad display(mail, Message\_headers) \wedge \Diamond user\_input(mail,'q'). \qquad (2)\end{aligned}$$

The same approach can be used to represent the intended behaviour of the mail tool and the more browser in standard UNIX systems. The following clause specifies that the user quits mail from inside the more browser if the prompt is displayed by the browser and the user eventually types 'q' and in the next interval the system displays the message headers to indicate that the user has left the mail application. Eventually, the user types 'q' to quit the mail tool:

$$\begin{aligned}&quit\_mail\_from\_more \Leftarrow\\&\quad display(more, prompt) \wedge \Diamond user\_input(more,'q') \wedge\\&\quad \bigcirc(display(mail, Message\_headers) \wedge \Diamond user\_input(mail,'q')). \qquad (3)\end{aligned}$$

A similar approach can be used to represent normal behaviour involving the transfer of mail between remote sites. The user is presented with the *Message_headers* by the mail tool. This indicates that the system is ready to receive input. This time, instead of quitting the tool, the user eventually composes a message by typing a command line that is formed by the word 'mail' and the recipient's address. The text of the message is typed in and sent. Eventually, under normal operation a reply will be received and can be identified by the sender indicated in the From: line. The Re: line might also be used to indicate a reply to a particular message. This additional requirement could be introduced using extra clauses but is omitted for the sake of brevity:

$$\begin{aligned}&request\_reply\_using\_mail \Leftarrow display(mail, Message\_headers) \wedge\\&\quad \Diamond(user\_input(mail,'mail' \frown User\_name \frown Message\_Text) \wedge\\&\quad \Diamond(display(mail, Message\_headers) \wedge\\&\quad element(Message\_headers,'From:' \frown User\_name))). \qquad (4)\end{aligned}$$

The previous clauses illustrate how temporal logic can be used to represent interface requirements at a high level of abstraction. The following section argues that significant problems arise when attempting to use this formalism to represent the system failures and user 'errors' that characterise breakdown.

2.2 Using Temporal Logic To Analyse Abnormal Behaviour

The first of the case studies, described above, suggests that breakdown can occur when users fail to recognise that they are no longer within a particular mode. In the case of the mail tool the use of the browser, more, can be considered as a mode within which the 'q' input has a different effect than if it were issued from the mail tool itself:

$$\begin{aligned} &typeahead\_quit \Leftarrow \\ &\quad display(more, prompt) \wedge \Diamond user\_input(more, 'q') \wedge \\ &\quad \bigcirc(user\_input(mail, 'q') \wedge not \;\; display(mail, Message\_headers)). \end{aligned} \tag{5}$$

In this clause, the user types the second 'q' in the next interval after they have issued their input to exit the more application. This second quit command may be issued before the display is updated to reflect the fact that the user has successfully exited from more. In other words, the user either intentionally or accidentally presses the 'q' key when the *Message_headers* have not yet been presented. If the second 'q' was intentional then this is characteristic of the type ahead exploited by expert users. If the second 'q' was accidental then this reflects an instance of the breakdown referred to in the introduction. Here we see an illustration of one of the most important weaknesses in the use of logic and other 'formal methods' for interface design. As they were originally proposed to support software engineering, they provide little or no means of distinguishing between the many different operator behaviours that lead to breakdown. The second case study highlights further weaknesses in the use of temporal logic as a means of representing and reasoning about breakdown. Here, the failure of local or remote systems may prevent a user from receiving a reply to their original mail message. The following clause specifies that a reply is received or a transmission failure occurs if the user sends a mail message and eventually mail presents the list of message headers and that list either contains a message from the original recipient or the mail demon that reports delivery failures for the recipient's machine:

$$\begin{aligned} &reply\_or\_transmit\_failure \Leftarrow display(mail, Message\_headers) \wedge \\ &\quad \Diamond(user\_input(mail, 'mail' \frown User \frown '@' \frown Address \frown Message\_Text) \wedge \\ &\quad \Diamond(display(mail, Message\_headers) \wedge \\ &\quad (element(Message\_headers, 'From :' \frown User \frown '@' \frown Address)) \vee \\ &\quad element(Message\_headers, 'From : deamon@' \frown Address)). \end{aligned} \tag{6}$$

The problem with this clause is that it is difficult for designers to specify the remedial actions that either the user or the system might take in response to the abnormal condition. For example, if eventually a message is received from the remote demon it is perfectly possible for the mail tool to automatically re-transmit the message. It is technically difficult to represent such a requirement because we do not know the exact interval in which the $\Diamond$ (read as 'eventually') operator will be evaluated as true. Informally, it might be true now or in the next interval or in the next, next interval. This means that designers can often be surprised by the behaviour of an interface that has been specified using temporal logics if these imprecise temporal requirements are satisfied at unforeseen moments during future traces of interaction. For example, if a large number of e-mail messages have been generated for a common failure point then their failure may eventually be reported in the same interval by a correspondingly large influx of error messages to a local machine.

The previous clauses illustrate some of the problems that designers must address when using interface specification notations to analyse breakdown in human computer interfaces. Firstly, temporal logic provides few techniques for describing problems that can potentially occur at several different points during an interaction. Secondly, the logic provides little or no notational support for representing and reasoning about the internal state of the system operator. In some cases, this makes it difficult to distinguish between operator errors and expert behaviour. In the more example, a novice might inadvertently type an extra 'q' to exit the application at an inopportune moment. An expert might issue the same input to exploit type ahead mechanisms. These limitations reflect the fact that temporal logic was originally applied to the software engineering of complex applications. Cognitive and perceptual requirements can be introduced to distinguish between expert and novice behaviours but this is not an intrinsic part of the notation. In contrast, the next section explores whether notations that are specifically designed to support the specification of user interfaces are better equipped to deal with potential breakdown in human computer interaction.

3 Analysing Breakdown Using UAN

In contrast to temporal logic, the UAN formalism was specifically developed to support user interface design. The following section examines whether this notation provides designers with better means of representing and reasoning about the impact of breakdown.

3.1 Using UAN To Analyse Normative Behaviour

The User Action Notation or UAN [11] is a language for describing the observable aspects of interaction plus related unobservable system consequences. It uses a tabular layout to categorise interaction events into user actions, feedback, local system state and consequent computational operations. A simple action language is used to capture user actions and system feedback. An important benefit of this notation, in view of our previous comments about the importance of time in breakdown, is that UAN has been extended to handle complex temporal relationships among tasks, including real-time constraints on actions [6].

The reading and filtering of mail messages, described in our first scenario, can best be described in UAN as a sequence of 'scan message' tasks which continue until the desired message is found. Figure 2 illustrates how each message scan task offers alternative actions, with specified conditions of viability. For clarity's sake, the internal system columns are omitted as they do not contribute to the description in this case [1]. The user continues to scan messages while she has not

Task scan messages

User Action	Feedback
while not at the desired message: (scan message)*	

Fig. 2. UAN for Scan Messages

found the one that she is looking for. Figure 3 illustrates how the underlying scan message task can be represented in UAN. In our second case study, the user's

Task scan message

User Action	Feedback
IF *full message displayed*: press <apace> ELSE press <q>	previous message or headers

Fig. 3. UAN for Scan Message

task is to send the mail. In the normative case, she should receive feedback from the recipient that the message has successfully reached its destination. This expected trace is shown in figure 4.

3.2 Using UAN To Analyse Abnormal Behaviour

Scanning Mail. A simple analysis of the scanning mail problem would identify the cause of the breakdown with the user's failure to correctly identify the

[1] The level of detail in the UAN description depends on the purpose to which it is put. The UAN action language includes as its terminal actions, single device input events (e.g., key presses and mouse clicks) and display-based output. For our purposes we need not finish the analysis to this level of detail.

Task send message

User Action	Feedback	Connection to Computation
despatch message	confirmation of despatch	message received
		reply sent
	confirmation of receipt	

Fig. 4. UAN for Send Message

current mode. That is, the user acts as if the first condition in the scan message task is viable, when it is not. This can be shown in UAN without difficulty, see Figure 5. Some UAN users make a practice of checking that there is a visible indication of any condition of viability. However, the error state is not part of

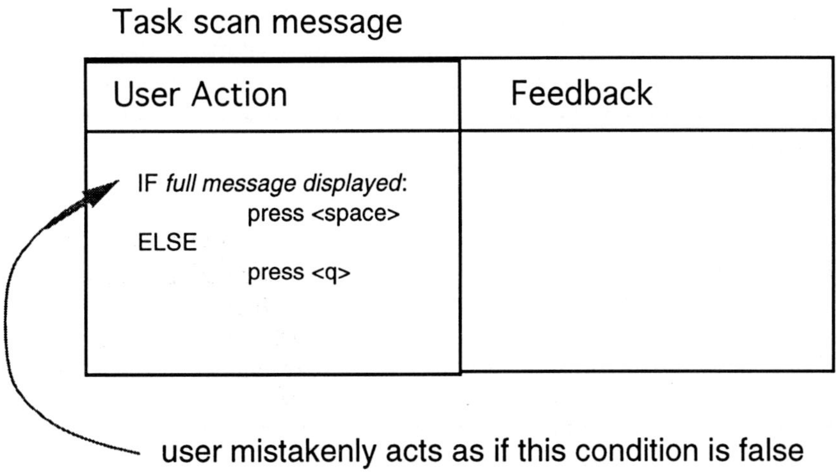

Task scan message

User Action	Feedback
IF *full message displayed*: press <space> ELSE press <q>	

user mistakenly acts as if this condition is false

Fig. 5. UAN for Scanning Mistake

the UAN description but an annotation of it. The UAN has no structures for allowing such information to be included in the description itself. Figure 6 shows how designers can develop alternative UAN descriptions to capture the user's 'erroneous' view of their interaction. However, the problem arises again in that we can only indicate that one is the normative and the other the abnormal version of the same task by means of annotations. Additionally, the problem in our first case study arose because the user was embedded in a long sequence of simple, repetitive actions. In such cases, users will tend to switch to a strategy

Task abnormal scan message

User Action	Feedback	Connection to comp
press <q>		mail system quits
	operating system prompt	

Fig. 6. UAN for Abnormal Scan Message Task

of action without checking feedback. Such strategies are so powerful (i.e., cognitively efficient) that, in the absence of errors, users have a strong inclination to adopt them [19, 22]. How can we deal with this more sophisticated analysis using the UAN? Almost all of the relevant structural features are present, viz., the UAN description includes both the reference to the repetition and to the fact that there are alternative strategies available for each repeated act. However, the description of the potentially problematic situation should include the fact that the there are a large number of repetitions and that the alternatives are "inexpensive" in terms of cognitive and/or articulatory effort (so that they can be carried out without reference to the visual feedback "cue"). This additional information can again only be added as annotations and not as a part of the UAN description itself.

Email communication breakdown The second case illustrates again the fact that the UAN is designed for describing normative behaviour. If a confirmation of receipt is not received, then the user is not sure what has occurred as a consequence of her original action. That is, one could say that she is not sure what task has been performed. Subsequent repair activity is largely a matter of discovering the actual, as opposed to assumed, user and system actions.

If the breakdown occurred because the message was sent to the wrong person due to the use of an incorrect alias, then the user performed 'despatch message to P2' while believing she had performed 'despatch message to P1'. Since the UAN does not include reference to the user's belief state, there is no way of specifying this mismatch. And, even if one could describe these actual and intended tasks, the notation has no mechanism for relating them.

If the breakdown occurred because of a network failure, then the problem resides in a mismatch between actual and believed actions in the Connection to Computation column. The UAN does not provide a way of asserting that some invisible state or action was believed to have occurred but did not.

Abnormal task states highlight the crucial and complex relationships between user's beliefs or assumptions and the actions and states which form the behavioural aspect of task performance. A focus on task breakdown emphasises the inadequacies of task description languages which, like the UAN, are limited to behavioural features of normal task performance.

It is important to remember that the UAN was conceived to assist designers in specifying a design. In our examples we have left behind simple design description and now concerned with the analysis of the interaction domain so that a successful (or more successful) design can be produced. As soon as we think of the notation as meant to help the process of design, rather than merely the specification of a particular design solution, the requirements for the notation change.

4 Notations For Error Tolerant Design

The previous sections have analysed the limitations that prevent designers from using notations, such as temporal logic and UAN, to assess the impact of breakdown during human computer interaction. In contrast, this section identifies techniques that can be used to avoid these limitations. In particular, we extend temporal logic to include epistemic operators. That is, operators that represent assertions about the knowledge possessed by operators during critical periods of interaction. This provides means of distinguishing between instances of breakdown and the more optimal behaviours exhibited by expert users. Unfortunately, this approach provides limited means of representing the asynchronous failures that occur during breakdown. This problems can be avoided by extending the application of exception handling techniques from the field of real-time and safety-critical software engineering to provide a means of representing and reasoning about system failure and operator 'error'.

4.1 Errors And Epsitemics

Breakdown is defined to occur when either operator 'error' or machine failure leads to sub-optimal traces of interaction. One of the reasons why interface notations, such as temporal logic and UAN, are so poorly equipped to deal with breakdown is that they provide little means of analysing the causes of operator error. They provide precise and concise means of representing observable behaviours by the system and the operator. Unfortunately, the use is left as a 'black box'. There is no means of representing or reasoning about inappropriate plans, slips and mistakes. Such limitations might be avoided if designers could incorporate elements of cognitive models into formal and semi-formal models of interactive systems. For example, Figure 7 presents a high level view of Barnard's Interacting Cognitive Subsystems (ICS) [1]. Visual and acoustic information is interpreted to produce object and morphonolexical representations. These help to form propositional representations using the bottom-up information from the sensory sub-systems and top-down information from an implicational encoding that helps to decide the "schematic" meaning.

Although the ICS model provides a coherent framework for analysing elements of users' cognitive processes, it does not provide designers with concrete means of analysing the contents of individual subsystems. There is no notation

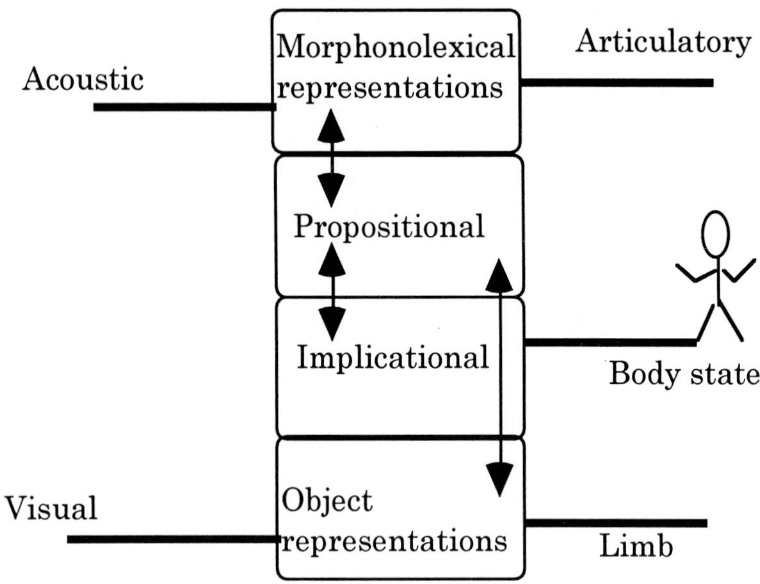

Fig. 7. Main Features of Barnard's ICS model

or design technique associated with the framework. Fortunately, a number of notations have been independently developed that might be linked with such a view of operator cognition. For instance, epistemic logics provide a convenient means of representing designers' expectations about a users' propositional and implicational subsystems. Epistemology, or the study of knowledge, has a history stretching back through Philosophy to the Ancient Greeks [8]. The modern history, however, dates back to von Wright [23], Hintikka [10] and Barwise and Perry [2]. Rather than present the detailed syntax and semantics of epistemic logic, the remainder of this paper exploits a simple propositional extension of temporal logic. The *kn* prefix is used to denote a fact about which the operator is certain. Elsewhere, we provide more detailed explanations of probabilistic alternatives to this epistemic approach [14]. Much of this work has modelled the ways in which an individual's knowledge changes over time. This has critical importance for analysing the impact of breakdown upon human computer interaction. For example, the following clauses illustrates how epistemic logic might be integrated into previous clauses to distinguish between expert and novice behaviour with the mail tool. In the case of expert interaction, the user knows at each stage that they are typing their input into the more application and then to the mail tool. The user exploits type-ahead techniques to issue the second 'q' before the mail-headers return. In the case of breakdown, the user does not know that the second 'q' is directed at the mail tool and, instead, believes that it is intended for the more browser:

$$expert\_typeahead\_quit \Leftarrow display(more, prompt)\wedge$$

$$\Diamond(user\_input(more,'q') \wedge kn(user\_input(more,'q')) \wedge$$
$$\bigcirc(user\_input(mail,'q') \wedge kn(user\_input(mail,'q') \wedge$$
$$not\ \ display(mail, Message\_headers)). \tag{7}$$

$$break\_down \Leftarrow display(more, prompt) \wedge$$
$$\Diamond(user\_input(more,'q') \wedge kn(user\_input(more,'q')) \wedge$$
$$\bigcirc(user\_input(mail,'q') \wedge not(kn(user\_input(mail,'q')) \wedge$$
$$kn(user\_input(more,'q') \wedge not\ \ display(mail, Message\_headers)). \tag{8}$$

It is important to re-emphasise that the use of the epistemic propositions is not intended to be a substitute for cognitive psychology. The aim is to explicitly record assumptions about operator knowledge under error conditions. Our intention is to alter the current focus of interface notations which is biased towards the normative behaviour of expert users.

Epistemics can also be applied to reason about the causes of breakdown in our second scenario. The following clause states that there is an inefficient repair if the user sends a message using a mail alias and there is no reply from that user or a mail daemon response to indicate a transmission failure until the user knows that the mail alias is incorrect. This is inefficient because the potential delay described by the $\mathcal{U}$ operator between the transmission and the check on the alias may be many hours or even days:

$$inefficient\_repair \Leftarrow$$
$$\Diamond(user\_input(mail,'mail' \frown Alias \frown Message\_Text) \wedge$$
$$not(element(Message\_headers,'From:' \frown User \frown '@' \frown Address) \vee$$
$$element(Message\_headers,'From: deamon@' \frown Address))\mathcal{U}$$
$$kn(not\ valid(Alias, User \frown '@' \frown Address))) \tag{9}$$

The previous clauses are intended to demonstrate that designers must have some means of representing and reasoning about elements of operator knowledge if they are to anticipate and respond to the potential for breakdown.

Thus far, we have concentrated upon extensions to the temporal logic notation. A similar approach can be adopted for the UAN formalism [2]. The user believed that they were in the more mode when they typed 'q'. As a result, they exited the mail system and the operating system prompt was re-displayed. Figure 8 illustrates how the user now knows at this point that they were not originally using the more browser of the mail tool. It is important to emphasise that our proposed use of epistemics represents a relatively simple extension of UAN and temporal logic. Clearly, the proposed annotations do not capture the

[2] Sharratt's Memory-Cognition-Action tables [21] included cognitive information in a task description notation and are somewhat similar to our enhancement of UAN. However, Memory-Cognition-Action tables were not designed to deal with breakdowns and in any case do not have the wide currency of use that is enjoyed by UAN.

Task 'break-down' in message scan

User knowledge	User Action	Feedback	Connection to comp
In 'more' mode	press <q>		mail system quits
Previously not in 'more' mode		operating system prompt	

Fig. 8. UAN for User Breakdown in Message Task

longer term tactics and strategies that users evolve to combat the uncertainties associated with breakdown. For instance, we have not shown how users may adapt their mail behaviour in the light of failures at remote sites. However, such extensions are completely feasible given the range of timing operators that are currently available in both temporal logic and the XUAN extension to UAN. This is the subject of continuing research. Rather than pursue this approach in greater detail, however, the following section focuses on the more pragmatic problem of how to represent exceptions during human-computer interaction.

4.2 Errors And Exceptions

The previous section has shown how elements of epistemic logic can be used to express observations about operator knowledge under breakdown. Unfortunately, this approach only addresses one of the causes that we have identified for 'failure' during human computer interaction. In particular, it does not show how notations, such as temporal logic and UAN, can be used to describe the human factors problems that arise from periodic machine failures, such as on remote sites during e-mail delivery. The following section shows how this weakness can be resolved by extending user interface notations with explicit exception handlers.

Previous sections have argued that an important problem in analysing breakdown is that it can have complex temporal characteristics. In our email example, this is illustrated by the delay that can arise between the user sending the message and a reply being sent from a remote daemon to indicate that the delivery has failed. This causes problems because the arrival of the return message is asynchronous. It occurs at some undefined time after the initial message has been sent. The user may have moved on to complete a range of many different tasks oblivious that there is still the chance that their email will fail. For the interface designers, this means that they would have to include an additional UAN table entry or a temporal logic clause to illustrate what would happen if the subsequent task were interrupted at any point by the arrival of the daemon's message. The following tables illustrate this problem, tasks can be interrupted if users detect the arrival of anticipated mail. In order to model this potential interruption to normative behaviour in UAN, designers must continually introduce the condition that specifies what should happen 'IF new message displayed'. Figure 9 illustrates how they must also repeat the epistemic constraint that urgent mail is anticipated. It is important to emphasise that almost all interface specification

Task read news

User Knowledge	User Action	Feedback
urgent mail is anticipated	IF *new message displayed*: start mail tool ELSE start news tool while not finished: (read newsgroup)*	draw mail tool window display message headers draw news reader window

Task read newsgroup

User Knowledge	User Action	Feedback
urgent mail is anticipated	IF *new message displayed*: start mail tool ELSE select newsgroup while not finished: (read items)*	draw mail tool window display message headers display news items

Fig. 9. UAN for Read News task.

notations suffer from problems when representing the effects of interruptions during breakdown. Production systems provide an important exception to this rule, see for instance Johnson [13]. In order to model this in Petri Nets, designers would have to introduce links at every place where the arrival of mail could alter the subsequent state transitions. In temporal logic, designers are forced to introduce additional clauses:

$$interrupt\_news\_reading \Leftarrow user_input(operating\_system, 'nn') \wedge$$
$$\Diamond(display(nn, newsgroups) \wedge \bigcirc(user\_input(nn, select\_group) \wedge$$
$$(display(operating\_system, new\_mail) \wedge user\_input(mail, reload)))). \quad (10)$$

At each stage of interaction after the initial mail message has been sent, the designer would be forced to specify that the user has the opportunity of interrupting their current task to check whether or not their mail has failed. This seems to be a heavy burden that cannot be excused considering the increasing opportunities that interface designers have to exploit the benefits of multi-tasking and asynchronous processing. Whether one looks at the multi-tasking being introduced through Windows95 and MacOS or the development of distributed information retrieval systems, such as the Web, it seems clear that the inability to handle asynchronous interrupts will be a critical requirement of future interface design notations. In practice, the proponents of previous formalisms have either ignored this issue or have suggested that designers ought to annotate the descriptions to take account of such problems [16]. In contrast, the remainder of this paper argues that such asynchronous events ought to be explicitly included within future generations of interface specification notations.

Exception handling is an important feature of many existing programming languages. For example, the following code illustrates how programmers can use the Ada language to deal with unexpected memory problems:

```
procedure MAIN is
Simple example of exception handling.
Declarations removed to simplify exposition
begin
      INITIALISE_BOOK_DATABASE;
      DO_OPTIONS;
      CLOSE_BOOK_DATABASE;

exception
      when OVERFLOW | MEMORY_ERROR =>
      PUT(''System Error: out of memory, cannot store book'');
      CLOSE_BOOK_DATABASE;
end MAIN;
```

When executed this code will open a database of books, it will offer a menu of options and will continue to execute the user's commands until they have finished. However given a finite amount of memory and fallible technology, it is possible that at some point during interaction the system will not be able to store an additional book. The user can be protected against this problem by placing an additional guard at every point where it might have arisen. This is good programming practice but it is also error prone. Programmers frequently do not anticipate all of the places where failure can occur. It can be difficult for all of the members of concurrent development teams to understand when and where these errors can occur. By explicitly flagging the exceptions that can disrupt normative behaviour, designers place important sign-posts both for themselves and their colleagues. In the previous example, if an `OVERFLOW` or a `MEMORY_ERROR` exception did occur at any point during interaction then the current procedure would be abandoned and control would pass to the nearest handler. In this case, the user would be presented with an appropriate warning and the system would be closed down in a relatively 'clean' fashion.

It seems strange that programmers can call upon a considerable degree of language and design support for the problems that arise during asynchronous system failures and yet interface designers have almost no means of addressing this problem. One means of resolving this inadequacy would be to introduce exception handlers into the tabular form of UAN: As with the Ada example, if the exception occurs in the read news task or anywhere in the read newsgroup sub-task then the task is interrupted by the user action immediately below the associated event; new message displayed. If the exception does not occur during

Task read news

User Knowledge	User Action	Feedback
	start mail tool start news tool while not finished: (read newsgroup)*	draw news reader window
EXCEPTION urgent mail is anticipated	*new message displayed* start mail tool	draw mail tool window display message headers

Task read newsgroup

User Action	Feedback
select newsgroup while not finished: (read items)*	display news items

Fig. 10. UAN with Exception in Read News Task.

the normal operation of the task then there is no need for the user to perform the 'start mail tool' action. This approach offers a number of benefits for the design of interactive systems. In particular, it avoids the repetition of guards that check for asynchronous events whenever they may interrupt a task. It also avoids the need to repeat the epistemic conditions that must be satisfied in order for the exception to take place. These are non-trivial benefits. By supporting the representation of such interruptions, the introduction of exceptions helps designers to consider the impact that breakdown can have upon the course of 'normative' interaction.

An important point here is that our extension to the UAN formalism should not increase the problems that designers must face when first learning how to exploit the notation. For this reason, we do not stipulate that every UAN table ought to have an exception handler. The point is that this techniques should be used sparingly and only for those infrequent conditions that interrupt the 'typical' course of interaction. If such failures are a regular problem then they ought to be explicitly considered within the main body of the UAN table.

A similar approach can be exploited within the temporal logic notation. Rather than introducing a separate tabular construct, the easiest approach here is to create a series of propositions that are quantified by the $\Diamond$ operator. For example, the following clause indicates that the normal task of reading the news can be interrupted if eventually the user is reading the news and eventually urgent mail is received. The interruption occurs if the user knows that there is urgent new mail and the operating system indicates that this mail has arrived and the user issues input to reload the new mail:

$$non\_normative\_interruption \Leftarrow \\ news\_reading \wedge \Diamond exception(urgent\_mail) \quad (11)$$

$$exception(urgent\_mail) \Leftarrow kn(urgent(new\_mail)) \wedge \\ display(operating\_system, new_m ail) \wedge user\_input(mail, reload). \quad (12)$$

The $\Diamond$ (read as 'eventually') operator specifies that the exception may or may not occur during any particular trace of interaction. If it does occur then the previous clauses do not state the effect which the interruption may have upon the news reading task. This is an important benefit of our approach. It is possible to represent and reason about the potential for breakdown without necessarily being forced to analyse all of the complex traces of interaction that may have to be considered during the later stages of development. The critical point here is that unless we have some means of analysing breakdown during the design of human-computer interfaces then we will continue to have systems that are designed only to support perfect users in perfect environments.

5 Conclusion and Further Work

This paper has argued that interface specification notations provide extremely poor means of representing and reasoning about breakdown during human computer interaction. Although existing notations provide good support for analysing normative behaviour, they cannot easily be used to analyse the problems that occur when users or systems fail to act as expected.

Our argument has been illustrated by temporal logic. This notation was originally extended to reason about software engineering but has subsequently been used to analyse complex human-computer interfaces. We have also examined the strengths and weaknesses of the User Action Notation. This was specifically developed to support the design of interactive systems. In both cases, we have identified two limitations that restrict the utility of these notations as means of analysing failure. Firstly, they provide no means of capturing the cognitive conditions that lead to operator 'error'. Secondly, they cannot easily be used to represent and reason about the impact that synchronous failures can have upon complex interaction.

The first of these problems has been addressed by the introduction of epistemic operators into both temporal logic and UAN. This provides means of distinguishing between behaviour that can either be characterised as an expert trace or a mistake by a novice depending on the knowledge that the users exploited to plan their interaction. This point was illustrated by the use of typeahead in the UNIX mail tool. If the user knew which more they were in then repeatedly issuing the command 'q' would have the desired effect of exiting the application. If, in contrast, the user was unaware of the mode then this same observable behaviour would have the undesirable affect of throwing them out of the mail tool and causing an unnecessary and inconvenient disk access. It would have been impossible to distinguish between these two behaviours using previous versions of interface specification notations.

The problem of asynchronous failures has been resolved by the introduction of exception handling routines into UAN and temporal logic. These routines describe what should happen when unexpected failures can potentially occur at many different points during interaction. It is important to emphasise that while these handlers can be introduced with relatively simple syntactic changes to the specification notations, they are intended to alter the existing pre-occupation with normative behaviour.

Potential interaction breakdowns are often very complex, both in terms of epistemological and temporal aspects, and the simple examples we have presented can only point the way towards the much richer modelling constructs and notations designers need to deal with breakdowns and their consequences. Epistemic operators and exception handlers are two notational enhancements which can help. Further research is needed both to elaborate them and to identify and develop additional analytic and descriptive tools.

Acknowledgements

Thanks go to the members of the Glasgow Interactive Systems group (GIST) who provided valuable advice and encouragement for this work. Johnson's research is supported by EPSRC grants GR/K55040 and GR/K69148.

References

1. P. Barnard. Interacting cognitive subsystems: A psycholinguistic approach to short-term memory. In A. Ellis, editor, *Progress in the Psychology of Language*, volume 2, pages 197–258. Lawrence Erlbaum, London, 1995.
2. J. Barwise and J. Perry. *Situations And Attitudes.* Bradford Books, Cambridge, United States of America, 1983.
3. R. Bastide and P. Palanque. Petri net objects for the design, validation and prototyping of user-driven interfaces. In D. Diaper, D. Gilmore, G. Cockton, and B. Shackel, editors, *Human-Computer Interaction—INTERACT'90*, pages 625–631. Elsevier Science Publications, North Holland, Netherlands, 1990.
4. J.M. Carroll. *The Nurnberg Funnel: Designing Minimalist Instruction For Practical Computer Skill.* MIT Press, Boston, United States of America, 1992.
5. D. Diaper and P. Johnson. Task analysis for knowledge description. In J. Long and A. Whitefield, editors, *Cognitive Ergonomics For Human Computer Interaction*, pages 191–224. Cambridge University Press, Cambridge, United Kingdom, 1989.
6. P. Gray, D. England, and S. McGowan. XUAN: Enhancing UAN to capture temporal relationships among actions. In G. Cockton, S. Draper, and G. Weir, editors, *People And Computers IX*, pages 301–312. Cambridge University Press, Cambridge, United Kingdom, 1984.
7. P.D. Gray and C.W. Johnson. A critical analysis of interface specification notations. In *The Design, Specification and Verification of Interactive Systems*, pages 113–133. Springer Verlag, Berlin, Germany, 1995.

8. J. Halpern. Reasoning about knowledge: A survey. In *Handbook Of Logic And Artificial Inteligence: Volume 4 - Epistemic And Temporal Reasoning*, pages 1–34. Clarendon Press, Oxford, United Kingdom, 1995.
9. T. Hewett. Importance of failure analysis for human-computer interface design. *Interacting With Computers*, 1(3):3–8, 1991.
10. J. Hintikka. *Knowledge And Belief.* Cornell University Press, Ithica, United States of America, 1962.
11. D. Hix and H.R. Hartson. *Developing User Interfaces.* John Wiley and Sons, London, 1993.
12. D. John and D. Kieras. The goms family of analysis techniques: Tools for design and evaluation. Technical Report CMU-CS-94-181, School of Computer Science, Carnegie Mellon University., Pittsburgh, PA, USA, 1994.
13. C.W. Johnson. *A Formal Approach To The Integration Of Human Factors And Systems Engineering.* PhD thesis, Department Of Computer Science, University of York, York, United Kingdom, 1992.
14. C.W. Johnson. A probabilistic logic for the development of safety-critical interactive systems. *International Journal Of Man-Machine Studies*, 39(2):333–351, 1993.
15. C.W. Johnson. Using Z to support the design of interactive, safety-critical systems. *IEE Software Engineering Journal*, 10(2):49–60, 1995.
16. C.W. Johnson. Literate specification: Using design rationale to support formal methods in the development of human-machine interfaces. *Human Computer Interaction Journal*, 1996. Acceped and to appear early in 1996.
17. S. Mitchell. The automatic filtering of electronic mail messages. Technical report, Dept of Computing Science, University of Glasgow, Scotland, 1996. Final year dissertation.
18. D.A. Norman. The 'problem' with automation : Inappropriate feedback and interaction not 'over-automation'. In D.E. Broadbent, J. Reason, and A. Baddeley, editors, *Human Factors In Hazardous Situations*, pages 137–145. Clarendon Press, Oxford, United Kingdom, 1990.
19. P. O'Donnell and S.Draper. How machine delays change user strategies. In C. Johnson, editor, *The Challenge Of Time.* Glasgow Interactive Systems Group, Glasgow, United Kingdom, 1995. G-95.1.
20. H. Petroski. *To Engineer Is Human: The Role Of Failure In Successful Design.* St. Martin's Press, New York, United States of America, 1986.
21. B. Sharratt. Memory-cognition-action tables: A pragmatic approach to analytical modelling. In *Interact T90*, pages 625–631. Elsevier Science, North Holland, 1990.
22. D. Taylor. The role of human action in man-machine system errors. In J. Rasmussen, K. Duncan, and J. Leplat, editors, *New Technology and Human Error.*, pages 287–292. John Wiley and Sons, London, United Kingdom, 1987.
23. G.H. von Wright. *An Essay In Modal Logic.* Elsevier, North Holland, Netherlands:, 1951.
24. T. Winograd and F. Flores. *Understanding Computers And Cognition.* Addison-Wesley, Reading, United States of America, 1987.

This article was processed using the LaTeX macro package with LLNCS style

Risk Analysis, Impact and Interaction Modelling

A. M. Dearden & M. D. Harrison

Department of Computer Science
University of York
York, YO1 5DD
UK
email: {andyd,mdh}@minster.cs.york.ac.uk

Abstract. Operator error has been blamed for many accidents and incidents in safety-critical systems. It is important that human-machine interface (HMI) designers are aware of the relationships between their design decisions, operator errors, and the hazards associated with a system. In this paper, we demonstrate how information from risk analysis can be combined with formal specification of the HMI, to support designers in exploring these relationships. We use the concept of *interactor* to model the human-machine interface (HMI); together with a concept of *impact*, which we define informally as: "the effect that an action or sequence of actions has on the safe and successful operation of a system." We show how interactors can be used as design representations for the HMI at the earliest stages of design, as well as providing a medium by which risk analysts can inform HMI designers about the impact of human-errors. To demonstrate the feasibility of this approach, we consider a simple, gas-fired, electricity generating plant as a case study. Our proposed approach is intended to complement, rather than compete with, existing design and analysis methods for the HMI. The method achieves this by making risk analysis information available in the early stages of HMI design.

Keywords

Impact, Safety-critical systems, Interactor, Risk analysis, Formal Models

1 Introduction

Operator error has been blamed for many accidents and incidents in safety-critical systems. It is important that human-machine interface (HMI) designers are aware of the relationships between their design decisions, operator errors, and the hazards associated with a system.

In this paper, we consider how formal models of the HMI can be used, together with information generated by risk analysts, to reason about the *impact* of operator actions and errors. Informally, we define *impact* as:

> the effect that an action or sequence of actions has on the safe and successful operation of a system.

We begin from the assumption that the operators of such systems develop their own perception of the impact of their actions, based on their experience of plant behaviour. We can refer to the operators' viewpoint as the *perceived impact*. Clearly, it is desirable that the perceived impact corresponds as closely as possible to the understanding of impact developed by formal risk analysis procedures. To assist operators in developing a correct perception of impact and actual risk: HMI designers should be aware of the relationship between their design decisions, human error and impact; and the completed design should reflect this understanding.

We use a simple case study to demonstrate how formal models, based on the concept of *interactor*, can facilitate communication between HMI designers and risk analysts during the early stages of the design process. By supporting such communication we hope to assist HMI designers in considering the impact of human error, whilst they develop the design. The case study involves the specification of an HMI for a simple gas-fired electricity generating plant.

1.1 Structure of the rest of this paper

In section 2, we briefly describe the background of the work in risk analysis, and interactor modelling. In section 3, we explore the notion of impact and develop a method for quantifying impact.

In section 4, we describe the power plant case study informally, and then develop an interactor specification for the plant. In section 5, we discuss how risk analysis information encoded in the form of fault trees can be used to reason about the impact of operator errors in performing procedures. Finally, in section 6, we discuss issues that will need to be considered in developing and applying this type of analysis to large scale, real world safety-critical systems.

2 Background

2.1 Risk Assessment

Risk assessment is the process of identifying, analysing and quantifying the risks associated with a system. Within risk assessment various terms, some of which are interchangeable in everyday use, are given very precise meanings. For instance, a clear distinction is drawn between hazard and risk. We list some basic definitions here, taken from Villemeur (1992a), to assist the reader who may be unfamiliar with this area.

Failure The termination of the ability to perform a required function.

Failure Mode Effect by which the failure of a component is observed.

Human Error The departure of a human operator's performance from what it should be, this departure exceeding allowable limits under given conditions.

Hazard Situation that is undesirable for man, society or the environment.

Risk Hazard measure combining a measure of the probability of the occurrence of undesirable events and a measure of the consequences of the events.

Probabilistic Risk Assessment Study aimed at evaluating the risks associated with a system based on probabilistic methods.

Preliminary Hazard and Risk Analysis Preliminary investigation of the hazards inherent in using a system and the risks of undesirable events occurring.

HMI designers and researchers may wish to sharpen the definition of human error as:

Human Error Deviation of a human operator's behaviour from the behaviour that was originally planned by the designer, and was used as the basis for risk analysis and system safety certification.

The measure of the consequences of an event is generally referred to as the event's *Severity*.

Typically, risk assessment attempts to identify and analyse risks from two complementary perspectives. The first perspective can be characterised as 'bottom-up', starting with the known failure modes of components, and investigating how these might lead to a hazard. These techniques offer qualitative methods for relating failures to hazards. The second perspective is 'top-down' beginning with a given hazard, and seeking to identify all the possible ways that the hazard might be caused. The existence of a previous bottom-up analysis can act as a significant aid to completeness checking when conducting a top-down analysis. Top-down analyses are often recorded in the form of fault trees, and can be used to support quantification in probabilistic risk assessments.

2.2 Existing methods of relating risk analysis and HMI design

A number of techniques are already available to enable HMI designers and risk analysts to exchange information. None of these techniques make use of a quantified concept of impact. Current techniques can be divided into two major classes.

Quantifying human error probabilities

The first set of techniques seek to quantify the probability that a human operator will commit a particular error when performing some procedure. This permits analysis of the risks associated with a given HMI design. Examples include: Technique for Human Error Rate Prediction (THERP) [Swain and Guttman, 1983], Technica Stima Errori Operatori (TESEO) [Bello and Colombari, 1980], Human Cognitive Reliability [Hannaman *et al.*, 1984] and the Success Likelihood Index Methodology (SLIM) [Embrey *et al.*, 1984]. Unfortunately, these techniques suffer from the difficulty associated with any attempt to construct quantitative, predictive models of human behaviour. Also, they can only be applied after the HMI design has been refined to quite a high level of detail.

Qualitative analysis using multi-disciplinary teams

These methods involve systematic investigation of the possible consequences of human errors or component failures, by multi-disciplined teams and discussion of possible design responses. The teams include risk analysts, human factors engineers, and domain engineers (i.e. those responsible for the design of the machine elements of the system). Kirwan (1992) provides a review of methods in this class. The major disadvantage of these techniques is that they are very resource intensive. Indeed, the process may be characterised as providing the HMI designer with the full time services of the domain engineer and the risk analyst, whilst the design is refined, and supplying the risk analyst with the full time services of an HMI designer and a domain engineer whilst the analysis is conducted. Alternatively, a team analysis could be conducted after the HMI design has been developed in some detail, in which case the early stages of HMI design would have to proceed unsupported by risk analysis information.

Our proposed study of impact aims to support these other techniques by providing the HMI designer with a way of interpreting some of the information generated by preliminary risk analysis. Conversely, the use of formal models of the evolving HMI design makes the design more accessible to risk analysts. We do not intend that our technique should replace existing methods, rather that it should be used to provide additional support to HMI designers during the early stages of design, where currently available methods are unsuitable.

2.3 Interactor models in HMI design

Many authors have argued for the use of formal modelling techniques in the development of safety-critical software. Using such techniques, analysis of the behaviour of a system can be conducted and the designer can investigate generalised claims about safety properties at an early stage of the design, rather than waiting for the construction of hi-fidelity working prototypes, and the results of simulations and testing which may never completely cover the space of possible conditions.

Given the suitability of formal modelling techniques for safety-critical software, we propose to support impact analysis for HMI designs by investigating formal models of the HMI using the concept of interactor [Duke and Harrison, 1993, Duke *et al.*, 1994]. An interactor is a component of a human-machine or human-computer interface, that incorporates state, behaviour and presentation information. For instance, a button or a menu can be modelled as an interactor. Complex interactors can be constructed from simple interactors, e.g. a scroll bar can be constructed from two buttons and a slider. The particular notation that we shall use for describing interactors is based on Duke & Harrison (1994).

2.4 Specifying procedures and identifying plausible errors

Given a specification of the behaviour of an interactive system as a collection of interactors, it is possible to specify operating procedures as hierarchical tasks where the leaf nodes of the hierarchy correspond to primitive operations of the interactor model.

Hollnagel (1993) argues that a number of generic error 'phenotypes' can be identified in human behaviour. These phenotypes include omission of a step in a procedure, commission of an incorrect step in a procedure, transposition of two steps of a procedure. Fields, Wright and Harrison (1995) demonstrate how Hollnagel's concept can be used to generate plausible error behaviours expressed as sequences of primitive actions on the interactor model. This is achieved by constructing a heirarchical task description, and then applying the phenotypical errors not only at the level of individual procedural steps, but also at the level of whole subtasks. Thus, given a hierarchical description of a procedure, and a set of phenotypical errors, a family of plausible erroneous human behaviours can be noted.

Having generated plausible error sequences, Fields *et al.* show how the interactor model can be used to investigate the state that would be reached if a given error sequence occurs. In practice, very few sequences are likely to lead directly to a hazard, since safety-critical systems are designed around the principle that no single point of failure should cause a hazard to occur. However, many sequences still have an impact in the sense that they could eventually lead to a hazard, if some other failures or human errors were to occur. The problem we address in this paper is how an HMI designer can obtain an evaluation of the significance of such error sequences, in order to guide design decisions. Such decisions need to consider how to prevent the sequence occurring and how to support the operator's perception of the impact of these interactions.

3 Impact

In section 1 we defined impact informally as:

> the effect that an action or sequence of actions has on the safe and successful operation of a system.

To make this concept usable by designers, and to relate it to the work of risk analysts, we require some way of quantifying impact or of ordering the relative impacts of different errors.

Within risk analysis, the concept of severity is used to quantify the degree to which some hazardous event is undesirable. Since safety-critical systems are designed with the philosophy that no single point of failure should lead to a hazard, human errors generally will not lead directly to hazards. Therefore their impact cannot be measured directly by reference to the severity of hazards. However, human errors can have 'an effect on the safe and successful operation of a system'. For instance, if an aircraft pilot accidentally fires the fire extinguishers for an engine when there is no fire, this does not lead to a hazard (there is not a fire). However, there is an effect on safe and successful operation, because if a fire occurs later in the flight, then the pilot will not be able to extinguish it.

The change that has occurred as a result of firing the extinguisher is an increase in the probability of major damage to the engine (and possibly loss of the aircraft) during the flight. This leads us to the conclusion that impact should be measured in terms of the change in risk, i.e. the change in probability of the hazard, and the relative severity of the hazard itself. The problem we address in this paper is how impact can be measured, and how information about impact measurements can be made available to HMI designers.

The procedure we suggest makes use of interactor models of the system state together with risk analysis information in the form of fault trees. We seek to avoid the difficulties inherent in attempting to quantify probabilities of human-error by concentrating instead on quantifying the consequences of hypothesised human-errors in terms of statistically-based estimates of the reliability of physical components. The interactor model will be used to determine the state reached by the machine components as a result of interaction sequences that arise from plausible human errors. The fault trees will be used to reason about the way that the risk of a hazard is changed when the probabilities associated with leaf nodes of the fault tree are changed to certainties (probabilities of 1 or 0) arising from erroneous interactions. We describe our procedure in detail in section 5.

A similar procedure is used within the Systematic Hazard Analysis Reduction Procedure (SHARP) as described by Villemeur (1992b) to select particular human errors for detailed probabalistic analysis. However, Villemeur's procedure is applied after the HMI design has been defined in detail, does not take advantage of the possibilities offered by recent developments in modelling interactive systems, and does not support the early stages of HMI design.

Our approach makes two fundamental assumptions:

1. That preliminary risk analysis activities for the plant can be carried out before the HMI design is developed in detail. This assumption seems valid for human-machine systems. In a human-machine system such as a chemical processing plant or an aircraft, the functional capabilities of the system are primarily determined by the machine elements of the system, with the human controlling the machine elements to achieve the system goals. The primary constraint on the design must always be the ability of the machine elements of the system to support the necessary functions. Thus, the design of the HMI will tend to follow the design of the machine elements of the system. This lag offers the opportunity for information gained from preliminary risk and hazard analysis to be fed into the HMI design process.

2. That the preliminary risk analysis considers states that the components can reach as a result of human intervention as possible failure modes of the components, even if those components cannot fail into those particular states. For instance, for a pump the only normal failure mode may be in the off state. We must assume that the risk analyst also considers how the pump being in the on state could contribute to a hazard. This assumption may imply some changes to current practices.

Given these two assumptions, the impact of an action can be measured by reference to the fault trees constructed during preliminary risk analysis, and this measure can be used to guide the process of HMI design.

4 Specifying an example system using interactors

In this section, we develop an interactor specification for a simple gas fired electricity generating system.

4.1 An informal description of the plant

We begin by describing the design of the physical plant, which is shown in figure 1. In the plant two pumps are arranged in parallel. These pumps circulate heating fluid through

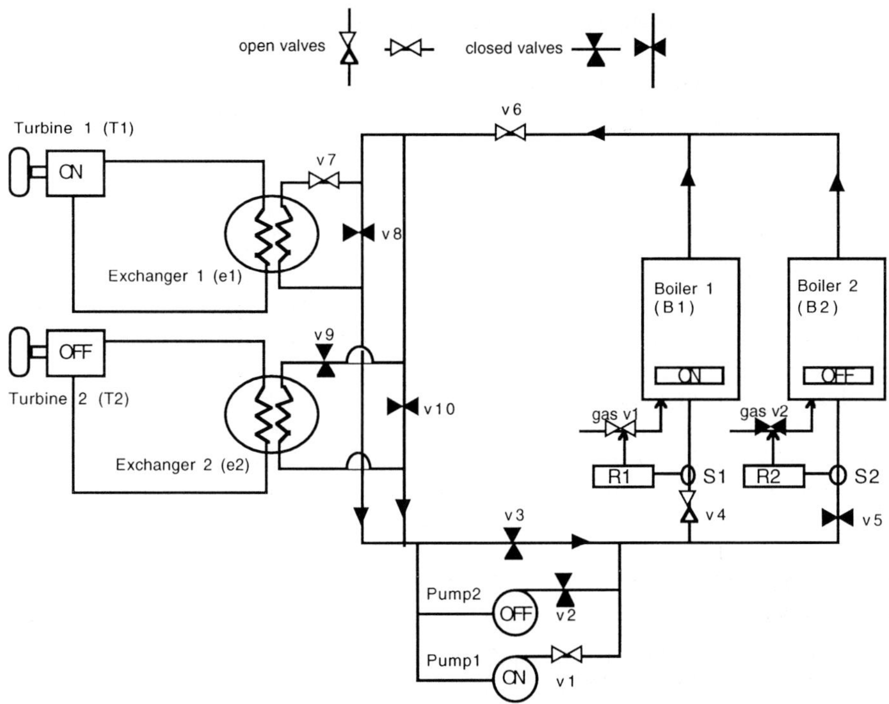

Fig. 1. *A simple, two turbine, power plant*

(one or two) gas fired boilers and then through (one or two) heat exchangers which transfer the heat to (one or two) secondary circuits in which steam driven turbines generate electricity.

The plant has three major operating modes, namely:

- low power A - using boiler 1, pump 1, exchanger 1, and requiring valves v1, v4, v6 and v7 to be open with all other valves closed as in figure 1;
- low power B - using boiler 2, pump 2, exchanger 2 and requiring valves v2, v5, v6, v9 to be open with all others closed; and
- high power - using both pumps, both boilers, both exchangers and requiring valves v1, v2, v4, v5, v6, v7, v9 open and all others closed.

The major hazard associated with the plant is overheating of the boilers, which might occur if they are left on with no flow through them. To guard against this, sensors monitor the flow of heating fluid into the boilers at points (s1, s2), and regulators (r1, r2) can be used to cut off the flow of gas into the boilers through the gas valves (gv1, gv2).

4.2 Developing an Interactor Specification

To begin the specification of the plant we break down the overall design into groups of components. First we identify the two secondary circuits, each of which consists of one turbine and one heat exchanger. In the primary circuit there are ten main valves, two pumps, two boilers and two regulator subsystems that provide the cutout function for the boilers to avoid overheating. For an interaction analysis, we do not decompose the regulatory subsystems, but model them as single units with a relatively complex function. This choice is made because the interactor model and our analysis is only concerned with components at the level at which the operator can interact with them. Since the operator cannot interact with the sensor or control component of the regulatory subsystem, we need only model it as a special case of a valve.

Each type of component will be modelled by an interactor which describes its state and behaviour. Presentation of the interactor is left as an open decision during the early stages of design. The interactors that describe the components will then be used as attribute types when describing the overall behaviour of the plant.

4.3 Describing plant components

Each interactor is described by a name, a set of attributes that comprise the state of the interactor, a set of actions that the interactor is able to engage in, and a set of axioms that constrain the possible behaviour of the interactor. The axioms are expressed in modal action logic (MAL) [Ryan *et al.*, 1991]. MAL includes all the usual logical operators, connectives and quantifiers $(\wedge, \vee, \neg, \Rightarrow, \forall, \exists)$ together with an additional action operator [A] where [A]P means that the action A is available, and that immediately after the action A the predicate P holds. The statement P $\Rightarrow$ [A]Q is read as, when the pre-condition P holds, the action A is possible, and it's post-condition is Q. Given a state, if an action is not mentioned by at least one axiom for which the pre-condition holds, or one axiom with no pre-condition, then that action is not available in that state.

The first interactor describes a valve. The valve is modelled in terms of two attributes, its state, opened or closed, and its mode, working or failed. The valve has actions that permit it to be opened or closed, and allow it to fail or be repaired.

interactor Valve - a description of a valve
attributes
  mode : {failed, working} - the mode of the valve, working or failed
  state : {opened, closed} - the state of the valve, open or closed

actions
  open, close - opening or closing the valve
  fail, repair - the failure, or repair of the valve

axioms

1 $\text{mode} = \text{working} \wedge \text{state} = \text{opened} \Rightarrow [\text{close}]\ \text{state} = \text{closed}$

2 $\text{mode} = \text{failed} \wedge \text{state} = s \Rightarrow [\text{close}]\ \text{mode} = \text{failed} \wedge \text{state} = s$

3 $\text{mode} = \text{working} \wedge \text{state} = \text{closed} \Rightarrow [\text{open}]\ \text{state} = \text{opened}$

4 $\text{mode} = \text{failed} \wedge \text{state} = s \Rightarrow [\text{open}]\ \text{mode} = \text{failed} \wedge \text{state} = s$

5 $[\text{fail}]\ \text{mode} = \text{failed}$

6 $[\text{repair}]\ \text{mode} = \text{working}$

For the valve, the first axiom states that if the valve is in the working mode, and is in the opened state, then the close action is available, and results in the state being closed. The second axiom states that if the valve is has already failed then the close action will not change the state. Axioms 3 and 4 are similar but refer to the open action. The final two axioms state that the fail and repair actions are always available. Also that, immediately after the fail action, the mode is failed, and immediately after the repair action, the mode is working. Note that the axioms are interpreted as defining all the conditions under which the actions are possible. The axioms 5 and 6 imply that the fail and repair actions will always be available. The rest of the axioms imply that close is only available when the state is opened, or the mode is failed, and the open action is only available if the state is closed or the mode is failed.

This completes the specification of the valve. The specification can be used to reason about the behaviour of the valve in response to given events or sequences of events.

The pumps can similarly be modelled, however the pumps can only fail into the off state. The pumps could be modelled by using a new type of attribute that takes values from the set {start, stop}. However, a more elegant alternative is to exploit the similarity between Valves and Pumps by using inheritance, and equating: start with open, stop with close, on with opened, and off with closed. Exploiting this similarity the Pump can be modelled as inheriting all the behaviour of a valve, but having an additional axiom that ensures that a pump can only fail in the off state.

interactor Pump - a description of a pump
Valve
attributes
actions
axioms

7 $\text{mode} = \text{failed} \Rightarrow \text{state} = \text{off}$

The Boiler and turbines are similar to the pump in having start and stop actions, on and off states, and failed or working modes. Also the boilers and turbines fail safe by always failing into the off state. For this reason they can be modelled as equivalent to the pump.

Boiler == Pump

Turbine == Pump

The heat exchangers do not have on, off switches, only failed (blocked) or working (open) states.

interactor Exchanger - the heat exchangers
attributes
state : {blocked, opened} - the state is either blocked or open
actions
fail, repair -
axioms
1 [fail] state = blocked
2 [repair] state = opened

The behaviour of the gas supply regulators for the boilers can be modelled by inheriting the general behaviour of a valve, and adding the behaviour of a cut off switch. Notice that this model treats the sensor and regulator as a single component, and abstracts away from the triggering behaviour of the regulator, merely recording that the regulator can cause the valve to close.

interactor Regulator - a description of a regulator valve as used for the gas suppl
Valve

attributes
conmode : {failed, working} - The mode of the regulator control compone
- working or failed. The mode attribute is inh
- from the valve describes the valve mode
constate : {closed, opened} - The state of the control component.
- The state attribute inherited from
- the Valve describes the valve state

actions
reset, cutout - the cutout and resetting the regulator control
confail, conrepair - the failure or repair of the regulator control
- fail, repair, open and close are inherited from Valve

axioms
1 conmode $\neq$ failed $\wedge$ mode $\neq$ failed $\Rightarrow$ [cutout] state = closed $\wedge$ constate = close
2 state = s $\wedge$ conmode = failed $\Rightarrow$ [cutout]state = s
3 conmode = failed $\Rightarrow$ constate = opened
4 constate = closed $\Rightarrow$ [reset] constate = opened
5 [conrepair]conmode = working
6 [confail]conmode = failed $\wedge$ constate = opened

4.4 Describing plant behaviour

To model the plant we can now utilise a collection of instances of the interactors introduced above. At this stage of the modelling there are no additional actions that can be applied to the Plant other than those actions that can be applied to the individual components. The only additional axioms needed to describe the behaviour of the plant state that each gas supply regulator must be in an opened state in order for the associated boiler to be on.

interactor Plant		- a model for the complete plant
attributes		
b_1, b_2	: Boiler	- the two boilers
p_1, p_2	: Pump	- the two pumps
x_1, x_2	: Exchanger	- the heat exchangers
$v_1, v_2, v_3, v_4, v_5,$		
$v_6, v_7, v_8, v_9, v_{10}$	: Valve	- the valves on the fluid circuit
reg_1, reg_2	: Regulator	- the regulator for the gas supply to the boilers
t_1, t_2	: Turbine	- the two turbines

actions

axioms

1 $b_1.state = on \Rightarrow reg_1.state = opened$

2 $b_2.state = on \Rightarrow reg_2.state = opened$

Additional axioms might be introduced at a later stage of design if more detail about the internal functioning of the system is required, or if higher level attributes are introduced. For instance, if the designer chose to introduce an attribute to describe whether there is a flow in a particular part of the circuit (perhaps to enable a display to report this to the operator), then axioms could be introduced that relate the state of the other components to the value of this attribute. For the analysis that we are conducting in this paper, additional axioms are not required.

This model can now be used as a starting point for the development of the human-machine interface. The notation provided by Duke & Harrison allows the designer to note which parts of the system state are to be perceivable directly by the operator, and what sensory modality will be used to carry such state information. At the level of design that we are concerned with, decisions about sensory modality might still be open, and the choices made do not affect the assessment of impact directly (although such decisions may well affect the probability of errors occurring). The reader is referred to Duke & Harrison (1993) for a more detailed discussion of the role of interactor models in design.

4.5 Specifying operating procedures

The interactor notation can also be used to describe the individual operating modes, and to specify intended operating procedures for a system.

For instance, in our power plant example, the designer may describe the idle mode (where all valves are closed, and all boilers, pumps and turbines are off) by introducing an initialisation axiom:

axioms

3 $[](\forall i : [1..10] v_i.state = closed \wedge v_i.mode = working) \wedge$
$(\forall i : \{1, 2\} p_i.state = off \wedge p_i.mode = working) \wedge$
$(\forall i : \{1, 2\} b_i.state = off \wedge b_i.mode = working) \wedge$
$(\forall i : \{1, 2\} t_i.state = off \wedge t_i.mode = working) \wedge$
$(\forall i : \{1, 2\} x_i.state = opened) \wedge$
$(\forall i : \{1, 2\} reg_i.state = closed \wedge reg_i.mode = working) \wedge$

(the [] annotation in MAL indicating that the predicate that follows shall hold before any events have occurred)

The designer's intended operating procedure to bring the plant to the low power A mode can be described hierarchically.

$$\begin{aligned}
&\text{Start_LowPwr_A} == \langle \text{Open_valves_A}, \text{Start_Pump_A}, \\
&\qquad\qquad\qquad\qquad \text{Start_Turbine_A}, \text{Start_Boiler_A} \rangle \\
&\text{Open_valves_A} == \langle \text{v1.open}, \text{v4.open}, \text{v6.open}, \text{v7.open} \rangle \\
&\text{Start_Pump_A} == \langle \text{p1.start} \rangle \\
&\text{Start_Turbine_A} == \langle \text{t1.start} \rangle \\
&\text{Start_Boiler_A} == \langle \text{b1.start} \rangle
\end{aligned}$$

Using this procedural specification together with the interactor model and the initialisation axiom a designer can verify that the procedure does achieve the goal of moving the plant to the low power A mode.

4.6 Identifying plausible errors

Having specified the operating procedures, the techniques used by Fields *et al.* (1995) can be applied to generate plausible error sequences. For instance, if we consider the heirarchical specification of the procedure above, together with the omission error phenotype, we generate a set of plausible error sequences including the sequence:

$$\langle \text{v1.open}, \text{v6.open}, \text{v7.open}, \text{p1.start}, \text{t1.start}, \text{b1.start} \rangle$$

Alternatively we might use the commission phenotype to generate a sequence that includes the opening of valve 5, i.e.:

$$\langle \text{v1.open}, \text{v4.open}, \text{v5.open}, \text{v6.open}, \text{v7.open}, \text{p1.start}, \text{t1.start}, \text{b1.start} \rangle$$

Using the interactor model we can see that the consequence of the first sequence would be to reach a state similar to low power A, but with the valve v4 still closed. The second sequence leads to the low power A mode but with v5 open.

Neither of these errors leads directly to the boiler overheating. This is what we would expect for a design that respects the principle that no single point of failure results in a hazard. However, if we now consider the possibility of the event reg1.fail happening in conjunction with the human errors above, the first sequence would lead to the boiler overheating (unless the operator took further action by shutting down the boiler manually), whereas the second sequence would not (unless a further failure occurred). A quantitative analysis of impact should capture this distinction.

5 Quantifying impact

As we noted in the previous section, the interactor model provides a mechanism by which identified errors can be mapped to consequences in terms of system state. In this section, we shall concentrate on how the impact can be quantified.

5.1 Using fault trees

We now consider how fault trees, generated by the risk analysts studying the design of the physical components of the plant, could provide information about the impacts of the errors to an HMI designer.

Fault trees describe how combinations of failures might lead to hazardous events occurring. Figure 2 shows a fault tree. At the root of the fault tree, the hazard and the

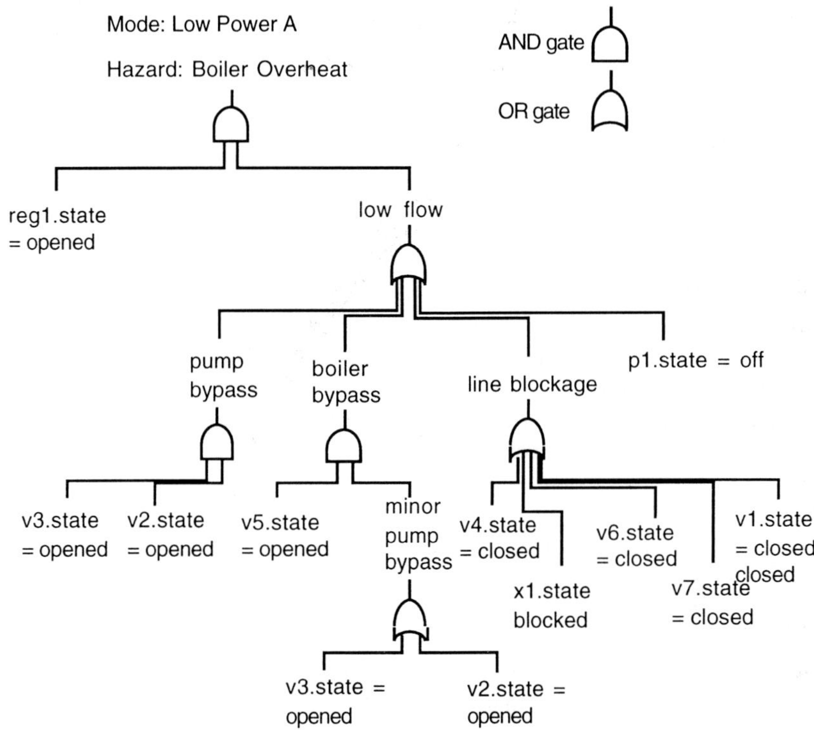

Fig. 2. *A fault tree for the hazard of boiler 1 overheating whilst in low power A mode*

mode for which the fault tree applies are noted. In figure 2, this is the overheating of boiler 1 when the system is in the low power A mode. Branching points of the fault tree are marked by logic gates. Starting from the root, the boiler overheating can occur if there is no flow and the gas regulator for the boiler fails. Low flow, in turn, can be caused by the pumps being bypassed, the boiler being bypassed, a blockage of the circuit or the failure of the pump. The leaf nodes of the fault tree are labelled by possible failure states of individual components, such as the fact that a particular valve is in the open or closed state, or the pump has failed off. Internal nodes of the tree can be given abstract labels to assist the reader in understanding the way that the faults at the leaves lead to the hazard at the root. These internal labels do not affect the formal semantics of the tree.

The fault tree in figure 2 has been integrated with the interactor model. The leaf nodes of the tree are labelled by component states from the model. Also, the gas regulatory system has been treated as a single component, as in the interactor model. Notice that the fault tree does not include any claims about possible failures arising from human error. This is a deliberate choice, since we cannot expect to obtain reliable estimates of human error at such an early stage of design, and because the notion of phenotypical error has already been applied to restrict our attention to plausible error sequences. When human error is ignored, and the probability of failure of individual components is known[1], then the fault tree can be used to compute the probability of the root event occurring due to a component failure. Such calculations provide a baseline from which questions of impact may be considered.

5.2 Calculating impact values

Our quantification of impact is based on measuring the change in the probability of a hazard occurring as a result of a given human error. This can be calculated by considering the probability of the root event of the fault tree under normal circumstances, and the probability of the root event given that the error is known to have occurred (i.e. the post conditions of the action are given a probability of 1). The difference between these two values is a measure of the impact of the action. Below, we illustrate this computation in the context of our example.

To simplify the calculation of probabilities we can transform the fault tree using boolean algebra to obtain the expression:

$$\begin{aligned}
&\text{reg1.state} = \text{opened} \wedge \\
&(\text{v1.state} = \text{closed} \vee \text{v4.state} = \text{closed} \vee \text{v6.state} = \text{closed} \vee \text{v7.state} = \text{closed} \\
&\vee \text{x1.state} = \text{blocked} \vee \text{p1.state} = \text{off} \\
&\vee (\text{v3.state} = \text{opened} \wedge \text{v2.state} = \text{opened}) \\
&\vee (\text{v3.state} = \text{opened} \wedge \text{v5.state} = \text{opened}) \\
&\vee (\text{v5.state} = \text{opened} \wedge \text{v2.state} = \text{opened}))
\end{aligned}$$

For the purposes of the example we assign component failure probabilities as follows:

- Probability of failure for regulator as 0.01 failures per unit time;
- for any valve to fail open 0.01 per unit time;
- for any valve to fail closed 0.001 per unit time;
- for pump to fail off 0.02 per unit time;
- for the exchanger to be blocked 0.0001 per unit time.

[1] Safety databases are used to make statistical data about the failure behaviour of existing standard components available throughout safety related industries. The failure behaviour of new components is typically extrapolated from information about similar existing components.

Using these figures the overall probability of the boiler overheating as a result of spontaneous component failure is (approximately[2]):

$$0.01 \times (4 \times 0.001 + 0.0001 + 0.02 + 3 \times 0.01 \times 0.01) = 0.000244$$

Now we consider the impact of sequence 1. The result of this sequence compared to low power A mode is that v4 is known to be closed, in which case the above boolean expression reduces to:

$$\text{reg1.state} = \text{opened} \vee \text{true}$$

i.e. it is sufficient that the regulator fails, this event having a probability of 0.01. The impact of the sequence is a function of the change in probability and the severity. The change in hazard probability for sequence 1 is $0.01 - 0.000244 = 0.09756$.

A similar calculation can be performed for the second error. In this case the relevant boolean expression is:

$$\begin{aligned}&\text{reg1.state} = \text{opened} \wedge\\ &(\text{v1.state} = \text{closed} \vee \text{v4.state} = \text{closed} \vee \text{v6.state} = \text{closed} \vee \text{v7.state} = \text{closed}\\ &\vee \text{x1.state} = \text{blocked} \vee \text{p1.state} = \text{off}\\ &\vee \text{v3.state} = \text{opened} \vee \text{v2.state} = \text{opened})\end{aligned}$$

This gives a new (approximate) hazard probability of:

$$0.01 \times (4 \times 0.001 + 0.0001 + 0.02 + 0.01 + 0.01) = 0.000441$$

Hence, the change in hazard probability is $0.000441 - 0.000244 = 0.000197$.

These two calculations show how the error sequences increase the probability of the hazard occurring. Since neither are close to 1, neither error on its own leads to the hazard. However, for the first sequence only one defense remains to prevent the hazard (namely the regulator), whereas for the second sequence the system is still relatively well protected, requiring two further errors or failures before the hazard occurs. This difference is reflected by a difference of 2 orders of magnitude in our calculation of impact.

5.3 Impact with respect to multiple hazards

Any general assessment of the impact of actions needs to compare actions in terms of their effects on multiple hazards. In an analysis of interaction sequences with respect to multiple hazards some sequences may reduce the probability of one hazard, whilst increasing the probability of another hazard. For example, in our case study, we might consider as a secondary hazard the failure of the plant to start generating electricity at a specified time. If we take the sequence:

$$\langle \text{v1.open}, \text{v4.open}, \text{v6.open}, \text{v7.open}, \text{p1.start}, \text{t1.start} \rangle$$

[2] Hazard probability calculations generally use the equation $P(A \vee B) = P(A) + P(B) - P(A \wedge B)$. The calculations in this section treat $P(A \vee B)$ as equal to $P(A) + P(B)$ ignoring the second order correction term

this will result in no electricity being generated, but it reduces the probability of the boiler overheating to zero. To evaluate the overall impact of this error on the system the HMI designer may need to know something about the relative severity of the hazards.

We have conducted some work investigating techniques from multi-criteria decision theory and utility theory to elicit and codify information about relative hazard severities in a form that can help HMI designers. This work is continuing.

5.4 Summary

In this section we have shown how fault trees constructed by risk analysts on the basis of the design of the physical components of a system, and referring only to the (known) failure characteristics of those components, can be used to quantify a notion of impact in terms of a change in the probability of a hazard occurring. The technique uses the formal model of the HMI to reason about the state that the system will reach after a given interaction sequence.

6 Discussion

In this paper we have demonstrated an approach to the integration of risk analysis information within HMI design by means of a formal interface modelling technique. A number of questions need to be addressed before applying such a technique in a safety-critical system. We consider a few of these below.

- Does current risk analysis practice provide sufficient data to support impact analysis?

  An important assumption was that the preliminary risk analysis considers all the modes that system components could reach as a result of human intervention, even if these states are not normal failure modes for the components. Ensuring the completeness of a risk analysis is a difficult task, and is often supported by bottom-up techniques such as failure modes and effects analysis where each possible failure mode of a component is considered in turn to identify how it might lead to a hazard. The introduction of impact analyses may require that additional failure modes, corresponding to states that a component can reach, but which cannot be reached as a result of failure, are considered. For instance, in our example the pump was assumed only to fail in the off state. However, in a risk analysis to support impact analysis we should require that the possibility of it being in the on state is considered as one possible failure mode. The current practice of risk analysis within an organisation will need to be considered before introducing impact analysis into the design process.

- What features does an interaction modelling technique require in order to support impact analysis?

  Three aspects of interactor modelling make it particularly suitable for impact analysis. These are: its formality, this is necessary in order that the state reached by a given interaction sequence can be clearly identified; its state-based model of the

HMI, which corresponds closely to the state based approach that is used by risk analysts; and its ability to specify the system in terms of sets of components, which is convenient when annotating fault trees which are also expressed in terms of component states. Other techniques that share these features are likely to be equally adaptable to support impact analysis.

- What is the scope of impact analysis?

  Any technique to support HMI design for safety critical systems will have some limitations in terms of its scope, i.e. the class of errors and failures that it addresses. Impact analysis, as we have described it, seeks to support reasoning about situations where human-errors combine with the failure of physical components. This class of errors may be more or less significant within different domains. However, HMI designers must recognise that there are other important sources of error that may affect the safety of complex human-machine systems, e.g. the conflict between different safety and production goals that may be faced by an operator, or problems of attentional dynamics in rapidly evolving situations. Such problems may require analysis of more than just the physical system and the HMI, as Woods (1994) suggests, for these problems

  > "the proper unit of analysis is the distributed cognitive system"[p123]

  We are currently planning further case studies, particularly in the medical physics domain, to test the scope of impact analysis.

- Can the technique be extended to handle large scale industrial design problems?

  It is certainly the case that our computation of impact is within the capabilities of existing fault tree management tools used for industrial risk analysis. Many existing computer based tools already permit the calculation of a similar measure known as the Birnbaum Importance Factor for a component [Henley and Kumamoto, 1981]. Interactor based specification does not as yet have tool support, but progress in the development of user-interface design environments [Foley and Sukaviriya, 1995] suggests that tool support is a realistic prospect.

- How might late changes in the design of physical components affect impact analysis and HMI designs informed by that analysis?

  Johnson (to appear) considers how risk analysis information could be incorporated within a design rationale. It may be that the relationship between design decisions and the results of impact analysis could be treated in a similar fashion.

- What features in an HMI design can be used to assist the operator in developing a correct perception of impact?

  This is perhaps the most challenging problem raised by this work. In this paper, we have concentrated on identifying high impact interactions, rather than on supporting the operator's perception of impact. We have made some small inroads in identifying design features that assist the operator in forming a correct perception of impact [Dearden and Harrison, 1996], but much work remains to be done.

7 Conclusion

We have presented a technique for integrating models of interactive systems with information derived from risk analysis. The techniques allows risk analysts and HMI designers to explore the impact of specific human-errors. The technique extends current methods because it can be used during the early stages of HMI design, and because it does not rely on questionable estimates of the probabilities of human-errors.

Acknowledgements

This work was funded by EPSRC research grant number GR/JO7686. We should also like to thank our anonymous reviewers for their helpful comments on earlier drafts of this paper.

References

[Bello and Colombari, 1980] Bello, G. C. and Colombari, V. (1980). Empirical technique to estimate operator errors (TESEO). *Reliability Engineering*, 1(3).

[Dearden and Harrison, 1996] Dearden, A. and Harrison, M. (1996). Impact as a human factor in interactive system design. In Redmill, F. and Anderson, T., editors, *Safety-critical Systems: The Convergence of High Tech and Human Factors*, 184 – 199. Springer.

[Duke *et al.*, 1994] Duke, D., Faconti, G., Harrison, M., and Paterno', F. (1994). Unifying views of interactors. In *Proc. International Workshop on Advanced Visual Interfaces*, 143 – 152. ACM Press.

[Duke and Harrison, 1994] Duke, D. and Harrison, M. (1994). *FSM: Overview and Worked Examples*. Technical Report SM/WP44, AMODEUS II project, ESPRIT Basic Research Action 7040.

[Duke and Harrison, 1993] Duke, D. J. and Harrison, M. D. (1993). Abstract interaction objects. *Computer Graphics Forum*, 12(3):25 – 36.

[Embrey *et al.*, 1984] Embrey, D. E., Humphreys, P. C., Rosa, E. A., Kirwan, B., and Rea, K. (1984). *SLIM-MAUD: An Approach to Assessing Human Error Probabilities Using Structured Expert Judgement*. Technical Report NUREG / CR 3518, Brookhaven National Laboratory.

[Fields *et al.*, 1995] Fields, R., Wright, P., and Harrison, M. (1995). A task centred approach to analysing human error tolerance requirements. In Zave, P., editor, *Proceedings, RE'95 The Second IEEE International Symposium on Requirements Engineering, York, UK*, 18–26. IEEE, New York.

[Foley and Sukaviriya, 1995] Foley, J. D. and Sukaviriya, P. N. (1995). History, Results and Bibliography of the User Interface Design Environment (UIDE), an Early Model-Based System for User Interface Design and Development. In Paterno', F., editor, *Interactive Systems: Design, Specification and Verification*, 3 – 14. Springer Verlag.

[Hannaman *et al.*, 1984] Hannaman, G. W., Spurgin, A. J., and Lukic, Y. D. (1984). *A model for assessing human cognitive reliability in PRA studies*. Technical Report NUS 4531, Electrical Power Research Institute.

[Henley and Kumamoto, 1981] Henley, E. J. and Kumamoto, H. (1981). *Reliability Engineering and Risk Assessment.* Prentice Hall.

[Hollnagel, 1993] Hollnagel, E. (1993). The phenotype of erroneous actions. *Int. Journal of Man-Machine Studies*, 39(1):1 – 32.

[Johnson, pear] Johnson, C. W. (to appear). Documenting the design of safety-critical user interfaces. *Interacting with Computers.*

[Kirwan, 1992] Kirwan, B. (1992). Human error identification in human reliability assessment. Part I: Overview of approaches. *Applied Ergonomics*, 23(5):299 – 318.

[Ryan *et al.*, 1991] Ryan, M., Fiadeiro, J., and Maibaum, T. (1991). Sharing actions and attributes in modal action logic. In Ito, T. and Meyer, A., editors, *Theoretical Aspects of Computer Software*, volume 526 of *Lecture notes in computer science*, 569–593. Springer Verlag.

[Swain and Guttman, 1983] Swain, A. D. and Guttman, H. E. (1983). *Handbook of Human Reliability Analysis with Emphasis on Nuclear Power Plant Applications, Final Report.* Technical Report NUREG/CR-1278 SAND80-0200 RX, AN, U. S. Nuclear Regulatory Commission.

[Villemeur, 1992a] Villemeur, A. (1992a). *Reliability, Availability, Maintainability and Safety Assessment*, volume 1. John Wiley.

[Villemeur, 1992b] Villemeur, A. (1992b). *Reliability, Availability, Maintainability and Safety Assessment*, volume 2. John Wiley.

[Woods *et al.*, 1994] Woods, D. D., Johannesen, L. J., Cook, R. I., and Sarter, N. B. (1994). *Behind Human Error: Cognitive Systems, Computers, and Hindsight.* Technical report, CSE-RIAC, Ohio State University.

This article was processed using the LaTeX macro package with LLNCS style

Design assistance for user-adapted interaction

D. Akoumianakis, A. Savidis and C. Stephanidis

Institute of Computer Science
Foundation for Research and Technology-Hellas
Science and Technology Park of Crete
P.O. Box 1385, GR-71110, Heraklion, Crete, Greece
Tel.: + 30 81 391741, Fax : +30 81 391740
E-mail: {demosthe, as, cs }@ics.forth.gr

Abstract This paper discusses current approaches to user interface adaptation and describes how some of their shortcomings can be overcome by supporting the articulation of user interface adaptation constituents during the early phases of design and development of a user interface. It is claimed that this type of adaptation support is required to ensure accessibility of a user interface by different user groups with varying abilities, requirements and preferences. Additionally, the paper describes the components of a prototype design environment called USE-IT which has been developed to support the automatic derivation of such adaptation decisions so as to ensure that the resulting user interface will be accessible by the target user (group). The tool is part of a novel user interface development platform which integrates design assistance and development support to provide a unifying basis for constructing high quality user interfaces that are accessible by different user groups, including disabled and elderly people.

Keywords user interface adaptation, design assistance, interaction metaphors, abstract interaction objects, adaptability rules

1 Introduction and Rationale

The need for adaptation of the user interface of computer-based applications and services is evident when one considers the diverse abilities and requirements of different user groups, the wide availability of alternative and non-conventional input/output technologies, the compelling need for more user friendly products as well as the recent trends towards universal accessibility and greater usability of interactive applications. It follows that user interface adaptation is a technique that may potentially serve three key requirements of the information technology industry, namely *product specialisation* or *individualisation*, *universal accessibility* and *high quality interaction*.

The requirement for more specialised products has been a driving force of developments in the information technology industry and has been comprehensively addressed in the past [1]. It has also provided the primary motivating factor for recent research in the domain of adaptive user interfaces. On the other hand, universal accessibility and high quality of interaction have both surfaced as requirements for a number of years, but have not been properly addressed until recently. Universal accessibility reflects the principle that all citizens should be provided with equal access opportunities in the emerging Information Society, the forthcoming applications and services in the context of developing a new telecommunications infrastructure. However, currently available tools for user interface development do not provide the tools to support universal access [2]. This is due to the underlying assumptions characterizing such tools which limit the type and style of interaction that can be effectively supported. For example, the assumption that all users are familiar with the currently prevailing visual desktop embodiment of the computer is clearly insufficient to allow access to interactive applications by certain user groups such as people with disabilities.

Consequently, given these shortcomings of the present paradigm of user interface software and technology, it follows that there is a compelling need to consider the issues involved in developing a new generation of development tools that will ease the tasks of the designer and the developer in adapting an interface to the specific individual user requirements as well as target technological platforms. Such a unifying basis for building user interfaces accessible by users with different requirements, abilities and preferences, also strives for higher quality of interaction for all users; by developing new software environments for designing as opposed to programming interaction, and supporting user interface adaptation with respect to different development platforms and user groups, it is claimed that the resulting user interfaces are inherently user-centered and more usable.

The objective of this work is to investigate the requirements of user interface software components which provide designers and developers with the required support for articulating adaptation constituents and designing, developing and maintaining the user interface, as opposed to arriving at dedicated and programming-intensive solutions. Thus, the normative perspective adopted in the present work is that support for user interface adaptation should be systematically embedded in high level user interface environments and unified development platforms, in order to empower designers and developers to articulate adaptation scenarios and corresponding contexts of use so as to facilitate accessible and more usable interactive systems. It follows, therefore, that there is a need to investigate the architectural components of user interface design environments and how these can be enhanced so as to support the articulation of plausible adaptations at the semantic, syntactic and lexical levels of interaction in order to allow for the automatic realisation of alternative scenarios of use, given the different user groups and the tasks that they need to perform with the user interface.

2 Towards A Software Architecture For Adaptable And Adaptive User Interfaces

There have been several studies aiming to investigate the numerous dimensions of adaptation in interactive software systems (for a review see [3]), namely, what constitutes an adaptation constituent, the level and timing of adaptation, the controlling agent, the type of knowledge that is required to arrive at meaningful adaptations, etc. Despite the substantial contributions of these efforts to the study of adaptation, there are still several issues that need attention, if user interface adaptation is to be adequately served by designers and developers of interactive software applications. More importantly, however, it is necessary to investigate and extract a reference model for user interface architectures in order to understand the range of adaptable and adaptive behaviour that is needed and the way in which it can be effectively supported throughout the user interface life cycle (i.e. from the early design phases to implementation and maintenance). In the past, there have been several attempts to extract a reference model from concrete user interface architectures in order to classify existing prototypes and to guide the construction of user interface software. The best known architectural abstractions of user interface software include the *Seeheim* [4], *PAC* [5], *ALV*, *MVC* and *Arch/Slinky* models [6]. However, they do not consider user interface adaptation aspects. Moreover, in the literature on User Interface Adaptation, architectural abstractions are almost missing. Mostly, specific prototype architectures are reported (see [7, 8, 9, 10, 11] etc.), instead of abstract models. Even those however, limit their scope to address adaptivity which constitutes only one dimension of user interface adaptation in interactive software systems.

It is important to note that both adaptability and adaptivity of user interfaces may be hardcoded, which implies that the user interface code has a pre-set structure. Typically, in such cases, adaptability and adaptivity are both built into the user interface code through rules which are local to the user interface and predetermined. This means that in case that the system's run-time behaviour requires enhancements, either in the form of additional adaptability or adaptivity rules, the user interface code would have to be upgraded and recompiled. Of course, modifications of the target application may also be required. It follows, therefore, that such architectures lead to *monolithic* systems which are likely to be large (in lines of code) and not easily modifiable.

An alternative approach would be to introduce the adaptability and adaptivity rules as "orthogonal" to the user interface, but not part of it. This is to say that such rules are not embedded in the user interface code, but they can be collected as supplementary information (in files) which can be consulted by the run-time libraries of the user interface development system (see Figure 1). The architecture described in the diagram of Figure 1 is clearly more flexible in the sense that the rules determining adaptable and adaptive behaviour of the user interface are not part of the user interface code. Instead, the user interface development toolkit, in addition to the other functions that it carries out, it also acts as an interpreter of adaptation decisions established either manually or by an external module (see below). This requires that the user interface development toolkit should support an explicit model of the adaptable and adaptive constituents which can be determined by the rules. Such abstractions are always desirable in user interface software and have been integral components of high level user interface development environments.

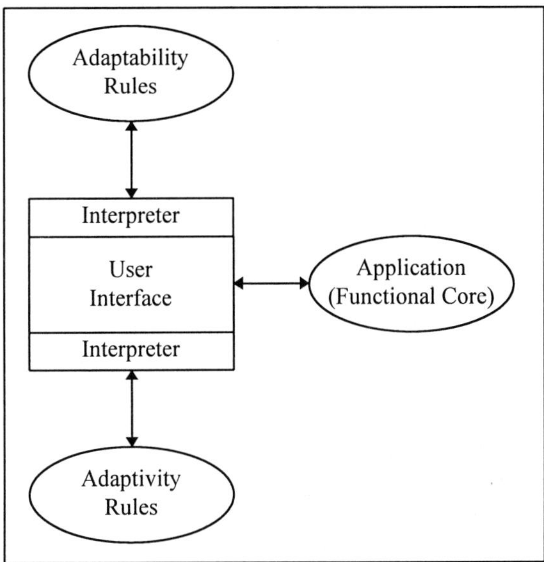

Fig. 1 User interface architecture for supporting adaptation

The only concern regarding the architectural abstraction depicted in the diagram of Figure 1 is the way in which the adaptability and adaptivity rule sets are produced. Figure 1 assumes that such rules may be hard-coded or editable through templates, but this may lead to ad hoc and non-systematic user interface designs. Alternatively, tools can be developed to support the automatic construction of appropriate adaptability and adaptivity rules according to the user's abilities, knowledge, interests as well as any preferences of specific adaptation constituents (i.e. interaction style, dialogue syntax, input/output devices, interaction techniques, etc). This observation leads to an enhanced architectural abstraction which is depicted in the diagram of Figure 2. In this revised architecture for adaptable and adaptive user interfaces, three tools have been introduced, namely, two design assistants that support the automatic generation of adaptability and adaptivity rules respectively, and a user interface development toolkit which is responsible for realising the adaptable and adaptive user interface on a target platform (i.e. MS-Windows, X-Windowing system, etc).

In the following two sections, the latter architectural abstraction for adaptable and adaptive user interfaces is further elaborated by exemplifying the communication protocols between the user interface development system and the two modules for adaptability and adaptivity.

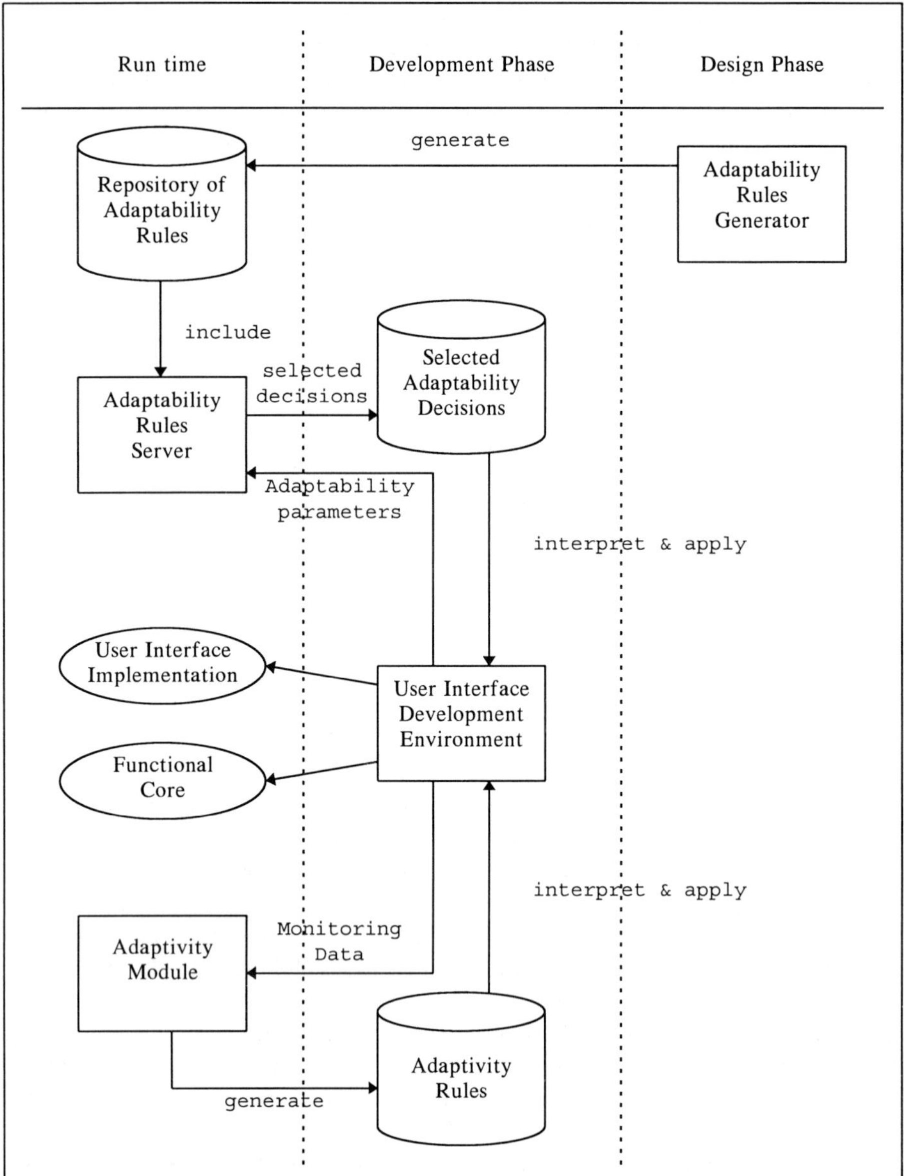

Fig. 2 Proposed architecture for adaptable and adaptive user interfaces

2.1 Communication Protocol Between The User Interface Development Tool And The Adaptability Module

In order to support adaptability, the underlying user interface development system requires knowledge about the adaptable user interface constituents. This knowledge allows the tool to properly instantiate an abstract physical interaction object into a concrete interaction object (e.g. an X Windows system menu). Consequently, the role of the adaptability module is to supply as much as possible of this knowledge. To achieve this, the adaptability module may be executed by the designer of the interface, before the development of the user interface commences, to compile an adaptation design scenario according to the user's abilities, requirements and preferences. During user interface development, the run-time libraries of a user interface development toolkit may consult the decisions comprising the adaptation scenario, in order to properly realise the adaptable properties of abstract interaction object classes on a target platform. In this respect, the user interface development process is separated from the design phase (i.e. we introduce "orthogonality"), since the user interface development system may be used by the user interface developer after the completion of the task of the adaptability module. As an example, let us consider the development of a simple user interface which involves the construction of a menu. Before the developer implements the interface with the user interface development toolkit, the adaptability module is used to compile a file containing maximally preferred interface adaptability rules. Such rules may follow the format depicted in Table 1. During user interface development, the developer uses the user interface development toolkit to implement the user interface (i.e. select the abstract physical interaction object classes which will be used). The run-time libraries of the UI toolkit utilise the information of Table 1 to adapt accordingly the details of the interaction and implement the user interface on the target platform. In this way, the menu is automatically adapted according to the user's requirements, abilities and preferences. The communication protocol described above is summarised in the diagram of Figure 3.

```
Menu.input_device=keyboard
Menu.inputTechnique=indirectPick2D
Menu.output_device=braille&speechSynthesiser
Menu.outTechnique=tactileTechnique
Menu.on_BrailleLines=2
Menu.on_BrailleCells=80
Menu.interim_feedback=speech
Menu.on_audioVoice=male
Menu.on_audioVolume=4
Menu.on_audioPitch=99
Menu.fontFamily=helvetica
Menu.topology=horizontal
Menu.access_policy=byKeyboard
```

Table 1 Hypothetical lexical adaptability rules

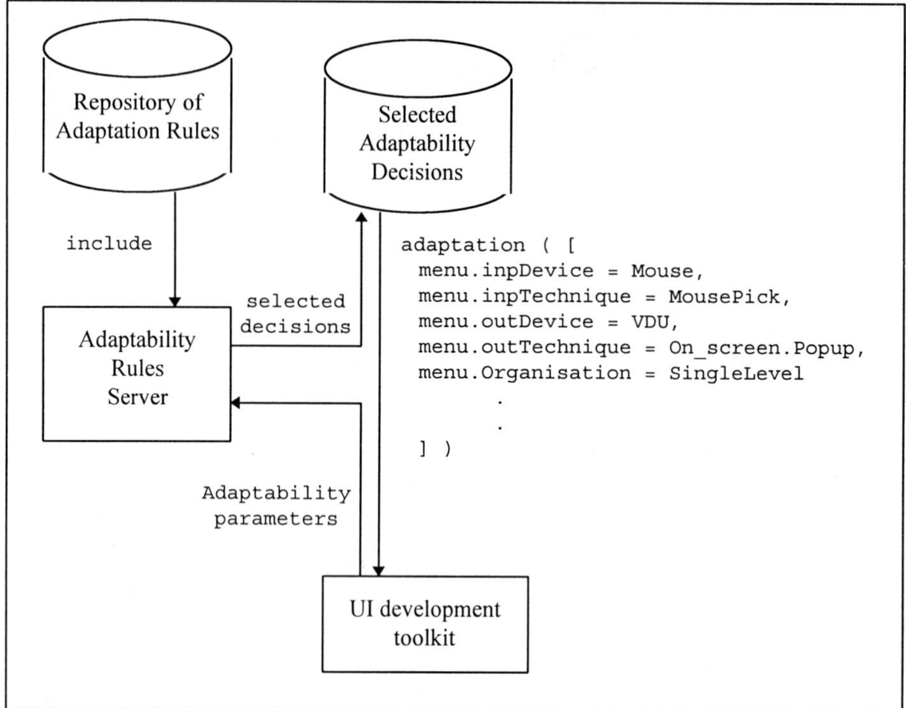

Fig. 3 Communication protocol for lexical user interface adaptability

2.2 Communication Protocol Between The User Interface Development Toolkit And The Adaptivity Module

Having briefly reviewed the communication protocol between the user interface development toolkit and the adaptability module, this section will briefly introduce a possible communication protocol between the user interface development toolkit and the adaptivity module. However, it is perhaps appropriate to consider first the meaning of adaptivity in this context. Let us assume that the user interface of a public information system is to be designed in such a way so as to exhibit adaptive behaviour in certain contexts of use. This means that the user interface should be provided with a mechanism to adapt dialogue characteristics depending on the current context of use. For purposes of simplicity we characterise context of use as being dependent on the task being performed and the current user. Thus, for instance, let us assume a simple file manager consisting of a listbox indicating the files available and three commands, namely, Open File, Delete file and Run. Run-time adaptation (i.e. adaptivity) may account for:

- the syntax of the dialogue (i.e. Function-Object versus Object-Function - see (i) and (ii) in Figure 4, respectively)
- the lexical properties of interaction objects; for instance, if interaction has been interrupted for some reason, the interface may prompt the user towards a likely option see (iii) in Figure 4.

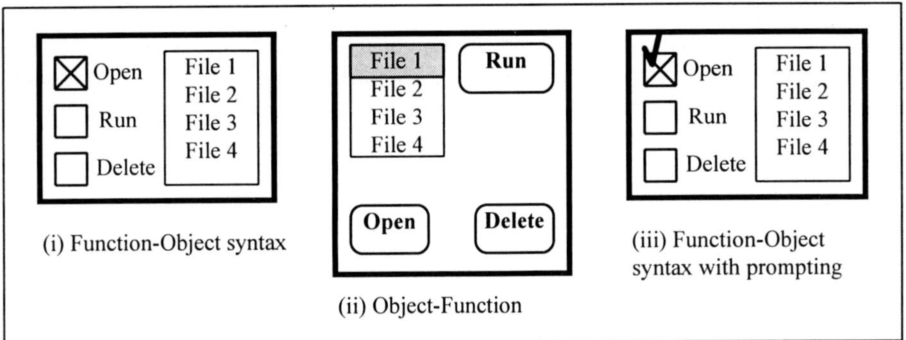

Fig. 4 Examples of adaptive behaviour

To support the type of adaptation described above, the communication protocol introduced previously needs to be slightly revised. More specifically, whereas in the case of adaptability the user interface development toolkit was merely interpreting the adaptability rules before instantiating an interaction object, in this case it should feed the adaptivity module with data so as to enable the latter to fire the appropriate adaptivity rule. This revised communication protocol is depicted in the diagram of Figure 5.

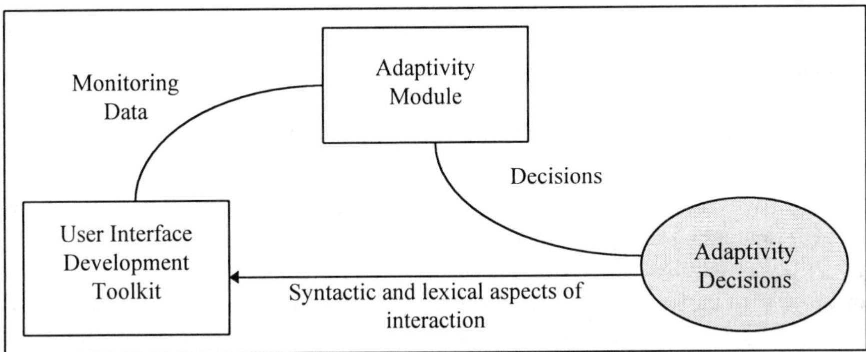

Fig. 5 Communication protocol between user interface development toolkit and adaptivity module

3 Overview of the design environment

To support the adaptability of user interface constituents during the early design and development phases, a new user interface development platform has been developed, comprising several tools which determine and apply adaptations at the lexical level of interaction [12, 13, 14].

In this context, a tool called USE-IT has been implemented which develops a semantics of adaptation at the lexical level and automatically constructs a lexical specification scenario depicting maximally preferred assignments to lexical attributes of abstract interaction objects (i.e. lexical adaptability rules), so as to ensure accessibility of the target user interface by the intended user group. The architecture of USE-IT is depicted in the diagram of Figure 6. In addition, new user interface development toolkits have been constructed supporting the development of visual, non-visual and unified interaction, while at the same time, providing developers with the required support to interpret and apply the lexical adaptability rules to implement a user-adapted interface.

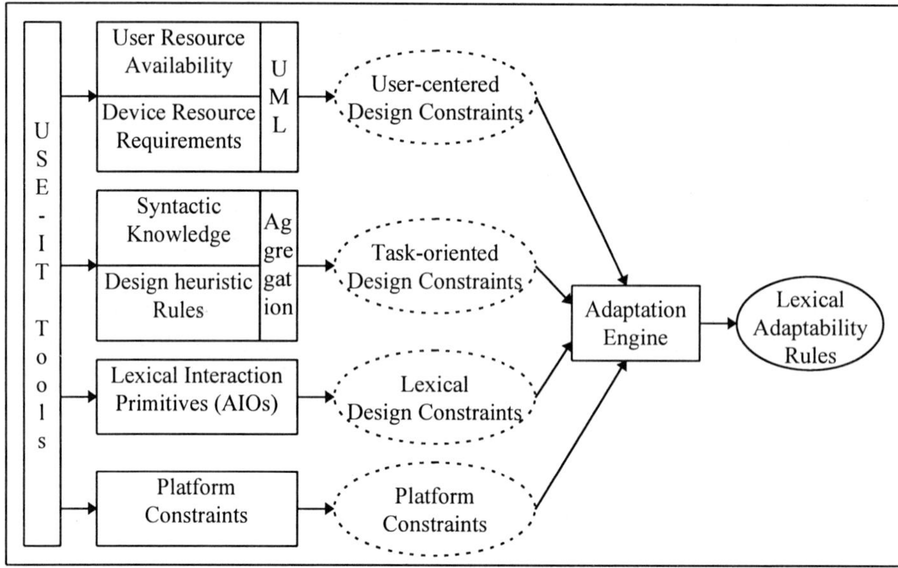

Fig. 6 Overview of the USE-IT architecture

The reason why our current efforts have concentrated on providing support for this type of lexical adaptability during the development phase is due to the fact that such adaptability is required to ensure the accessibility of the user interface by different user groups. In particular, it is claimed that if the development tools can support the level of abstraction required to enable designers and developers to adapt non-trivial attributes of lexical interaction (such as input device, input interaction technique, output device, output interaction techniques, access policy, navigation policy, topology, feedback, etc), then it is possible to develop interfaces which are automatically adapted to the individual user requirements, abilities and preferences.

To this effect, the tools developed thus far provide support for visual and non-visual interaction in two graphical environments, namely MS-Windows and X-Windowing system by means of integrating toolkits with enhanced or alternative (in the case of non-visual interaction) interaction capabilities [12].

Consequently, the approach to supporting adaptations, which has been followed thus far, can be summarised as follows:

I. User interface adaptation rules are compiled by the USE-IT tool.
II. User interface adaptation rules compiled by USE-IT are subsequently interpreted by the run-time libraries of the underlying high level user interface development

toolkit with which the user interface is to be implemented so that adaptations are instantiated onto a target technology platform.

In this manner, adaptation of the lexical layer of interaction is automatically supported during the initial design and development phases of the user interface. In the following sub-sections, an attempt is made to describe some of the design details that have characterised the implementation of the USE-IT tool so as to provide a better insight into what is being adapted, when and how.

3.1 Adaptation constituents

USE-IT compiles adaptation decisions for attributes of physical interaction object classes. These are instances of abstract interaction object classes binded to a particular interaction technology. In the recent literature [15, 16], the term abstract interaction object (AIO) has been associated with several properties, briefly summarised as follows: (a) AIOs are application domain independent; (b) they encapsulate all the necessary interaction properties (i.e. appearance, placement, behaviour, state, etc) by means of attributes (i.e. size, width, colour, and methods such as selection, activation, state change, etc); (c) they preserve a degree of independence from particular windowing systems and environments (i.e. they are platform independent).

In the context of the present work, the term is used in a broader sense to include additional properties such as the following: (a) AIOs are adaptable to the end user (i.e. their attributes can be adapted through reasoning); and (b) AIOs are metaphor independent (e.g. an AIO, such as the Button object, can be applicable for both the Desktop and Rooms metaphor, through perhaps different realisations). An abstract interaction object when binded to a particular interaction metaphor or lexical technology inherits additional attributes specific to that interaction metaphor. Thus, for instance, the implementation of an abstract `Button` on a visual lexical technology supporting the visual desktop metaphor could be conceived as follows:

```
DesTop Button
   [  method      Selected;
      method      FocusedIn;
      method      FocusedOut;
      string      Font = 'Courier';
      int         FontSize=12;
      string      Foreground='white';
      string      Background='black';
      Button.inputDevice = 2ButtonScan;
      Button.inputTechnique   =  Scanning2D;
      Button.ScanMode    =  TimeScan;
      Button.ScanTime    =  500;
   ]
```

An adaptable physical interaction object class AIO_c is a relation defined as the Cartesian product of a set of attributes $A_1,...,A_n$. An adapted physical interaction object class is any unary subset of AIO_c. It is important to note that these definitions are *logical* definitions and, as such, they are bound by the underlying user interface development tool. In other words, the range of attributes of an object as well as the possible assignments to these attributes may differ depending on the tool used to implement the interface. Consequently, the power of such a tool lies in the level of abstraction that it supports to facilitate access to non-trivial attributes (i.e. topology, accessPolicy, navigationPolicy, initiationFeedback, interimFeedback, completionFeedback, etc).

Example: Consider an AIO_c called `Button` and defined as follows:

```
Button :- TP x AP x ID x OD
```

where `TP`, `AP`, `ID` and `OD` are attributes defined as follows:

```
TP(TopologyPolicy) = {horizontal ,vertical}
AP(AccessPolicy)   = {byKeyboard ,bySpeech}
ID(InputDevice) = {1switch ,2switch ,5switch}
OD(OutputDevice)   = {Crt , Braille}
```

An adaptation decision is an assignment of a value to an attribute of the abstract interaction object class `Button`. A collection of adaptation decisions define a unary relation which is a subset of

```
Button :- TP x AP x ID x OD
```

and which is the adapted abstract interaction object class `Button`.

3.2 Binding Adaptation Of Abstract Interaction Objects To Interaction Metaphors and Task Contexts

Two important primitive concepts which determine the derivation of an adaptation decision are the prevailing metaphor of interaction and the application-specific task context. In general, interaction metaphors may be either embedded in the User Interface (e.g. menus as interaction objects follow the "restaurant" metaphor) or may characterise the properties and the attitude of the overall interaction environment (e.g. the desktop metaphor presents the user with an interaction environment based on sheets of papers called windows, folders, etc). In the present work, it is assumed that each development platform (i.e. OSF/Motif, MS-Windows, etc) serves one interaction metaphor (e.g. the visual desktop). Consequently, each of those platforms provides the implementational support that is required for the interactive environment of the metaphor. Different interaction metaphors may be facilitated either through the enhancement of existing development platforms or by developing new ones. An example is CommonKit [17] which supports non-visual interaction based on the non-visual Rooms interaction metaphor. Consequently, depending on the interaction metaphor (e.g. visual desktop, NonVisualRooms), typically two things may change, namely the associated interaction object classes and the attribute value range or attribute adaptability range (i.e. set of constants from which the value of the attribute may be drawn).

Task awareness is another critical technical requirement that needs to be supported explicitly. The principle behind task awareness is based on the belief that the user interface may be required to exhibit different interactive behaviour during different interaction contexts. For instance, to practically support the development of an interface which combines Desk-Top and Book behaviour, the developer needs constructs which are sufficiently expressive to describe when properties of each metaphor are to be applied and how. It follows, therefore, that task awareness reflects the requirement for differentiating lexical and syntactic properties of the dialogue depending on what the user is trying to accomplish or what the interface is attempting to convey. The work to support task-aware interface design and development involves a number of issues, which can be briefly summarised as follows. Designers and developers need to be provided with explicit constructs to specify different (desirable) states in the dialogue. Additionally, there is a need for mechanisms and techniques for binding interaction objects to dialogue states. Through such a localisation of the interactive behaviour of an object, it is made possible to practically support fusion of metaphors, provided that the underlying toolkits preserve the principle of toolkit inter-operability [2]. Thus, it will be possible to derive and assign different interactive properties to interaction objects based on the requirements of the dialogue state and the metaphor that is deemed as appropriate.

To support the contextual binding of adaptations decisions to specific interaction metaphors and contexts of use, USE-IT supports the explicit representation of metaphor and task oriented design knowledge. This means that the designer is able to declare the interaction objects available in a particular interaction metaphor, the attributes of these objects and the interaction techniques which are supported as well as the interaction requirements that should determine any adaptation decisions in a particular task context. More specifically, the translation from abstract interaction objects to platform-specific objects is largely determined by the context assigned to the interaction task. Parameters of the context of an interaction include the range of objects which are available, their attributes, the sequence of tasks that the designer has declared, as well as the metaphor of interaction during this task.

3.3 User Model Driven Reasoning Of Design Issues

In order to provide the required support for user-centered design of an interactive application, a technique for describing the characteristics of a user (group) which are critical for the design of the user interface has also been developed. This is an area were several developments have been undertaken in the past. One strand has explored the definition of models of the user from psychological and/or cognitive perspective. Typical examples include Goals Operators Methods and Selection Rules (GOMS), Programmable User Models (PUM), Runnable User Models (RUM), but also the SANE toolkit for cognitive modeling and user-centered design, etc. Another approach has been concerned with the construction of models of the users interests, knowledge, goals and beliefs and have provided various classes of representation techniques and tools for acquiring, representing and embedding such user models in interactive software systems. Examples of this category of tools include user modeling shell systems such as GUMS [18], UM [19], UMT [20], BGP-MS [21], PROTUM [22], and user modeling servers [23]. Unfortunately, the latter cluster of developments have followed a path which has not addressed their potential integration into high level user interface development platforms. The only effort known in this direction has been the results of recent work in the UIDE system where an overlay of UIDE's knowledge is used to compile a user model [11]. Consequently, there is a need to embed user-modeling facilities and components into user interface design and development environments so as to allow user model driven reasoning when designing the interface, but also at run-time in order to determine adaptivity-oriented behaviour.

USE-IT provides a tool, as part of the integrated design environment, which allows the elicitation of user characteristics and the interpretation of the implications of these user characteristics on the lexical and syntactic levels of the user interface. The user models constructed by the tool are visible, transparent and modifiable by the designer. In addition, "what-if" type of exploration of any tentative user model is supported. More specifically, the designer will be able to change parameters of the user model and assess tentative interface design considerations for a particular interaction task. In this manner, the implication of a particular user model on the design of the interface becomes directly visible to the designer during the early phases of interface development.

3.4 Reasoning About Lexical And Syntactic Design

The user-computer interaction process is realised at three different levels [24]:

(i) The *semantic* level, which concerns the general functionality which needs to be made accessible to the user (e.g. editing files, storing/retrieving records);

(ii) The *syntactic* level, which concerns the structure and the syntax of the dialogue with which the semantics are made accessible to the user (e.g. steps taken by the user, tasks to be accomplished and decomposition of tasks);

(iii) The *lexical* level, which concerns the structure of the actual input/output items which are used to physically realise the dialogue (e.g. interaction objects as well as input devices, and interaction techniques).

Tools to empower developers when designing user interfaces, should ideally, address the wide range of issues related to the adaptations that may be required at any of the three levels of user-computer interaction. At the semantic level, it is important to support the adaptation of the information content that the user interface communicates. At the syntactic level, tools are required to facilitate identification of appropriate task sequences and determination of dialogue syntax (interaction style, sequence, task composition and order) through which the semantics of the information to be communicated to the user, is conveyed. Finally, at the lexical level, adaptation may be required to support alternative interactive behaviours of the user interface constituents and the way that the user-computer interaction is to be accomplished (i.e. input/output devices, interaction techniques, attributes of interaction objects, etc). To provide the required support for reasoning towards primarily lexical details, USE-IT combines user modelling techniques (depicting user-centred design constraints), task specifications (depicting task requirements and interaction / interface / dialogue constraints), and platform oriented details (i.e. interaction metaphors, objects, presentation alternatives, devices, etc).

4 Design assistance: Towards the automatic derivation of adaptation decisions

From the discussion thus far, it becomes apparent that the semantics of adaptation decisions, as introduced in the previous sections, can be encapsulated in a relational data model such as that depicted in the diagram of Figure 7. Normalizing such a data model (see Figure 8) results in the introduction of an additional entity type namely `adaptation` which is defined by the following relation:

```
adaptation(MetaphorId,TskContextId,ObjClsId,Atttr,Value)
```

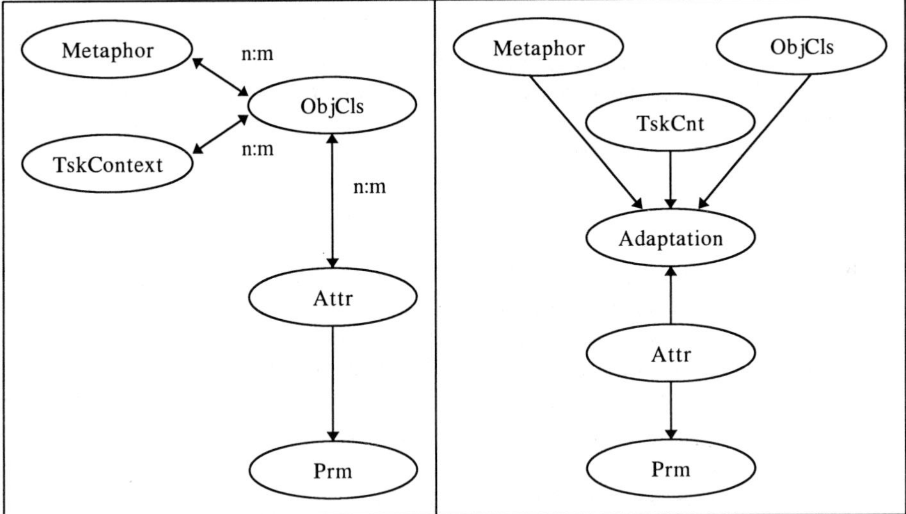

Fig. 7 Structure of oblecst classes **Fig. 8** The normalised data model

In the context of the present work, the task of adapting abstract interaction object classes involves the derivation of the *extension* of the `adaptation` relation, given a particular design scenario. To facilitate this, the designer is provided with various tools (knowledge editors, rule generators, etc) for managing the knowledge components of the overall environment. Thus, prior to running the adaptation engine to compile the adaptation decisions, the designer develops a precise representation of the platform constraints (i.e. input/output devices, available interaction object classes per interaction metaphor, their attributes, etc), the user constraints (i.e. user capabilities and preferences), task constraints (i.e. task contexts and their requirements). Regarding the task contexts defined by the designer, it is important to note that at any point in time there is an abstract task context called `anyOther` which is used to solve two main problems. The first is to enable the tool to handle negative information. Thus, the availability of a primitive task context such as `anyOther` allows the designer to encode design knowledge such as the following:

```
IF the user has colour blindness
   THEN inputDevice='2 switchScanning with time scan' and
      IF the user is engaged in the task context message_editing
         THEN timeScan=500 and
            ScanningHighLigther=Gray
      ELSE timeScan=1500 and
         ScanningHighLigther=Gray
ELSE inputDevice='2 switchScanning with no time scan'
   and ScanningHighLigther=Red
```

The above rule requires that, if the user has colour blindness and is engaged in `message_editing`, then `timeScan` should be allowed and set to 500 and the colour of the highlighter should be set to gray. If the user has colour blindness, but is NOT engaged in `message_editing`, then `timeScan` should be allowed and set to 1500 and the colour of the highlighter should be set to gray. If, on the other hand, the user does not have colour blindness, then irrespective of the current task context, time scan should be disallowed, while the colour of the highlighter should be set to red.

The second problem that is addressed by the introduction of the `anyOther` task context relates to the requirement to support *incremental* design and derivation of additional knowledge based on partial descriptions of the design domain. The first issue, namely incremental design, relates to the ability to introduce new design parameters such as for example new task contexts, as additional knowledge about the user or the interaction context become available, and thus improve the design scenario and the respective implementation. The second issue, namely supporting partial design descriptions, reflects a technical requirement of the tool which is critical, if design assistance is to be provided. More specifically, one important issue is to allow the designer to input partial knowledge about a design problem and to employ a design assistant tool, which would derive and subsequently use in the inference process additional (missing) knowledge immediately deducible from the initial specification.

For example, consider a 2D interactive manipulation task, such as selecting and changing the colour of an object in a three-dimensional space. It is known that in 2D interactive tasks, indirect and continuous pointing devices such as the mouse, the trackball and the joystick are better than direct devices due to the awkward position of the arm. Moreover, it is also known that when the task to be performed is conceived by the user as a separable (as opposed to integral) task, then a device which fits better this perceptual structure of the task should be used [25]. Given such a hypothetical problem description, the designer is asked to select the most appropriate device. There are several problems that may result in a sub-optimal decision on behalf of the designer. First of all, he may not be aware of the notion of separable versus integral tasks and devices which may be better suited in each case. Assuming that this not a problem, however, the next issue is that the designer faces a dilemma. This is due to the fact that, if the designer follows the former criterion, he should select one indirect and continuous pointing device, which could be anyone of those mentioned above, as there is no evidence of preference of one over the others. If, on the other hand, the designer follows the second criterion, then the task when performed in 2D space is considered as separable, while the same task when performed in 3D it constitutes an integral task. Each consideration has implications on the selection of the device. The situation is worse when both criteria are considered as equally valid and should be satisfied.

A typical session with USE-IT results in the compilation of adaptation decisions for specific attributes of physical interaction objects. To demonstrate this, let us consider a hypothetical design scenario involving the construction of a communicator for a user community that is distinguished by the following characteristics:

```
P: User has only one hand
Q: Fingertips can be used for key selection
R: User has relative colour blindness
S: User is able to respond in timed patterns
```

Given the above, let us assume a user model containing the following factual information[1]:

```
evalue("Ability to initiate movement")
evalue("Ability to control movement")
evalue("Ability to reproduce the movement on demand")
evalue("Ability for gross temporal control")
evalue("Ability to read standard text")
evalue("Ability to read enlarged text")
evalue("Ability for visual tracking")
control_act("Movement of one hand")
control_act("Movement of head")
control_act("Directed eye-gaze")
contact_site("Fist")
```

Let us also assume that differentiation of lexical assignments should occur under three distinctive task contexts, namely `visual-keyboard`, `message-editor` and `anyOther`. We also assume a visual interaction metaphor implemented by enhancement of MS-WINDOWS™ to accommodate scanning of interaction objects. Finally, for the purpose of simplicity, it is assumed that the design team has provided the tool with sufficient design knowledge (in the form of the heuristic rule cited earlier), as well as that the consistency and integrity of the knowledge base has been verified[2]. The outcome[3] of USE-IT for the above example is as follows:

---

[1] It is important to mention that a user model also includes derivation rules which are, however, irrelevant for the present example.

[2] All these facilities can be interactively accommodated using the USE-IT's knowledge editors and rule generators (Akoumianakis et. al., 1995c).

[3] This output is a formatted version compiled from the occurences of the predicate **`adaptation(M,T,O,A,Value)`** introduced earlier in this section.

```
visual_desktop 384
   Visual-keyboard
      listBox
         input_device = 1 Switch/TimeScan
         inputTechnique = scanning2D
         scanmode = 1
         timescan = 1500
         dm_bordercolor.red = 0
         dm_bordercolor.green = 255
         dm_bordercolor.blue = 0
         dm_borderstyle = 1
         dm_borderwidth = 2
         em_borderwidth = 2
         em_bordercolor.red = 255
         em_bordercolor.green = 0
         em_bordercolor.blue = 0
         em_borderstyle = 2
         output_device = vdu
         outputTechnique = displayTechnique
```

The above sample outcome depicts adaptations for the interaction object `listbox` during the task context `Visual_keyboard` and the visual desktop interaction metaphor. Such adaptations are produced for all abstract interaction objects and all task contexts and interaction metaphors. As shown in the example, USE-IT provides the development tool with a total count of adaptation decisions derived per interaction metaphor.

5 Adaptability rules as a type of design assistance

One important task during the initial design phase of a user interface is the availability of assistance to the designer. Design assistance can take various forms including the provision of training modules, styleguides, guidelines, etc. The primary medium through which such assistance has been provided to the designers up until now is printed-paper. Only very recently there has been a shift of perspective towards more interactive means and in some cases there have been efforts towards the development of knowledge-based systems and tools for working with guidelines. One important shortcoming of these efforts has been the lack of support at the designer's finger tips. In other words, there are no tools which can provide the designer with direct and context-specific feedback while designing the interface. Moreover, the few cases that this is achieved offers limited value as the support that is offered is sometimes trivial (for instance, existing tools can detect that the designer has omitted the title in a list box, or that the selected font in the title is not consistent with the font in the dialogue box). However, there are a broad range of issues on which a designer is likely to appreciate meaningful assistance. For instance, it would be useful if a designer could query a tool to find out the preferred feedback option in a target acquisition task; or to obtain guidance on whether to use a 3D tracker as opposed to a mouse for the interactive manipulation of an object in 3D space, the available options in terms of interaction techniques (i.e. in case that the task involved is the change in colour of the object then the use of the conventional mouse would require a mode change parameter to complement the selection of the object); etc.

Until now, design support has been merely addressed through general high-level design suggestions (such as, for example, design guidelines and heuristics, like use of forms with explicit confirmation for field values, which could be targeted either to specific or non-specific application domains). The problem with such approaches for practically supporting the interface design process can be summarised as follows:

I. Guidelines are usually too specialised and their applicability is often limited to particular contexts.
II. Guidelines that are general, lose power and value when applied to specific contexts.
III. The relevant topic had small or no relevance with the target application domain.
IV. The initial interaction objectives of guidelines may be different, or even contradictory with the target application domain and consequently the guidelines may not be applicable at all.
V. The topic and objectives may have strong relevance; however, applying the guidelines directly may be problematic, while an "adaptation" of the original guidelines would be much less than practically trivial.
VI. Different design guidelines for the same topic and objectives may result in incompatible instructions.
VII. Large number of guidelines.

Guidelines are not sufficiently structured in such a way so as to reflect a comprehensive design process.The above problems may limit considerably the practical integration of such generated design suggestions during the design process. Moreover, automating the application of such design decisions through an interface development system cannot be realised. Currently, there is no support for explicitly incorporating design decisions into the interface development process by means of automatic interpretation and realisation of decisions within the resulting interface implementation.

The USE-IT tool has over-passed this difficulty by extending design suggestions to more concrete interface design scenarios. The USE-IT tool has the ability to generate different rules concerning attributes of interaction object classes according to the particular interaction contexts. These decisions can be interpreted and applied automatically during user-computer interaction by interactive applications, which are built through specific interface development systems (see Figure 9). Consequently, such interactive applications practically implement the design decisions generated by the USE-IT tool. This behaviour is achieved by the proper synergy of the USE-IT tool with the interface development platform which has to understand and apply decisions provided by the USE-IT tool. This strategy of implementationally separating systems with different roles during the development process has been realised in the current generation of the unified user interface development platform, while it brings about several advantages:

- o *Modification independence*, since modifications in one system do not affect the other. In our approach, the communication between the systems is reduced to the file which is produced by the USE-IT tool and read by the interface implementation, while the protocol is mainly the syntax of that file.
- o *Implementation independence*, since different programming languages can be employed for different systems. For instance, the USE-IT tool has been developed via the Prolog language, while the interface implementation is provided in the C++ language.
- o *Design role resolution*, since the USE-IT tool is targeted to interface designers and human-factors specialists, while the interface development systems concern interface implementation experts. It should be noted that existing interface development environments usually require that the designer also deals with implementation notations.
- o *Knowledge reusability*, since design decisions for a specific domain can be directly re-used for interactive applications within the same domain. This is possible since the design decisions can be easily transferred to the new interface implementation (the interface implementation will automatically apply the rules).

The interface development systems which have been implemented support powerful methods for abstraction of interaction objects and interaction techniques. This has been an important feature for practically supporting the design decisions generated by the USE-IT tool which relies upon a sophisticated model of the lexical layer of interaction that is not supported by existing toolkits of interaction objects. It should be noted that with more sophisticated and well structured models of the lexical layer, it is possible to accomplish high quality of adaptability. Consequently, it is critical to have interface development systems which practically provide better organisation and decomposition of lexical interaction elements. We have utilised the object abstraction methods of the interface

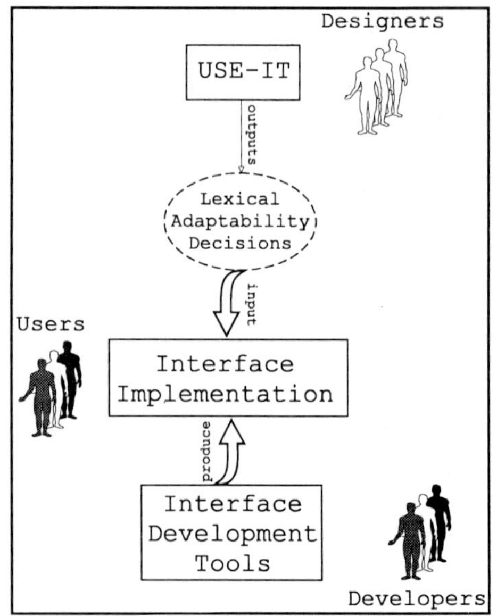

Fig 9 Synergy of USE-IT with interface development tools.

development framework so as to match the structural aspects of the lexical level of interaction as they are realized from the USE-IT tool point of view.

6 Status Of The System And Plans For Future Work

The first prototype of the USE-IT system was implemented using the LPA Prolog Programming Environment and the knowledge representation and facilities of the ESTA Expert System Shell. Subsequently, due to implementational limitations the system was ported into PDC Visual Prolog. Currently, USE-IT runs under MS-Windows and facilitates the derivation of lexical adaptability rules. This is to say that all the modules of Figure 6 have been implemented and can be used to compile adaptability rules for two interaction metaphors, namely the visual desktop as supported by MS-Windows and the non-visual rooms as originally supported in the COMMONKIT [17]. It should also be mentioned that USE-IT, as indeed the remaining user interface development tools of the unified user interface development platform [12] will become public domain software early in 1997. Finally, it is foreseen that a number of additional features will be incorporated into the current system in order to support syntactic adaptability of interaction elements as well as the propagation of human factors knowledge into user interface design and development phases.

7 Summary and conclusions

This paper has reviewed user interface adaptation techniques, given certain trends within the information technology industry, namely increased product specialization, universal accessibility in the emerging information society and demand for higher quality of interaction with computer-based interactive applications and services. It is claimed that the way in which user interface adaptation has been conceived and supported by recent research efforts does not provide the means for addressing comprehensively the above new requirements. The main reason for this is that at present the techniques of user interface adaptability and adaptivity are implemented on top of a class of user interface development tools which does not support the development of high quality user interfaces for diverse user groups with different abilities, requirements and preferences. This implies that the supported adaptations, irrespective of their timing or agent of initiation, are bound by the limitations of the currently prevailing visual desktop embodiment of the computer and the way that this is implemented in the various user interface development tools and systems (i.e. range of interaction objects, attributes, interaction techniques offerred by these development tools). Additionally, the range of adaptation constituents considered is inherently limited to either pre-determined dialogue syntaxes or some presentation attributes, thus failing to account for basic design concerns such as the the details of the way in which interaction is managed at the physical level.

In order to eliminate some of these shortcomings, the work presented in this paper described the rationale and technical ground for another approach to user interface adaptation, which is initiated by the developer of the user interface and takes place during the early design and development phases of an interactive application. This type of adaptation is currently not supported by any of the existing tools, systems or environments for user interface development, as it requires certain qualities that are commonly missing from user interface development tools of the present paradigm. The paper has discussed the most important of these qualities, and presented one technical approach towards their implementation. This approach is currently being elaborated in the context of the ACCESS (TP 1001) project of the TIDE Programme of the European Commission which has developed a novel platform for *unified* interface development.

The current version of USE-IT is utilised to provide lexical adaptability for user interfaces of interactive applications in the domains of interpersonal communication for speech/motor and cognitive/language impaired users, and hypermedia applications accessible by both blind and sighted users, in the context of the ACCESS project. Extensions are already underway to support syntactic adaptability; this is important as there are application domains in which a dialogue sequencing differs depending on the target user groups. For instance, in the domain of hypermedia applications, it is highly likely that the dialogue of a user interface will require support for adaptability if it is to integrate both sighted and blind users.

Acknowledgments

The work presented in this paper is carried out in the context of the ACCESS project (TP1001) funded by the TIDE Programme of the Commission of the European Union. Partners in this consortium are: CNR-IROE (Prime Contractor), Italy; ICS-FORTH, Greece; University of Athens, Greece; RNIB, U.K.; SELECO, Italy; MA Systems Ltd., U.K.; Hereward College, U.K.; National Research and Development Centre for Welfare and Health, Finland; VTT, Finland; PIKO Systems, Finland; University of Hertfordshire, U.K.

References

1. Galer, M., Harker, S., Ziegler, J. (1992): Methods and Tools in User-Centered Design for Information Technology, Human Factors in Information Technology 9, North-Holland.

2. Savidis, A., Stephanidis, C. (1995): Towards multimedia interfaces for all: a new generation of tools supporting integration of design-time and run-time adaptivity methods. NSF/Multimedia'95 Workshop on Adaptive Multimedia Technologies for People with disabilities, November 10, San Francisco.

3. Dieterich, H., Malinowski, U., Kuhme, T., Schneider-Hufschmidt, M.(1993): State of the Art in Adaptive User Interfaces. Adaptive User Interfaces: Principles and Practice, M. Schneider-Hufschmidt, T Kuhme and U. Malinowski (Eds.), Adaptive User Interfaces, Amsterdam: Elsevier Science Publishers B.V, North-Holland, pp. 13-48.

4. Ten Hagen, P.J.W. (1990): Critique of the Seeheim model. In User Interface Management and Design, Duce, D., A., Gomes, M., R., Hopgood, F., R., A., and Lee, J., R. (Eds), Eurographics Seminars, Springer-Verlag.

5. Coutaz, J. (1990): Architecture models for interactive software: Failures and trends, In Engineering for Human-Computer Interaction, G. Cocton (Ed), North-Holland, pp: 473-490.

6. The UIMS Developers Workshop (1992): A Metamodel for the run-time architecture of an interactive system, SIGCHI Bulletin 24, 1.

7. Cote-Munoz, A., H. (1993): AIDA: An Adaptive System for Interactive Drafting and CAD Applications, in M. Schneider-Hufschmidt, T. Kuhme and U. Mallinowski (Eds.), Adaptive User Interfaces, Amsterdam: Elsevier Science Publishers B.V, North-Holland, pp. 225-240.

8. Sherman, H., E., Shortliffe, H., E. (1993): A User-Adaptable Interface to predict Users' Needs, in M. Schneider-Hufschmidt, T. Kuhme and U. Mallinowski (Eds.), Adaptive User Interfaces, Amsterdam: Elsevier Science Publishers B.V, North-Holland, pp. 285-315

9. Zimek (1991): Design of an adaptable/adaptive UIMS in production, in Human Aspects in Computing - Design and Use of Interactive Systems and work with terminals, Bullinger (Editor), pp. 748-752, Elsevier.

10. Ancieri, F., Dell'Ommo, P., Nardelli, E., Vocca, P (1991): A user Modeling System, in Human Aspects in Computing - Design and Use of Interactive Systems and work with terminals, Bullinger (Editor), pp. 440-447, Elservier.

11. Sukaviriya, P., Foley, J (1993): Supporting Adaptive Interfaces in a knowledge-based user Interface Environment, In W. D. Gray, W. E. Hefley, and D. Murray (Eds.), Proceedings of the 1993 International Workshop on Intelligent User Interfaces (pp. 107-114), Orlando, FL. New York: ACM Press.

12. Stephanidis, C. (1995): Towards User Interfaces for All: Some Critical Issues, in Proceedings of HCI International '95 Conference on Human Computer Interaction, pp. 137-143, Elsevier.

13. Stephanidis, C., Savidis, A., Akoumianakis, D. (1995): Towards user interfaces for all, Conference Proceedings of 2nd TIDE Congress, pp. 167-170.

14. Stephanidis, C., Savidis, A., Akoumianakis, D. (1996): Development tools towards User Interfaces for All, to appear in International Journal of Human-Computer Interaction.

15. Myers, A., B. (1990): A new Model for Handling Input, ACM Transactions on Information Systems, 8(3), pp. 289-320.

16. Bodard, F., Hennebert, A-M., Leheureux, J-M, Provot, I., Vanderdonckt, J. (1994): A model-based Approach to Presentation: A Continuum from Task Analysis

to Prototype, in Proceedings of Eurographics Workshop on Design, Specification and Verification of Interactive Systems, pp. 25-39.

17. Savidis, A., Stephanidis, C. (1995): Developing Non-Visual Interaction on the basis of the Rooms metaphor, in Companion of CHI'95 Conference on Human Factors in Computing Systems, pp. 146-147, ACM Press.

18. Finin, T. (1989): GUMS: A General User Modelling Shell, in User Models in Dialogue Systems, A. Kobsa, W. Wahlster (editors), pp. 411-430.

19. Kay, J. (1995): The um toolkit for reusable, long-term user models, User Modelling and User-adapted Interaction, 4(3).

20. Brajnik, G., Tasso, C. (1994): A Shell for Developing Non-Monotonic User Modelling Systems, International Journal of Human Computer-Studies, 40, pp. 31-62.

21. Kobsa, A., Pohl, W. (1995): The user modelling shell system BGP-MS, in User Modelling and User-adapted interaction 4(2), pp. 59-106.

22. Vergara, H. (1994): PROTUM - A Prolog based Tool for User Modelling, Bericht Nr. 55/94 (WIS-Memo 10), University of Konstanz, Germany.

23. Orwant, L., J. (1995): Heterogeneous Learning in the Doppelganger User Modelling System, in User Modelling and User Adapted Interaction , 4(2), pp: 107-130.

24. Foley, J., D., Wallace, V., L., Chan, P (1984) : The human factors of Computer Graphics interaction techniques, IEEE Computer Graphics and Applications, Vol. 4(11), pp. 13-48.

25. Jacob, R., Sibert, L., Mcfarlane, D., Mullen, M (1994): Integrality and Separability of Input Devices, ACM Transactions on Computer-Human Interaction, 1(1), pp3-26.

26. Akoumianakis, D., Petrie, H., Morley, S., and Stephanidis C. (1995): Supporting user interface adaptability during the design and development process, Adjunct Proceedings of HCI-95 People and Computers Conference, University of Huddersfield, U.K.

27. Akoumianakis, D., Stephanidis C. (1995): User Modelling for adaptable interface design, in Proceedings of the 6th International Conference on Human Computer Interaction (HCI International'95), Tokyo, Japan 9-14 July, Volume 1, pp.1071-1076.

28. Akoumianakis, D., Savidis, A., Stephanidis C. (1995): An expert user interface design assistance for deriving lexical adaptability rules, Proceedings of the Third World Congress on Expert Systems (WCES-96), February 5-9, Seoul, Korea.

29. Benyon, D., Murray, D. (1993): Adaptive Systems: from intelligent tutoring to autonomous agents, Knowledge-Based Systems, 6(4), pp.197-219

30. Browne, P., D. (1993): Experiences from the AID project, in M. Schneider-Hufschmidt, T. Kuhme and U. Mallinowski (Eds.), Adaptive User Interfaces, pp: 69-78, Amsterdam: Elsevier Science Publishers B.V, North-Holland.

31. De Carolis, B., de Rises, F. (1994): Modelling Adaptive Interaction of OPADE by petri Nets, SIGCHI, Vol. 26, No. 2, pp. 48-52.

32. Koller, F. (1993): A demonstrator based investigation of adaptability, in M. Schneider-Hufschmidt, T. Kuhme and U. Mallinowski (Eds.), Adaptive User Interfaces, pp. 183-196, Amsterdam: Elsevier Science Publishers B.V, North-Holland.

33. Lai, K., Malone, T. (1988): Object Lens: A Spreedsheet for Cooperative Work, Proc. of the Conference on CSCW, ACM, New York, pp.115-124.

34. MacLean, A., Carter, K., Lovstrand, L., Moran, T, (1990): User-Tailorable Systems: Pressing the Issues with Buttons, CHI'90, ACM, New York, pp. 175-182.

35. Okada (1994): Adaptation by task intention identification, in FRIEND 21 Conf. Proc., Japan, 1995.

36. Robertson, G., Henderson, D., Card, S. (1991): Buttons as First Class Objects on an XDesktop, UIST '91, ACM, New Yoark, pp. 35-44.

37. Savidis, A., Stephanidis, C. (1995): Developing Dual User Interfaces for Integrating Blind and Sighted Users : The HOMER UIMS", in Proceedings of CHI'95 Conference on Human Factors in Computing Systems, pp:106-113, ACM Press.

GRALPLA: An Algebraic Specification Language For Interactive Graphic Systems

J.C. Torres, M. Gea, F.L. Gutierrez, M. Cabrera, M. Rodriguez

Dpt. Lenguajes y Sistemas Informáticos. Univ. Granada.
E.T.S.I. Informática. Av. Andalucía 38. E-18071 Granada. Spain
email: <jctorres,mgea,fgutierr,mcabrera,mlrogriguez>@ugr.es

Abstract. The specification of interactive graphic systems involves the use of formal methods to describe the synchronization restrictions and graphic information. Several proposals have been made trying to join formalisms for the specification of concurrent systems with some methods to describe the graphic component. This paper presents an algebraic specification language which has been designed to specify interactive graphic systems. The language is founded on the use of a mathematical formalism to describe the graphic component, with an extension of algebraic specification language to describe synchronization, using guarded operations. The language is an extension of the previous Gralpla language, to which new features have been added to allow the specification of dynamic changes of the graphic representations and of processes. A prototyping tool has been defined for the language to generate C++ prototypes of the specifications.

1 Introduction

Formal methods have been used broadly in the specification of computer systems. Any specification must describe, with the necessary precision and without ambiguity, the system to be built, and it must allow us to perform formal reasoning. This approach is still more important for interactive systems, because the complexity of the dialogue component involves description of graphic objects and the specification of synchronization restrictions.

Several proposals have been made to apply formal methods to the specification of interactive graphic systems[1,2]. Algebraic specification languages have been used to specify graphic systems, mainly graphic standards[3].

This paper presents a new version of Gralpla, which is an algebraic specification language which uses a mathematical formalism to describe the graphic component[4,5,6], to which guarded functions have been incorporated to describe synchronization. This language is an extension of an earlier version, which was also developed at the University of Granada[7,8,9] and to which the following features have been incorporated: dynamic modification of the graphic representation, feedback control, processes, error management, type inheritance, asynchronous function calls and the use of function as parameter.

The language has been defined to incorporate an associated translation tool, to allow the automatic construction of prototypes in a high level language (C++). These prototypes can be used as a first version of the system to be developed.

The following section contains a comparison with some other specification languages used for the specification of interactive systems. Section 2 describes the main characteristic of the language proposed, and its syntactic and semantic structure. In section 3 and 4 we discuss with more detail the mechanism for describing the graphical and interactive components. Section 5 gives a simple, yet complete, example.

We conclude with a brief discussion on how to perform formal reasoning for verifying the specifications.

2 Comparison with other specification methods

Formal methods might be property-oriented or model-oriented. Model-oriented methods describe a model of the system, which is used as support to define the system behaviour. Property-oriented specification methods make a description of the system only indicating its external properties[2,10,11].

We are going to compare GRALPLA with some formal specification languages which have been previously used to specify interactive systems: Object-Z, OBJ, LOTOS and Petri nets. Our aim is not to do a complete study of the wide range of specification methods, but to show some of the differences and similarities of Gralpla with some other well known methods.

There are some differences between Object-Z[10] and GRALPLA. First, both specification languages have different natures (model vs property oriented), secondly ,they have different mathematical bases. Nevertheless, the two languages incorporate similar facilities for the specification of object oriented systems, like hierarchy of types and inheritance. Object-Z allows us to define new classes starting from the existent classes, and in GRALPLA it is possible to define types which are a specialization of a more general one, previously defined.

We can find many common characteristics between OBJ[11] and GRALPLA. Both follow a method guided by properties, and both are algebraic specification languages. Both languages allow us to define a type of error that will be returned for functions (OBJ allows to define any type of error and GRALPLA has three types of error: warning, severe and fatal), and to describe when a function will return a type of error. OBJ specification can be executed. And GRALPLA has an automatic tool for the construction of prototypes in a high level language.

LOTOS is a specification language developed for the specification and implementation of distributed systems[12]. It is a suitable method for the description of concurrent systems. LOTOS combines process algebra (inherited from CSP and CSS) with data algebra (using the specification language ACT ONE). In LOTOS, the main elements of the description are the processes (entities that can carry out internal actions not observable from the environment, and can communicate with other processes through events), this mechanism corresponds with process functions and asynchronous functions in GRALPLA. LOTOS describes the process behaviour using algebraic expressions, and LOTOS operators such as sequentiality, parallel composition and disabling. GRALPLA uses guards as synchronization mechanisms.

Petri nets are a suitable formalism for describing process behaviour and synchronization, they have been used to specify interactive systems and user interfaces[13].

Table one shows the comparison of some of the characteristic for the languages mentioned above. Some of these formalisms have a graphical notation to represent the system structure. This is the case for Lotos, for which there exists a graphical notation (G-Lotos), and for Petri Net. But any of them, except Gralpla, have use a graphic formalism as support, which may be used to specify the object rendering information.

Table 1. Comparison of specification languages.

	Object-Z	OBJ	LOTOS	Petri Net	Gralpla
Model vs property	model	property	property	model	property
Object oriented	X			X	X
Error management		X			X
Sync. mechanism			X	X	X
Process specification			X	X	X
Graphic notation			X	X	X
Graphic Formalism					X
Development Tool		X	X	X	X

3 The specification language

GRALPLA is an algebraic specification language. The language supports the definition of abstract data type hierarchy, error management, specification of graphic information and specification of processes and synchronization. A tool to check the structure and completeness of specifications and to generate a prototype in a high level language has been designed. These prototypes can be used as a first version of the system to be developed.

The specification of a system using GRALPLA consists of a collection of modules, each one specifying a type. The specification of every type consists of an interface, a collection of functions, and a set of conditions that hold the objects of the defined type, expressed as a set of axioms on the type functions. The text of a type definition contains the following sections: header, dependencies, constructors, functions, axioms and synchronization. A correct type specification is formed with a sequence of these components in the order described (see appendix 1). The remainder of this section describes the syntax and semantics of these components[9].

Header. This section contains overall characteristics of the type. It defines the type name (object_id), which is also the name of the carrier set, and an optional list of formal parameters, used to build generic modules. The parameter list is a sequence of identifiers, which are used to define generic types. We can restrict the valid substitution of parameters, assigning a type to them. For instance, parametric object list[T:point]. Functions can be treated as parameters; parameter function might link object with a predefined behaviour.

The header might contain a list of types from which the actual type inherits its behaviour. The type inherits all the functions defined in its supertypes. The behaviour of inherit functions can be redefined. These mechanisms (parametrized types and subtyping) let us manage *abstract data type hierarchy* using inheritance and generic types.

The keyword *graphic* means that the type implements a graphic entity, which has a graphic representation, and will be specified in the type using graphic objects.

Dependencies. This section gives a list of the types which are used by the actual one. That is, the actual type may use its carrier set and all its public functions.

Constructors. This section contains the constructor signature. Constructors are functions which allow us to build objects of the type specified. There must be at least one constructor. A type list can be optionally used to allow the definition of parameters when creating an object.

Functions. This section contains the signature of the type functions. Any function may return more than one result; in this case selectors must be included to identify every result. Any function definition has five components: function qualifiers, function identifier, an optional list of selector identifiers, a list of argument types, and a list of result types. Any function must have at least one argument and one result.

We treat known error situations as functions. Each error function has associated with it a type of error, describing the system tolerance to the error situation (robustness), and an optional explanation comment of the error situation. We can use error functions in the right hand side of the axioms (given an error result) and in the left hand side (providing error recovery). In this way it is possible to specify what happens when an error appears and what is the behaviour of the object in these circumstances.

Qualifiers are used to declare special characteristics of the function. *Private* functions define an auxiliary function which do not appear on the type interface. The *asynchronous* and *process* qualifiers are used to modify the function activation mechanism. They are explained in section 3.

Axioms. This section contains the axioms that hold for this type. Every axiom is an equation, whose left-hand side is a function. To write the equation it is possible to use auxiliary variables, whose only purpose is to establish relations within the axiom.

Table 2. Specification Language Sintax

```
<Module_espec> ::= <Header> [<Dependencies>] <Constructors>
                   <Functions> <Axioms>  [ <synchronization> ]

<Header> ::= [interface] [parametric] [graphic] object object_name
     [<Param_element> {,<Param_element> }] [:<object_id> {,<object_id>}];

<Param_element> ::= <Parameter> | <parametric_function>

<Parameter> ::= parameter_id [ : type_id ]

<Parametric_function> ::= func_name [ < selector_id { , selector_id } > ] :
          func ( Parameter { , Parameter } -> Parameter { , Parameter})

<Dependencies>::= [import {object_id [<Parameter_id> {,<Parameter_id>} ]};]

<Constructors> ::= [Constructors] { [private] function_id :
                   { type_id { , type_id } } -> object_id [ <Explicit_rep> ] ; }

<Explicit_rep>::=[where graphic_rep'('function_id [var_id{,var_id }')'=<Term>]

<Functions> ::=  Functions { [private] [process] [asynchronous]
          function_id [ < selector_id { , selector_id } > ]  :
          { type_id { , type_id } } -> type_id { , type_id } [ <error_description> ] ;}

<error_description> ::= := error [ ( <type_error> [, " comment " ]) ]

<type_error> ::= warning | fatal

<Axioms> ::= Axioms [ var { var_id { , var_id } : type_id } ; ]
          { function_id ( <Term> { , <Term> } ) [. selector_id]  = <Process>

<Process> ::= <Expresion>  |  ( <Process> )  |  <Process> '|' <Process>

<Expression> ::= <Term> | if ( <Term> ) <Expression>
                 [ else <Expression> ] endif

<Term>::=  <Term> <Operation> <Term>   | not <Term> | var_id |
          int_cte | real_cte | char_cte | true | false
          | function_id  [ ( <Term> { , <Term> } ) [ . selector_id] ]

<Operation> ::= + | - | * | / | and | or | < | >  | >= | <= | <>  | =

<Synchronization> ::= Synchronization
                 { do function_id ( var_id { , var_id } ) when <Expression> ; }
```

Each axiom specifies a condition on the functions behaviour, which must always be true, and which can be externally tested, without defining the internal structure of the system. An axiom is an equation specifying the value of one result of a function, when applied after another function. They are interpreted as directional equations: what appears on the left hand side can be carried out as explained on the right hand side. Both sides of the axioms must be of compatible data types.

To understand the system behaviour it is necessary to express, in some way, the system state, but algebraic specification provides no representation for it. Some algebraic specification formalisms reason directly on the sequence of functions performed on the system, using term rewriting. Our approach is to use the functional history as implicit representation of the state of the system[14]. This representation allows us to implement a prototype from the specification. Of course not every function call must be stored in the functional history. From a conceptual point of view, functions can be classified in three groups: constructors, generators (which generate new object states) and consultors (which do not modify the object state). The behaviour of constructor and generator functions can be defined using axioms. When this is not the case, the function is considered a basic function, whose behaviour can not be derived from others functions. Only basic functions are conceptually needed to keep track of the object state.

Synchronization. The synchronization section may impose execution restrictions on the functions defined in the type. An associated boolean expression controls when the function must be performed.

The specification is *complete* if there is at least one axiom for each derived function specifying its behaviour when it is called after every basic function. The specification is *consistent* if there is only one axiom for each basic function.

In order to avoid defining a large number of axioms, a default behaviour for the functions has been assumed. This allows us to not write axioms which match the default behaviour. More precisely we assume that functions commute. That is:

$$f(g(h)) = g(f(h));$$

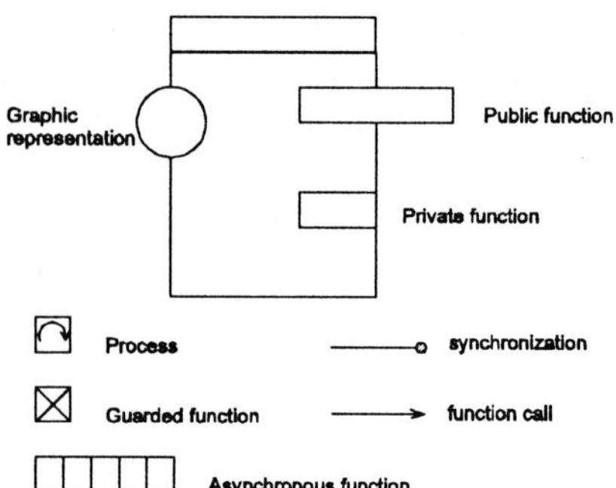

Fig. 1. Graphic notation to describe the system structure.

Previous work on implementing algebraic specifications has described methods to build an interpreter which accepts any expression formed with functions from the specification and gives its result. Such an implementation can be used only as a tester for the specification. GRALPLA has a one pass translator which accepts a specification module and generates its code in a high level language which can be used as an early system implementation. At present, a version generating C++ code for a preliminary version of the language is available.

Once we can ensure that the specification is correct, we can use our prototype as a first version for the design and implementation process. In this way, we can handle each class from the prototype and substitute it with a more efficient, hand coded, class. Using this method we can work with only one class at any time, carrying out the integration test for the new class with a full correct system, and thus ensuring that the next system version is also correct. In this way it is also easier to distribute the work in a team.

The overall structure of the system to be specified is described using the graphic notation of figure 1, where the box represents an object type. The notation shows the object graphic appearance, its public functions and functions involved on synchronization restrictions (feedback and interaction process).

4 Specifying interactive systems

The language incorporates the following notions to model interactive systems:

· *Asynchronous functions*, which are invoked following an asynchronous procedure call. This kind of functions allows us to represent the dynamic creation of processes.

· *Synchronization conditions*, using guarded operations expressed as preconditions.

· *Processes*, which are defined as functions that are processed forever.

We will now explain the semantics of these constructions. An *asynchronous* function uses an asynchronous calling mechanism. The caller process is not suspended when calling the function, it just provides the function input parameters to the function and continues its execution. The called module will perform the function at any moment in the future. Requests are attended in a sequential order. This kind of function must not have return parameters (except the own module type). The following signature:

asynchronous c-move: point, cursor -> cursor;

corresponds to an asynchronous function of the cursor object. When this function is called, its input parameter is stored in a pending request queue. There is only one queue for each module, in order to allow processing the event in temporal order. In some sense this is similar to a sampling method.

The *synchronization* section may impose execution restrictions to the functions defined in the module. A function appearing in the *do* part of a synchronization statement can be carried out only when the associated boolean expression is evaluated as true. Any request done while the expression is false will not be processed, until the condition becomes true. Note that this implies the suspension of the caller process if the called function is not asynchronous. When the condition becomes true the function is activated. Any assumption is done about the order in which activated functions are performed. For example, the sentence:

do cursor_pos(C) when is_moved(C);

declares that *is_moved* is the guard for *cursor_pos*. So any call to *cursor_pos* when *is_moved*(C) is not true will be suspended until the condition become true.

A *process* function expresses an independent activity within the module, which is performed without the direct intervention of any external agent. This kind of functions must have a precondition within the synchronization section, and can not have any input nor output parameter, except the module type. Process functions can be used to model operations which are performed as spontaneous actions within the object. For example:

process cursor_pos: cursor -> cursor;

declares that *cursor_pos* is a process. This function will be evaluated any time its guard is true.

A special operator (&) is used to mark the call to an asynchronous function within the specification. This is used to clarify the specification, as this call implies the creation of a new process.

The object state can now be represented by the story of basic functions performed on the object, with the sequence of pending operations.

5 Specifying graphic components

Graphic components play an important role in most interactive systems. Every kind of application needs different graphic renderings, which might change its appearance as a consequence of the interaction process. To use a formal approach for specifying an interactive system we need a formal framework for the specification of graphic objects.

We use a precise mathematical definition of graphic objects, defining object properties without imposing any representation. In this abstract view of graphic objects, we might change their representation (depending on the application structure) and the specification remains valid.

We can consider different abstract levels in the construction of objects. Graphic objects might be described by using two functions for inquiring their aspect and presence values. The *aspect* of a graphic object describes its visual appearance (colour, transparency, reflection coefficient, etc.), and other non-geometrical attributes (density, type of material, etc.). The *presence* of a graphic object describes its space occupance, that is the multiplicity with which an object occupies space. It is a countable property, and this allows us to contemplate the construction of a graphic object by the superposition or subtraction of other graphic objects.

A *graphic object*, in the universe $\mathcal{U}=(\pi,\delta,n)$, is a pair (μ,α), where:

μ is a function, called *presence function*, defined as μ: $\mathbb{R}^n \to \pi$

α is a function, called *aspect function*, defined as α: $\mathbb{R}^n \to \delta$

where π and δ are presence and aspect domain respectively[5].

We can use two different methods to describe the graphic representation of an object. The first one is to define its aspect and presence functions. For example, an arrow may be defined using these functions:

```
aspect(pt, arrow(p1,p2,a)) = if (is_in(p,arrow(p1,p2)) a else 0;

presence(pt, arrow(p1,p2,a)) = if (is_in(p,arrow(p1,p2)) 1 else 0;

is_in(pt,arrow(p1,p2,a)) =
    if (((pt.x = pt.y) and (p0.x ≤ p.x ≤ p1.x) and (p0.y ≤ p.y ≤ p1.y))
      or  ((p.y >= (p1.x - p.x) and ( 1/2(p1.x - p0.x)+p0.x ≤ p.x ≤ p1.x)
      and (1/2(p1.y-p0.y)+p0.y ≤ p.y ≤ p1.y))) true else false;
```

Note that in this example, aspect domain is irrelevant for specification details (indexed colour, grayscale, etc.). Graphic visualization is explicit, but we do not impose any restrictions on lower level details (rasterization, aspect ratio, etc).

Graphic object arrow;

arrow: point, point -> arrow;

Functions

move:	point, arrow	-> arrow;
highlight:	arrow	-> arrow;
hide:	arrow	-> arrow;
selected:	arrow	-> arrow;

Axioms

p1,p2, p: point
k: aspect;

is_in(p, move(p1,A)) = is_in(p-p1,A);
is_in(hide(A)) = false;

aspect(p, move(p1,A)) = aspect(p-p1,A);
aspect(highlight(A)) = k*aspect(A); // an arrow brighter, k>1
aspect(hide(A) = 0;

presence(p, move(p1,A)) = presence(p-p1,A);
presence(hide(A)) = 0;

show(hide(A)) = A;

Specification 1. Arrow.

When using this definition style, changes in the graphic component can be described adding axioms for aspect or presence functions (see specification 1).

On the other hand, the description of the graphic representation might be done using a graphic object as an abstract description. Graphic objects can be operated to define complex representations from simpler ones (graphic operator include addition, product by scalar, union, intersection, complement, object product, geometrical transformation[5]). Representation is denoted using the **graphic_rep** clause. Changes in the graphic representation are specified, in this case, as changes over the graphic representation used. **Graphic_rep** function is a process that performs a change in the representation when some other functions are called. In the cursor specification, we can impose that the graphic representation moves when the cursor moves. For instance, the single cursor object use the graphic representation of the arrow (specification 2).

This method for defining the graphic representation allows us to change the graphic representation in a simple way. For example, in the previous specification we can change the where clause

where graphic_rep(cursor(p1)) = cross_hair(p1);

Note that the remainder of the specification must not be changed. It is also possible to change the behaviour of graphic representation. For example when the object is highlighted

graphic_rep(highlight(C)) = F[yellow](graphic_rep(C));

where F[yellow](O) is a function assigning colour yellow to object O[6].

Graphic object with **dynamic behaviours** might control their visualization. They have an inherent function (*feedback*) describing *when* it is necessary to update the object visualization. **Feedback** is a process function that controls *when* the object representation is updated on the display, using a boolean expression as a guard. The visualization is updated whenever the condition has just changed or it is true. In the previous example, there are functions for hiding the cursor, so the visualization must be done only when it is visible.

Synchronization
do feedback(C) when not hide(C);

This approach characterizes graphic objects using only graphic functions and abstract representations, and it does not impose any concrete representation. In this way we define *how* it is represented (*graphic_rep*), and *when* it is actualized (*feedback*), this lends us greater flexibility in further stages of development.

6 Example

This section contains an example of the specification of a simple interactive graphic system. The example is a simple window system, which allows us to carry out *drag and drop* operation of an icon over a window using a cursor. Behaviour of the different elements are easily described, considering all the facets of the interaction process: change of visualization, synchronization and dialogue component.

Briefly, the behaviour of each element is as follows:

Cursor. The cursor follows the locator device movement. The activation trigger is the movement. The user might hide or show it, and inquire about its position.

Icon. The icon is a representation of an application object. Actions performed on the icon are reflected in the associated object. Its appearance might change when it is selected (triggering the mouse while locating the cursor onto it), and it might be dragged.

Window. A window represents a window with an associated command (delete, move, save, etc). The action is performed when the selected object is put on the window.

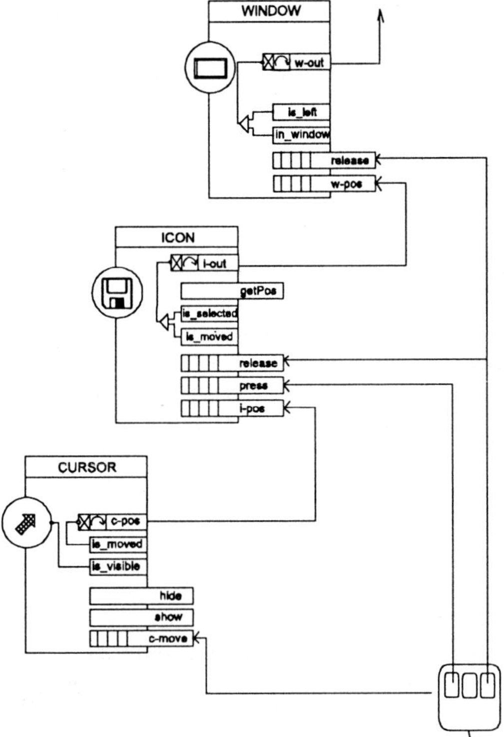

Fig. 2. Example system structure.

The overall structure of the example system is represented in figure 2 where connections between objects are also shown. The cursor follows the locating device movement; it does not change its rendering. Additional functionality is added to manage its appearance (show, hide), and inquiring about its actual measure (getPosition).

Asynchronous functions are used to receive events from the device. In the specification of the cursor type, one asynchronous function *cursor_move* has been defined which receives cursor displacement from the device. This function is a derived function, as it behaviour is defined with some axioms.

The process cursor_pos is responsible of sending the icon position to the upper level, whenever the cursor has been moved. Note that the process has a guard which implies that function is performed when the input device has been moved.

The rendering of the object has been defined as *arrow(P)* when the object is created, but it is updated using axioms 13 and 14. Axiom 13 specify that graphic representation when an addmove operation is the last one stored in the functional history is the translation by the given displacement of the original arrow. Axiom 14 establish that the graphic representation is hidden when the last operation performed is hide.

```
Parametric Graphic Object cursor [ A: graphic_object, f: func ( A, point -> A) ];

Import arrow;      // arrow is a graphic object describing the cursor graphic representation

cursor : Point -> cursor
        where graphic_rep(cursor(P)) = arrow (P);

Functions
asynchronous  cursor_move:   cursor, Point -> cursor;   // Receives locator inputs
process       cursor_pos:    cursor        -> cursor;   // Dispatches position
              hide:          cursor        -> cursor;
              show:          cursor        -> cursor;   // Set visibility on
              getPosition:   cursor        -> Point;    // gets the actual measure
private       addMove:       cursor, Point -> cursor;   // Records cursor movements
private       is_moved:      cursor        -> bool;     // Has it been moved
private       is_visible:    cursor        -> bool;     // Is it visible?
private       call:          cursor, graphic_object -> cursor; // To the upper level

Axioms
  var    pt, p:    Point
         C:   cursor;

cursor_move(cursor(pt),p)           = addMove(cursor(pt), p);   // Get device position
cursor_move(addMove(C,pt),p)        = addMove(C,pt+p);
cursor_move(hide(C))                = hide(C);                  // Don't change if hiden

cursor_pos(addMove(C,pt)) = call(cursor(pt+getPosition(C)), &f(A, pt));

show(hide(C))      = show(C);              // The effect of hide is reversed with show.
show(cursor(pt))   = cursor(pt);

is_visible(hide(C))          = true;
is_visible(cursor(pt))       = false;

is_moved(cursor(p))          = false;
is_moved(addMove(C,pt))      = true;       // There are pending moves to treat.

getPosition(cursor(pt))      = pt;
getPosition(addMove(C,pt))   = pt + getPosition(C);

graphic_rep(addMove(C,pt)) = move(graphic_rep(C),pt);
graphic_rep(hide(C)) = hide(graphic_rep(C));

synchronization
  do feedback(C) when is_visible(C);
  do cursor_pos(C) when is_moved(C);
```

Specification 2. Type Cursor.

```
Parametric Graphic Object Icon [ A: graphic_object, g: func ( A, point -> A)];

import bitmap;

   Icon : Point  -> Icon;

Functions
asynchronous icon_pos:     Icon, Point  -> Icon;   // Receives cursor inputs.
asynchronous press:        Icon         -> Icon;   // Receives button press events.
asynchronous release:      Icon         -> Icon;   // Receives button release events.
process      icon_out:     Icon         -> Icon;   // Dispatches position to the upper level.
             getPos:       Icon         -> Point;  // Retrieves current cursor position.
private      setPos:       Icon, Point  -> Icon;   // Changes the current cursor position.
private      putAt:        Icon, Point  -> Icon;   // Places the Icon at a point.
private      is_selected:  Icon         -> bool;   // Has it been pressed?
private      is_moved:     Icon         -> bool;   // Has it been moved?
private      active:       Icon         -> Icon;   // Attach the icon to the cursor
             call:         Icon, graphic_object -> Icon;   // Send data to the upper level

Axioms
   var   p,pt:    Point
         I:    Icon;

icon_pos(setPos(I,pt),p) =  setPos(I,p);               // Updates internal position
icon_pos(Icon(pt),p) =      setPos(Icon(pt),p);

press(setPos(I,p)) = if (is_in(I,p)) active(setPos(I,p));  // The icon is activated

release(active(I)) = I;                                // The icon is released

icon_out(setPos(I,pt)) = call(putAt(I,pt), & g(A,pt));  // Dispatches the icon movement

getPos(setPos(I,pt)) =      pt;
getPos(Icon(pt)) =          pt;

putAt(Icon(pt),p) = Icon(p);

is_selected(active(I)) =    true;
is_selected(Icon(pt)) =     false;

is_moved(setPos(I,pt)) =    true;
is_moved(Icon(pt)) =        false;

graphic_rep(Icon(pt)) = bitmap(pt);                    // Appearance described by the bitmap
graphic_rep(active(I)) = inverse(graphic_rep(I));
graphic_rep(setPos(p,I)) = if(is_in(p,I)) move(graphic_rep(I),p);
```

Specification 3. Type Icon.

```
Parametric Graphic Object Icon2 [ A: graphic_object, g: func ( A, point -> A)] :
                          Icon [ A: graphic_object, g: func ( A, point -> A)]:
                          // It inherits from Icon.

  Icon2 : Point  -> Icon2;

Axioms
  var    I:        Icon2;

press(active(I)) = I;                 // The press operation is used to release.

release(I) = I;                       // Release has no effect now.
```

Specification 4. Derived Icon type.

The feedback condition specifies that the graphic representation of the cursor on the output device must be updated whenever the object state has changed while the *is_visible* function is evaluated to true and when the value of this guard has changed. It can also be seen that some axioms have not been included, for instance:

is_visible(AddMove(pt,C)) = AddMove(pt,is_visible(C));

The definition of the type Icon is shown in specification 2. The icon has two renderings (normal and selected) which are changed when the cursor is over it and the locator device is triggered. It is affected by the cursor movement when it is selected and the cursor moves.

We can use inheritance to define a subtype of Icon. This subtype can change it behaviour, or extend it interface. As an example let us define a subtype of Icon for which the behaviour of the drag operation has been changed in the following way: the Icon is selected by pressing and is dragged until a new press operation is performed. For this case, it is necessary to redefine the behaviour of the release operation and to add a new axiom for the press function (see specification 3).

In order to handle error situations, we can include error functions. For example, suppose that we want to consider as an error the case when a press operation is performed been the cursor outside the Icon. To do this we can include the following error function in the Icon signature:

NotActivable: Icon -> Icon := error(warning, "Nothing to select");

which declares *NotActivable* as a warning error function, whose user meaning is that there is *nothing to select*. This function must be call when the user try to select and the cursor is not on the icon, so the axiom for press must be changed by:

```
press(setPos(I,p)) =  if (is_in(I,p)) active(setPos(I,p))    // The icon is activated.
                      else NotActivable(setPos(I,p));        // It is not activable.
```

```
Graphic Object window [A: graphic_object, f: func(A -> A);

Import rectangle, icon[window,window_pos];

window: Point, Point -> window;

Functions
asynchronous  window_pos:  window, Point  -> window;   // Icon position input.
asynchronous  released:    window         -> window;   // Device release actions.
process   window_out:      window         -> window;   // Dispatches to the application
private       in:          window         -> window;   // The icon is on the window.
private       is_left      window         -> boolean;  // Has it been released?
private       in_window    window         -> boolean;  // Is it on the window?
private       cleanEvents: window         -> window;   // Take off all the previous events.
private       action:      window, graphic_object -> window;  // Send messages to the
application

Axioms
    var  W:         window
         p1,p2,p:   Point;

window_pos(window(p1,p2),p) =      if(is_in(window(p1,p2),p)) in(window(p1,p2))
                                   else window(p1,p2);
window_pos(released(W),p) =        window_pos(W,p);   // The icon has been moved

window_out(W) = action(cleanEvents(W), &f(A));        // Notifies to the application.

is_left(released(W)) =       true;                    // A release has been processed.
is_left(window(p1,p2)) =     false;

in_window(in(W)) =                 true;         // The icon is on the window
in_window(window(p1,p2)) =         false;

cleanEvents(released(W)) =  cleanEvents(W);
cleanEvents(in(W)) =               cleanEvents(W);
cleanEvents(window(p1,p2)) =       window(p1,p2);

graphic_rep(window(p1,p2)) =       rectangle(p1,p2);
graphic_rep(in(W)) =               highlight(graphic_rep(W)); //Highlight if on window.

Synchronization
```

Specification 5. Type Window.

It is also possible to add axioms for NotActivable; this allows the specification of error recovery mechanisms.

The definition of type window is shown in specification 4. Note that the connection between object has been established using parameter functions, which provide a convenient way to define static connections. In this example, the cursor parameter must be fixed when an object is defined from type cursor. Assigning a parameter to the object fixes the topology (see import clause).

The window has two interaction dialogues. If the icon is dragged over the window, it changes the visualization (highlight), and if the mouse is released when the icon is over the window, the associated command is activated.

7 Verification of the specification

The verification of algebraic specification can be carried out using structural induction over the axioms. Gralpla has two specific characteristics which do not make verification straightforward: the use of graphic representation, and the definition of process and synchronization mechanism.

Within this section we will give a brief discussion on how to reason about the specifications[9]. The graphic representation of an object can be handled using functions (either graphic-rep or aspect and presence), whose value is a consequence of the object state.

To prove properties about the concurrent components, it is necessary to work with the state representation illustrated in section 3. That is, the object state can be symbolized as three sequences: performed functions (FH), pending asynchronous functions (PAF) and blocked function (BF):

$$S = (FH, PAF, BF) = (f_0\ f_1\ f_2 ...., g_0\ g_1 ..., h_0\ h_1\ ...)$$

Process functions with synchronization restrictions belong to the blocked functions (BF) sequence of the state (they will be performed when the guarded condition is true). When the process operation is performed, it is added again to the BF sequence. At any moment the first function in PAF (g_0) can be performed. This implies that the system change to a new state

$$S' = (FH', PAF', BF) = (g_0(FH), g_1 ...., BF)$$

where $g_0(FH)$ denotes the new functional history obtained applying, in the conventional sense, g_0 to the previous one.

Whenever the state changes, the preconditions of all blocked functions are checked, and if any of them has become true the function is performed. In this case the state changes to:

$$S'' = (FH'', PAF', BF') = (h_i(FH'), PAF', BF-h_i)$$

We can use this ideas to reason about the specification. We can, for instance, prove that one state can be reached from other one, or that one property (expressed as an equation) hold for a given state. It is also possible to prove lifeness properties for the specification[4].

We will now show this on one example: proving that it is always possible to reach the state *"send a cursor movement to the icon"* from the cursor creation, that is to say that cursor will call its parameter function f. This is equivalent to proof that there exists a sequence of functions that lets us reach a state on which this function is called. The initial state is:

S_0 = ({cursor}, { }, {cursor_pos})

for which the only functions are the constructor and the process cursor_pos, which is initially blocked. We will prove that there exist a sequence of actions performed by the user that let us reach a state on which a position is sent to the icon.

Let us suppose that the user performs a movement on the input device, then a call to *cursor_move* is executed. The state of the object will be:

S_1 = ({cursor}, {cursor_move}, {cursor_pos})

Then, the object must eventually perform the cursor_move operation, and as the guard for cursor_pos is not true, the only possible operation for the object is to carry out the cursor_move. Doing this, the object reachs the state:

S_2 = ({cursor_move(cursor(pt),p)},{ }, {cursor_pos})

which, from axiom 1 is equivalent to:

S_2 = ({addMove(cursor(pt),p},{ }, {cursor_pos}) =

=({addMove,cursor},{ }, {cursor_pos})

as addMove is a basic function.

The guard of cursor_pos is *moved(addMove(cursor))* which is now true. This implies that the process cursor_pos will be eventually performed, being the next state:

S_3 = ({cursor_pos(addMove(cursor(pt),p)},{ }, { })

and, from axiom 4, the new state will be

S_4 = ({call(cursor(pt+getPosition(C)), &f(A, pt))},{ }, { })

which implies calling to the upper level function *f*.

8 Conclusion

In this paper we have presented an algebraic specification language for specifying interactive graphic systems.

Our approach allows us to describe all the features involved in an interactive system in a unified way using only one specification method. Algebraic specification allows formal reasoning about properties, at different levels of abstraction.

The features of the language concerning the specification of the graphic and interaction components have been explained, and a simple, but complete, example has been specified.

It is possible to use a prototyping tool to generate code in a high level language which can be useful to evaluate the system, and as a design and implementation prototype.

The basis for the verification of specifications has been shown, and a proof example has been given.

The specifications obtained using Gralpla are more longest than those obtained with other specification languages (such as object-Z), but the mathematical concepts involved are simpler, and so the specification text is easier to read for a non expert. Besides this, the use of inheritance can used to derive simpler specification as subtypes of predefined ones. Future work will include the definition of library of types for user interface design.

References

1. Faconti, G.; Paterno, F.: *An Approach to the Formal Specification of the Components of an Interaction*. EUROGRAPHICS 1990
2. Paterno, Fabio; Faconti, G.: *On the use of LOTOS to describe Graphical Interaction*. HCI, 1992
3. Duce, D.A.: *Formal Specification of Graphics Software*. NATO ASI Series. Theoretical Foundations of Computer Graphics and CAD. Springer Verlag, 1988
4. J.C. Torres; B. Clares: *Using an Abstract Model for the Specification of Interactive Graphics Systems*. F. Paterno (Ed.): Design, Specification and Verification of Interactive Graphic Systems. Springer Verlag, 1994.
5. J.C. Torres; B. Clares: *Graphic Objects: A Mathematical Abstract Model For Computer Graphics*. Computer Graphic Forum, Vol. 12, N.5, 1993.
6. Torres, J.C.; Clares, B.: *A Formal Approach to the Specification of Graphic Object Functions*, Proceedings Eurographics 1994
7. M. Gea; J.C. Torres: *Object Oriented Prototyping of Graphic Application from Algebraic Specification*. Fourth Eurographics Workshop on Object Oriented Graphics, Sintra, Portugal, 9-11 May 1994.
8. Gutierrez, F.L.; Gea, M.: *Especificacion Formal de sistemas interactivos basados en interadores*. Proceedings CEIG, 1995

9. J.C. Torres; M. Gea; F.L. Gutierrez, M. Cabrera; M.Rodriguez: *The Gralpla Specification Language*. Dpt. Lenguajes y Sistemas Informáticos, Universidad de Granada, Spain. Report 96-1, 1996.
10. R. Duke; P. King; G. Rose; G. Smith: *The Object-Z Specification Language. Version 1*. Sotware Verification Research Centre. Deparment of Computer Science. University of Queensland. Australia. Technical Report N.91-1. May 1991.
11. J.A. Goguen; J.J. Tardo: *An introduction to OBJ: a language for writing and testing formal algebraic Program Specifications*. Software Specification Techniques. Addison Wesley, 1986.
12. G.P. Faconti; A. Fornari; N. Zani: *Visual Representation of Formal Specification: An Application to hierarchical Logical Input Devices*. DSVIS'94. Pisa. June 1994.
13. P.A. Palanque; R. Bastide: *Petri Net Based Design of User-Driven Interfaces Using Interactive Cooperative Objects Formalism*. DSVIS'94. Pisa. June 1994.
14. W. Mallgren: *Formal Specification of Interactive Graphic Programming Languages*. ACM Press. 1982.

Fusion Engines and Melting Pots

J.K. Hyde[1] and D.J. Duke[2]

[1]School of Computing Science, Middlesex University, Bounds Green Road, London, N11 2NQ, UK. email: joanne12@mdx.ac.uk
[2]Department of Computer Science, University of York, Heslington, York, YO1 5DD, UK. email: duke@minster.york.ac.uk

Abstract. Emerging multi-modal technology relies on innovative techniques for managing data, both at the user interface level, and at the internal application level. To assess design alternatives for this class of system, it is desirable to have models that focus attention on the critical features of these systems. Formal methods of software specification are known to provide this abstractive power in many contexts. This paper shows that an established specification technique can be used profitably to model, assess and improve the design of a generic kernel for multi-modal systems.

1 Introduction

Advances in technology are allowing new interfaces to be developed which can make use of novel input techniques, such as speech and gesture, and which can handle more than one modality of input at a time. Such systems are known as "multi-modal" [6]. However, these new forms of interaction require new models of processing of input. Coutaz and Nigay have developed a generic mechanism - called the "fusion engine" - for handling the input processing requirements of such systems. This has been used to implement a prototype application for multi-modal technology called the MATIS system [13, 12, 14]. In this paper we use formal specification techniques to examine and critique aspects of multi-modal fusion. The specification is based on published accounts of the fusion engine; however some simplifications have been made to keep the technical account simple and compact.

Formal methods are being increasingly used in software development in order to check the correctness and functionality of systems, but they have not yet been widely applied in checking system usability. One reason for this is that: "the concept of usability is too vague, founded on experimental psychology, to permit useful analysis or reasoning." [5]. Not only are usability criteria difficult to reason about in a formal context, but the actual concept of usability is still far from being clearly defined and is subject to much debate, and the work that has been done is not always immediately applicable to the design process. However, formal methods are potentially a very useful tool in examining system usability, since not only do they allow aspects of a system to be described in detail, and the implications to be examined whilst ignoring other issues through abstraction, they also allow a design to be changed and the modifications assessed. By using a formal notation to construct a model of a system, they allow the designer to gain an insight into the structures and relationships that are of importance, and to manipulate those relationships and examine the implications of change without needing an actual implementation of a system. This is not a claim that formal specifications are superior to algorithmic descriptions, but rather that they provide abstractions that are useful in representing and arguing about aspects of a system that ultimately affect the user interface.

To date, most work that has been done on the application of formal methods to multi-modal or other interactively sophisticated interfaces has focused on the interface or user task issues, for example [9]. Architectural support for modalities such as speech and gesture seem to require a model based explicitly on time, and so far the difficulty of using real-time techniques in this area has appeared to be incompatible with the insight that might be gained. However, the premise of this paper is that there are interesting generic aspects of multi-modal architectures that can be addressed independently of the complexities of real-time and of the details of a specific interface. In particular, we believe that an existing formal notation (Z - [15]) can provide a useful abstraction of the Fusion Engine, allowing examination of usability issues and design alternatives without the necessity for re-implementation.

It is not the purpose of this paper to debate the merits of particular specification languages. Z was chosen as it is well documented, and supports the kind of structured, abstract model of state that we believed would be most helpful in describing the behaviour of the fusion engine. Other notations, for example a process algebra, could also be used, but would highlight other aspects of the system. Section 2 of the paper examines the role and operation of the fusion engine implemented in MATIS. A Z specification of fusion, based on the the MATIS engine, is then developed in Section 3. This specification is assessed in Section 4 by means of certain usability criteria. A variation on the original specification, designed to solve some of the usability problems highlighted, is then outlined and assessed in Section 5.

2 The Matis System

Multi-modal systems make use of more than one device for the input of data to the computer by the user, and allow the use of modalities in parallel and sequential combinations. The wider range of modalities brings many benefits. It allows the system to have a better match with the abilities of the user, in that disabilities can be circumvented by the use of alternative modalities. It also frees the user from the artificial restrictions imposed by the use of single modalities, allowing the utilisation of more natural and expressive methods of communication, for instance pointing at parts of the display. Interaction can be made more precise and succinct, in that humans can use one modality to specify an object, and another modality to describe the operation to be performed. So not only do multi-modal systems allow for the differing abilities and needs of users, they also allow for a more effective communication via the interface. Indeed, they have great potential in many areas, and [5]: "...can make it feasible to use software in situations where more austere interface technology would be either unhelpful or unusual, for example during surgical procedures or for control of advanced aircraft." However, using more than one modality to input data can be problematic. The key issue is understanding how the computer should attempt to combine data from the different modalities. The two main concerns are temporal: specifically, the correct correlation of inputs against time; and contextual: the correlation of inputs in terms of their content. These are not distinct problems, since the assignment of contextually related information can be influenced by the time of input, and the assignment of temporally related information can be affected by the context.

MATIS is a system which allows users to retrieve information from a database about flights using speech, keyboard, mouse, and direct manipulation. These modalities can be used individually, for example by saying: "show me the USAir flights from Boston to Denver", or in combination, for example by saying: "show me the USAir flights from Boston to this city" whilst selecting Boston with the mouse at the same time. The mechanism which Nigay and Coutaz have developed to support MATIS is known as the Fusion Engine. This utilises "melting pots", matrices with domain information on one axis, and time along the other.

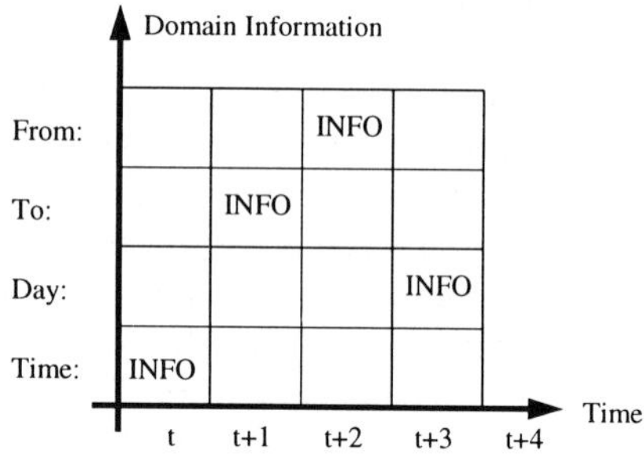

Fig. 1. *A Melting Pot in MATIS.*

When the user selects a destination, such as Denver, the information is mapped into a cell (or column) of the matrix or melting pot. The 'row' is determined by the field on the query that the selected information refers to, while the 'column' is defined by the time at which the information was accepted by the system. If the user then went on to select a departure point, such as Boston, a new melting pot would be created, and the two melting pots would be combined together to form a new melting pot which contained both departure and destination information.

Melting pots are combined, or fused, if they were produced in the same time segment and the contents of the pots do not clash, or where the contents of the melting pots are complementary (independently of time). Time and context provide a basis for the assignment of melting pots for fusion, and for resolving potentially redundant information. Nigay and Coutaz have identified three different types of fusion: microtemporal, macrotemporal, and contextual fusion. Microtemporal and macrotemporal fusion both rely on time values to give a framework for the fusion process. With microtemporal fusion, melting pots that are produced concurrently are fused together as belonging to the same temporal event horizon, provided that their contents do not preclude this. With macrotemporal fusion, melting pots created within a certain time span are fused, provided that their contents do not clash. Only with contextual fusion does time play no part, and there fusion is completed after the melting pots have been examined to make sure that their components are complementary and do not clash.

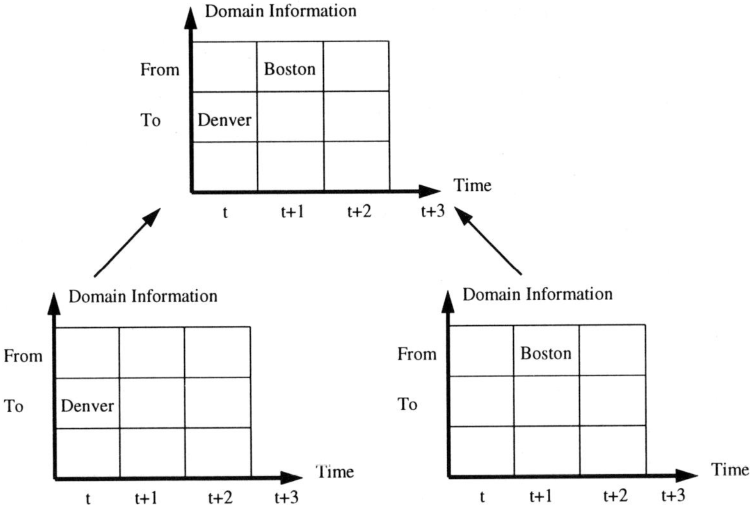

Fig. 2. *Fusion of two Melting Pots.*

The fusion engine utilises an eager fusion strategy, in that the system will immediately attempt fusion, regardless of whether or not there is any information still being processed. Information is occasionally assigned to a melting pot prematurely because of this, since other information input before, but processed later, means that a different context is placed on the fusion, for example a (temporally) later mouse input can often be processed before speech. In these cases, the original fusion, for example of the mouse input with existing melting pots may have to be undone, and the information re-assigned taking into account the newly processed speech.

When a user inputs the same information at the same time, but using different modalities, one of those inputs is said to be redundant, since the information is duplicated. The fusion engine will consider the duplicated input as redundant only in the case where the inputs were provided in parallel. Sequential redundancy would allow a duplicate input made later than the original to be considered redundant. The fusion engine as implemented in MATIS does not allow this, and effectively only supports parallel redundancy.

3 Specification of a Fusion Engine

The specification in this section describes a fusion engine based on the one implemented within MATIS. For clarity, some details of the system described by Nigay [12] are omitted; for example it is assumed that any input containing more than one piece of information will be decomposed into multiple melting pots. This has been done to simplify the description of the fusion engine, and does not reflect the structure of its implementation. Fission of inputs into columns is not covered by this specification. The section begins with a description of the basic types used by this specification, and gradually builds up the specification from a column to a melting pot to a fusion engine level. A description of the actual implementation of the MATIS engine can be found in [12].

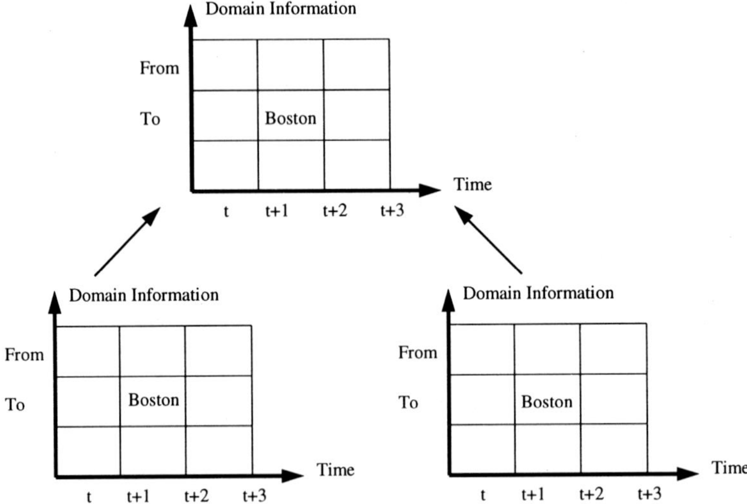

Fig. 3. *Redundancy in MATIS.*

3.1 Definition of given types and column

As discussed above, the MATIS fusion engine uses matrices known as melting pots as its basic unit for fusion. In this specification we have chosen to regard melting pots as collections of columns. A *column* is modelled as a schema with three variables: slotname, value and time.

- '*Slotname*' identifies the domain-dependent role of a piece of information recorded within the system, and is defined as being of the given type '*Field*'. For a given application, it is possible to define the extension of this type. In the case of MATIS, field would designate the attributes of a query, for example, destination, cost, and meal, etc.

- '*Value*' represents the specific information present in a field and is assigned the type '*Info*'. For MATIS, the extension of this type is the set of possible values any attribute of a query might hold, for example, 'Pittsburgh', 'Denver,' '$53.00' etc.

- '*Time*' in this specification represents the time at which the column was created within the fusion engine, rather that the time of entry by the user as in the MATIS implementation. Here we have chosen to model time using a given type '*Number*'. The type *Number* is not identified with the natural numbers, in order to avoid assumptions about how time is represented within an implementation.

The *Column* is therefore defined as a piece of information which has been placed in a designated field and associated with a time value, and is represented in the specification by the following schema:

$$
\begin{array}{|l}
\hline
\textit{Column} \\
\hline
slotname : Field \\
value : Info \\
time : number \\
\hline
\end{array}
$$

3.2 The Holding Pot

The 'holding pot' is an explicit rendering of part of the MATIS implementation that is usually implicit. It represents the buffer that exists between the various input devices, and the fusion engine itself. Use of this structure in the specification is intended to aid in reasoning about the workings of the fusion engine, and is discussed further in [11]. When a column is first produced, it is assigned to the holding pot, which is a sequence of *Columns* in time order, most recently produced last, that are awaiting fusion with the melting pots. The structure of the holding pot is introduced below, as a schema. Initially the holding pot is empty.

$$
\begin{array}{|l}
\hline
\textit{HoldingPot} \\
\hline
holdpot : \mathrm{seq}\ Column \\
\hline
\end{array}
\qquad
\begin{array}{|l}
\hline
\textit{HoldingPot}_{init} \\
\hline
HoldingPot \\
\hline
holdpot = \langle\rangle \\
\hline
\end{array}
$$

Before the column is added to the holding pot, the pot is checked to see that there are no other columns in the pot with exactly the same attributes. If there is, then the new column is not added to *holdpot*, as it is considered to be redundant information. This way columns can be checked for redundancy at the very start of the fusion process. If there is not an identical column already in the holding pot, then the column is placed in the holding pot in time order, and is now ready to be fused with an existing melting pot, or be assigned to a new melting pot. If multiple columns share the same time, then the column that is being added to the holding pot is placed in sequence behind those with which it shares the same time. This is an arbitrary choice, and the new column could just have easily been placed ahead of those columns. The operation of adding a column to the holding pot is defined below:

$$
\begin{array}{|l}
\hline
\textit{AddtoHoldingPot} \\
\hline
\Delta HoldingPot \\
newcol? : Column \\
first, second : \mathrm{seq}\ Column \\
\hline
newcol? \in \mathrm{ran}\ holdpot \\
first = holdpot \upharpoonright \{c : Column \mid c.time \geq newcol?.time\} \\
second = holdpot \upharpoonright \{c : Column \mid c.time < newcol?.time\} \\
holdpot' = first \frown \langle newcol? \rangle \frown second \\
\hline
\end{array}
$$

3.3 The Melting Pot

Now that column has been specified, it is possible to define the melting pot. In MATIS, a melting pot is defined as a matrix, with two axes, one representing the information recorded about a query, and the other representing time. In this specification it is represented as a set of columns, in which no column may share the same slotname with another column. The melting pot is therefore defined in the specification by the following schema:

$$
\begin{array}{|l}
\hline
\textit{MeltingPot} \\
\textit{meltpot} : \mathbb{P}\ \textit{Column} \\
\hline
\forall c_1, c_2 : \textit{meltpot} \bullet c_1 \neq c_2 \Rightarrow c_1.\textit{slotname} \neq c_2.\textit{slotname} \\
\hline
\end{array}
$$

The different types of fusion are given below, in terms of *HoldingPot* and an arbitrary melting pot. These 'localised' specifications will later be incorporated into the fusion engine as a whole by means of framing schemas.

3.4 Contextual Fusion

Contextual fusion is straightforward to represent, since it does not depend on any temporal considerations, but instead relies on examining the information present in the melting pot. The column at the front of the holding pot has its slotname compared against the slotnames of the columns in the melting pot. If there is no column present in the melting pot with that slotname, then the new column is removed from the holding pot and is added to the melting pot. This operation is represented in the specification by the schema *ComplementaryFusion*, in which the new column *newcol?* is the column at the front of the holding pot. Provided that its slotname does not match that of a column already in the meltpot, the new column is removed from the holding pot and placed into the melting pot.

$$
\begin{array}{|l}
\hline
\textit{ComplementaryFusion} \\
\Delta \textit{HoldingPot} \\
\Delta \textit{MeltingPot} \\
\textit{newcol?} : \textit{Column} \\
\textit{old} : \textit{Column} \\
\hline
\textit{newcol?} = \textit{head holdpot} \\
\textit{newcol?}.\textit{slotname} \notin \{c : \textit{meltpot} \bullet c.\textit{slotname}\} \\
\textit{meltpot}' = \textit{meltpot} \cup \{\textit{newcol?}\} \\
\textit{holdpot}' = \textit{tail holdpot} \\
\hline
\end{array}
$$

3.5 Redundancy

If the slotname of the new column is already matched by a column present in the melting pot, time becomes a factor in the behaviour of the engine. If the new column shares the same slotname and the same value as an existing column, and the time of the new column is the same as that of the old column, then the new column is considered redundant and is removed from the holding pot, but is not added to the melting pot.

Redundant

$\Delta HoldingPot$
$\Delta MeltingPot$
$newcol? : Column$
$old : Column$

$newcol? = head\ holdpot$
$(newcol?.slotname, newcol?.value) \in \{c : meltpot \bullet (c.slotname, c.value)\}$
$\{old\} = \{c : meltpot \mid c.slotname = newcol?.slotname\}$
$newcol?.time = old.time$
$holdpot' = tail\ holdpot$

The new column is checked to see if it shares slotname and value with a column already in the melting pot. If such a column exists in the melting pot, and the time associated with the 'old' column is the same as that of the new, then the new column is not added to the melting pot, but is instead removed from the holding pot as redundant information.

3.6 Non-Redundancy

Two cases need to be considered. First, if the new column shares the same slotname and the same value as an existing column, and the time of the new column is *later* than that of the existing column, then the new column is taken to be not associated with that melting pot and is instead assigned to a different melting pot, as forming part of another query. Therefore it is the same information but a new query, and is represented by the schema *SameButNew*. The crux of this schema is that although the new column shares the same attributes of slotname and value as an existing column, the new column has a greater time value than the existing column, therefore showing that the new column was produced later.

Secondly, if the new column shares the same slotname but *not* the same value as an existing column, and the time of the new column is later than or equal to that of the old column, then the new column is taken to be not associated with that melting pot, because of the difference in values, and is instead assigned to a different melting pot, as forming part of another query. This is represented by the *NewQuery* schema. The essential point of this schema is that although the new column shares the same slotname as an existing column, the value of the new column is different, so therefore must be assigned to a different melting pot, provided that the new column was not created earlier than the existing column. If a column had been created earlier than the existing column, that would necessitate the undoing of fusion already accomplished.

$$
\begin{array}{|l}
\hline
\textit{SameButNew} \\
\Delta HoldingPot \\
\Delta MeltingPot \\
newcol? : Column \\
old : Column \\
\hline
newcol? = head\ holdpot \\
(newcol?.slotname, newcol?.value) \in \{c : meltpot \bullet (c.slotname, c.value)\} \\
\{old\} = \{c : meltpot \mid c.slotname = newcol?.slotname\} \\
newcol?.time > old.time \\
holdpot' = tail\ holdpot \\
\hline
\end{array}
$$

$$
\begin{array}{|l}
\hline
\textit{NewQuery} \\
\Delta HoldingPot \\
\Delta MeltingPot \\
newcol? : Column \\
old : Column \\
\hline
newcol? = head\ holdpot \\
newcol?.slotname \in \{c : meltpot \bullet c.slotname\} \\
(newcol?.slotname, newcol?.value) \notin \{c : meltpot \bullet (c.slotname, c.value)\} \\
\{old\} = \{c : meltpot \mid c.slotname = newcol?.slotname\} \\
newcol?.time \geq old.time \\
holdpot' = tail\ holdpot \\
\hline
\end{array}
$$

Each of these operations involves the holding pot and some melting pot, and will shortly be placed into the overall context of the fusion engine.

3.7 Undoing fusion

Depending on which modality was used to input data, and because of the delays in processing columns, occasionally a column may be fused with a melting pot, and another column, which was created earlier but delayed due to processing problems, might actually be the one that should be fused there. If this happens, the fusion has to be undone. In this operation, the new column is checked to see if it has the same slotname as an existing column. If it has, and the existing column's time is later than that of the new column, then the old column is removed from the melting pot and reassigned to the holding pot. The new column is now fused with the melting pot. All the other columns can stay in the melting pot, until any new columns with the same slotname arrive with earlier times to challenge their position. This undoing of the eager fusion is represented in the specification by the *NewQuery* schema. The new column is checked to see if a column sharing it slotname and value already exists in the melting pot. If so, the new column is checked to see which column was produced earlier. If the new column was produced earlier than the old column, then the old column is removed from the melting pot and is placed in the holding pot. The new column is removed from the holding pot and is placed in the melting pot.

$$
\begin{array}{|l}
\textit{NewQuery} \\
\hline
\Delta \textit{HoldingPot} \\
\Delta \textit{MeltingPot} \\
\textit{newcol?} : \textit{Column} \\
\textit{old} : \textit{Column} \\
\textit{xmelt} : \mathbb{P}\ \textit{Column} \\
\hline
\textit{newcol?} = \textit{head holdpot} \\
\textit{newcol?.slotname} \in \{c : \textit{meltpot} \bullet c.\textit{slotname}\} \\
\{\textit{old}\} = \{c : \textit{meltpot} \mid c.\textit{slotname} = \textit{newcol?.slotname} \\
\qquad\qquad\qquad\qquad \wedge\ c.\textit{time} = \textit{newcol?.time}\} \\
\textit{newcol?.time} < \textit{old.time} \\
\textit{xmeltpot} = \textit{meltpot} \setminus \{\textit{old}\} \wedge \textit{meltpot}' = \textit{xmeltpot} \cup \{\textit{newcol?}\} \\
\textit{holdpot}' = \textit{tail holdpot} \frown \langle \textit{old} \rangle \\
\hline
\end{array}
$$

3.8 The Fusion Engine

The Fusion Engine is represented as a *HoldingPot* and a collection of 'pots', which are *MeltingPots* assigned names. The given type '*Handle*' represents the names of individual melting pots. The Fusion Engine is represented by the following schema:

$$
\begin{array}{|l}
\textit{FusionEngine} \\
\hline
\textit{HoldingPot} \\
\textit{pot} : \textit{Handle} \nrightarrow \textit{MeltingPot} \\
\hline
\end{array}
$$

Although melting pots have been defined and used, and the operations which affect them specified, the creation of a *MeltingPot* has not yet been defined. This is represented by the schema given below.

$$
\begin{array}{|l}
\textit{CreateMeltPot} \\
\hline
\Delta \textit{FusionEngine} \\
\textit{newcol?} : \textit{Column} \\
\hline
\exists\ \textit{MeltingPot};\ \textit{newpot} : \textit{Handle} \mid \\
\qquad \textit{newpot} \notin \operatorname{dom} \textit{pot} \\
\qquad \textit{meltpot} = \{\textit{newcol?}\} \\
\qquad \textit{pot}' = \textit{pot} \cup \{\textit{newpot} \mapsto \theta \textit{MeltingPot}\} \\
\hline
\end{array}
$$

There are two occasions when a new melting pot is needed. The first is where the *SameButNew* schema has been used, and the second is when the *NewQuery* schema has been used. These are equivalent to situations in MATIS such as a user entering data that is identical to an earlier input, and the user entering data which fills the same slotname, but has a different value, to an earlier input. The creation of a new melting pot is fully defined by the schema on the left, while the schema on the right is a 'framing' schema that is shortly used to promote the remaining operations.

$$\begin{array}{|l}
\hline
\textit{NewMeltPot} \\
\textit{CreateMeltPot} \\
\textit{NewQuery} \\
candidate : Handle \\
\hline
pot(candidate) = \theta Meltpot \\
pot'(candidate) = \theta Meltpot' \\
\hline
\end{array}$$

$$\begin{array}{|l}
\hline
\textit{NewMeltPot2} \\
\textit{NewMeltPot} \\
\hline
\{candidate\} \lhd pot \\
\quad = \{candidate\} \lhd pot' \\
\hline
\end{array}$$

3.9 Promotion Schemas

The $NewMeltPot2$ schema describes a generic operation that affects the holding pot and a specific melting pot. This is used to promote the operations defined in terms of one melting pot and the holding pot up to the level of the system as a whole. This is achieved by conjoining each of the operations $ComplementaryFusion$, $UndoFusion$, and $SameButNew$ with the frame. The final operation, $Redundant$, is slightly different as it does not change any melting pot.

$$FuseDate1 \mathrel{\widehat{=}} ComplementaryFusion \wedge NewMeltPot2$$
$$FuseDate2 \mathrel{\widehat{=}} SameButNew \wedge NewMeltPot2$$
$$FuseDate3 \mathrel{\widehat{=}} UndoFusion \wedge NewMeltPot2$$
$$FuseData4 \mathrel{\widehat{=}} Redundant \wedge [\, \Delta FusionEngine \mid pots' = pots \,]$$

As mentioned earlier, this model differs slightly from the Fusion Engine implemented in MATIS. Here for example it is assumed that there is only one pending melting pot with which fusion is attempted, identified by $candidate$. In the full implementation there may be several candidate pots with which fusion is attempted before a new melting pot is created. However, the key aspects of the Fusion Engine that are relevant to this paper have now been specified using Z. Subsequent sections now address the analysis of fusion engine mechanism via this specification.

4 Assessment of the Specification

The newness of the technology enabling multi-modal systems means that there has so far been little systematic research into the usability of multi-modal systems. Thus it is very difficult to reason about the usability of such systems. The most directly relevant usability research into multi-modal systems so far is embodied by the CARE properties [7], which were developed to be used as a starting point for determining the usability of multi-modal systems. CARE properties are ways of describing and assessing aspects of multi-modal interaction, in terms of the relationships between states and modalities. There are four properties: Equivalence, Assignment, Redundancy, and Complementarity. For a fuller discussion of the properties the reader is recommended to [7], but for the purposes of this paper a brief description is attached below:

Equivalence: this is when there is a full choice of modalities open to the user. No one modality is dominant, and any modality available can be used by the user without restriction.

Assignment: this is where there is no choice between which modality should be used for a specific task, either because there is no modality alternative, or the system imposes a specific choice.

Redundancy: This is where equivalent modalities can be used at the same time to convey the same information.

Complementarity: This is where modalities can be used together to achieve a goal, such as indicating a piece of information with a mouse, and specifying the operation to be performed on this information by speech.

Space precludes a detailed analysis of the specification in terms of these properties. However, a brief assessment is set out below.

Equivalence: In the definition of columns and in the operation of the fusion engine described in the specification, there is no restriction on the use of modalities; any modality can be used to input data into the system. Indeed, there is no explicit reference to modality within the specification, since the system makes no distinction between modalities. Therefore the system conforms to the equivalence criteria.

Assignment: As far as this specification is concerned, there is no restriction on which modalities must be used, other than that imposed by the user's preferences. As shown above in equivalence, the user can use any modality to enter information into the system, and since the system makes no distinction between modalities, it does not restrict the user to any specific modality for entering that information.

Redundancy: This is handled by this specification, but only within certain limits. In the *Redundant* schema, it can be seen that only if the time value of the new column ($newcol?.time$) is exactly equal to the time value of the existing column ($old.time$) is the new column considered as redundant and disregarded. Otherwise, if the new column has a lesser time value, (i.e. it was produced earlier than the column already in the melting pot) the existing column is removed from the melting pot and the new column added. If the new column has a later time value, then it is assigned to another melting pot as it is thought to be part of a different query.

An example of a problem linked to redundancy is that a user may have a tendency to re-iterate what they have already said whilst inputing new information. They therefore might inadvertently begin a new query in the system which they do not need. If the user does not realise this, it may result in confusion and frustration. For example, consider a user who has entered the information "Denver" by means of the keyboard into the destination field of the query. If the user then uses the speech input device and says: "Please give details of all flights to Denver from Pittsburgh", the Denver part of the input is redundant. However, the system would be unaware of this, since it was not created at exactly the same time as the other "Denver" input, and would therefore assign the whole of the speech input to a different melting pot. This is potentially a major problem. The fusion engine as it stands therefore can only handle parallel redundancy.

Complementarity: Two or more modalities can be used together without fusion restrictions, provided that they obey the temporal constraints as laid out in the specification and mentioned above in the discussion on redundancy.

With regards to the CARE properties, it seems that the fusion engine fulfils the requirements for Assignment and Equivalence. However, with regards to the properties of Redundancy and Complementarity, the fusion engine could be improved. One of the main problems concerns the size of the temporal window in the redundancy operation. If the window is too small, then users are not able to be natural in their use of modalities and input, and if it is too large, then data might be lost because the system believes them to be redundant when in reality they form part of new and unrelated queries. There is also the question as to whether or not the user might actually realise from any feedback on the display that incorrect fusion has taken place.

The other problem uncovered by CARE in this analysis is that of the fusion of complementary inputs, and again this is linked to how the fusion engine handles the time aspect of the system. Time is used to synchronise complementary inputs, so if an input happens outside the temporal window allowed by the system, it may not be assigned to a query in the way that the user expects. (In fact, the implementation of MATIS allows the temporal window to be adjusted to suit users' needs. We do not attempt to critique this mechanism here.)

Although the CARE properties highlight the fact that redundancy is important, they give little indication of how it might be managed in order to achieve a sensible fusion algorithm, when fusion should occur, and whether or not an "eager" fusion policy is of greater benefit than a "lazy" policy. Therefore there is a need to assess the specification using other usability criteria. It must be remembered that the fusion engine is not a self-contained artefact - it is part of an architecture that supports a multi-modal user interface. Although the engine is not explicitly part of this interface, some components of the system are reflected in the presentation. For example, the implementation of MATIS associates melting pots with the 'query' forms that the user manipulates. For this reason, properties of the fusion mechanism as described in the preceding sections can be perceived by the user. In [2], Abowd sets out a catalogue of usability properties linking internal and perceivable state, against which systems can be examined. Four of these properties - Observation, Presentation, Correspondence and Predictability - are selected here as a basis for further reasoning about the usability of the fusion engine.

Observation: This has been defined by Abowd [2] (pg 41) as: "...trying to determine how individuals understand that which they perceive." This is an elusive subject, but some work on this has been accomplished by Barnard and May in their development of the ICS model [4], as to how the user processes visually displayed information. The main point is that the output should be of a form that the user can easily understand, although it is still not entirely clear how that might be accomplished.

Presentation: This is the relationship between internal System representations and Output representations, the relationship between what is, and what is displayed on the screen. As Abowd states [2] (pg 59): "Lack of ambiguity between distinct *System* states and the *Output* is at least a necessary condition for overall predictability and proper goal assessment."

Correspondence: This is defined by Abowd [2] (pg 62), to mean: "a correspondence between their external stimuli and responses and the internal states to which they are linked in their agent representation."

Predictability: This measures the ease with which the user can predict how their actions will affect reaching their desired goal. As Abowd states [2] (pg 70): "A predictable system is one in which it would be possible for the user to internalise a model which would be of benefit to future interaction." Abowd models this as an absolute, as in the system either is or is not predictable, and therefore makes no allowance for semi-complete user models of the system.

We now apply these criteria to the fusion engine as described in the specification:

Observation: There is a problem with the usability of the system due to the use of eager fusion. Fusion must sometimes be undone, because columns have been wrongly assigned to melting pots. Although eager fusion has the advantage of providing immediate feedback, the direct correlation between the internal state and display can mislead and confuse the user. Therefore there is a trade-off between the user's desire for immediate feedback, and the user's desire for consistent feedback. In some situations, eager fusion would not be a problem, but in others, such as safety-critical situations it may be essential that feedback is always correct. The existence of this issue does not imply that the Fusion Engine mechanism is flawed, but simply that it may be useful to consider alternative approaches to presenting the state of the system, in order to highlight potential uncertainties.

Presentation: If each column represents one piece of information entered into the system by the user, and each melting pot represents a query displayed on the interface, there will be a one-to-one correlation between what is shown on the screen and what is in the system, because of this relationship between the fields in the query and the columns in the melting pot. The display will therefore accurately reflect the current status of the system.

Correspondence: Just as there is a correlation between what is shown on the screen and what is present in the system, so there is a correlation between what happens to the interface and what happens to the internal workings of the fusion engine. When new information arrives at the fusion engine, a new column is created in the system and is assigned to a melting pot, and information appears in a query in the interface. In other words, all changes in the presentation are reflected in the internal state, and vice versa.

Predictability: One problem here revolves around the issue of redundancy. Because of the way in which the specification currently handles redundancy, the user might not be aware that information input into the system only a very short time after other identical information may be assigned to a new melting pot/query. The other problem is that posed by the use of an eager fusion strategy. Because of the way that the system will attempt to fuse columns immediately without any time delay to allow other columns input earlier to catch up, misleading information may appear on the screen, reflecting the fact that a column has been wrongly assigned to a melting pot. Thus the user may find it difficult to predict how the system will respond in any given scenario, because of this capacity for erroneous output. Even if the system corrects itself within a very short period of time, using the *UndoFusion* operation, the user may become confused and may make incorrect assumptions about the internal workings of the system. At the very least, changes in the display brought about by the undo mechanism may make it difficult for the user to develop a coherent mental model of the operation of the system.

By applying a collection of usability criteria to an abstract model of the fusion engine, we have been able to argue that the fusion engine is not as user-friendly as might be hoped. The value of the specification in this process was that the important elements of the engine's behaviour could be captured, while ignoring other issues. For example, the specification describes the *potential* for the system to undo fusion, without resorting to operational details of precise timing or efficient implementation. From the analysis above we have identified a number of problems, including the way the Fusion Engine handles redundancy, and its use of an eager fusion strategy. Although the eager fusion strategy may be beneficial to the user in providing immediate feedback, the properties of Predictability and Observation are compromised by its use. The treatment of redundancy by the system is not ideal, and at present affects the Predictability, Redundancy and Complementarity properties.

5 Improving the Fusion Engine

The analysis in the previous section has identified a number of weaknesses in the fusion engine. Having identified these problems, we now consider means of eliminating or at least minimising their impact on the usability of the system. Critically, this variation can be described within the same specification framework as the original analysis. That is, we avoid the necessity for building a new specification (or worse, a new implementation) by phasing design changes into the existing description. By reusing the specification in this way, we both capitalise on the analyst's existing familiarity with the model, and amortise the cost of building the original model. The aim of the variation is to minimise the use of the *UndoFusion* schema by introducing a time delay before fusion was attempted, thus changing the "eager" fusion strategy to a "lazier" fusion strategy. It also attempts to resolve the issue of redundancy by making it possible for columns created after the original column to be considered redundant.

Before a column is fused in the *ComplementaryFusion* schema, it is checked to see if a certain time identified as 'interval' has passed. The inclusion of an interval gives other columns that might have been created earlier but were delayed by system processing the chance to 'catch up' and be allocated to melting pots earlier. The suggested 'delay interval' achieves the required effect by allowing columns to be placed into the correct order within the holding pot, before being assigned to a (potentially incorrect) melting pot. This results in less erroneous feedback to the user because there is less erroneous fusion to undo. Thus the system no longer uses a strictly "eager" fusion strategy and instead utilises a "lazy", or at least, "lazier" fusion strategy. This is represented in the new schema below:

$$\begin{array}{|l}
\hline
\textit{ComplementaryFusion}_{alt} \\
\textit{ComplementaryFusion} \\
\textit{interval} : \textit{Number} \\
\textit{currentTime} : \textit{Number} \\
\hline
\textit{currentTime} - \textit{newcol?.time} \geq \textit{interval} \\
\hline
\end{array}$$

The new column is checked to see what time it was created, and whether the interval time has passed. The time since the creation is checked by taking the *currentTime* of the system, and subtracting the column creation time (*newcol?.time*). If the remainder is equal to or bigger than the *interval* value, then the interval time has passed and the new column can be fused. If the interval time has not yet passed, then the column remains in the holding pot. If the interval time has passed, then the column is fused according to the usual criteria. Since there is little change to the original *ComplementaryFusion* schema except the addition of the variables *interval* and *currentTime*, and the line

$$currentTime - newcol?.time \geq interval,$$

the old schema is included within the new *AltComplementaryFusion* schema.

Although Z does not necessarily allow for the easy handling of time within schemas, this problem has been bypassed by using variables, and not defining how these variables acquire their values. This is something that can be extensionally defined at an implementational level. However, for the purposes of reasoning about the effect of a delay on the system the use of variables as above is quite adequate, and mirrors other approaches to time found in the specification literature, for example [1].

With the *Redundant* schema, the new column is checked to see if it shares the same slotname and value with an existing column in the melting pot. If this is so, and the new column was created within a specified duration of time, or *window*, then it is taken to be redundant, and merely a repetition of information rather than comprising information for a new query.

In the original *Redundant* operation, only columns that shared the same slotname, value and time would be considered redundant. The modified operation, given below, treats columns that share the same slotname and value, and were created during a specified *time window* as redundant. That is, the definition of redundant information has been expanded to include columns that were created not only at the same time as the original column but also shortly afterwards. Fusion of the new column is not attempted until a certain time *interval* has passed, as in the *AltComplementaryFusion* schema, thus allowing columns delayed by input processing to be placed in an appropriate position within the Holding Pot.

$$\begin{array}{|l}
\multicolumn{1}{l}{Redundant_{alt}} \\
\Delta MeltingPot \\
\Delta HoldingPot \\
old, newcol? : Column \\
interval, window : Number \\
\hline
newcol? = head\ holdpot \\
currentTime - newcol?.time \geq interval \\
(newcol?.slotname, newcol?.value) \in \{c : meltpot \bullet (c.slotname, c.value)\} \\
\{old\} = \{c : meltpot \mid c.slotname = newcol?.slotname\} \\
newcol?.time - old.time \leq window \\
holdpot' = tail\ holdpot \\
\hline
\end{array}$$

If the new column shares the same slotname and value as an existing column, but was created outside the 'window' of time, then it is taken to form part of a new query. This is in contrast to the original specification, where any column not produced at exactly the same time as a column already in the melting pot would be considered as forming part of a new query. The modified operation is given below.

$$\begin{array}{|l}
\multicolumn{1}{l}{SameButNew_{alt}} \\
\Delta HoldingPot \\
\Delta MeltingPot \\
old, newcol? : Column \\
interval, window : Number \\
\hline
newcol? = head\ holdpot \\
currentTime - newcol?.time \geq interval \\
(newcol?.slotname, newcol?.value) \in \{c : meltpot \bullet (c.slotname, c.value)\} \\
\{old\} = \{c : meltpot \mid c.slotname = newcol?.slotname\} \\
newcol?.time - old.time > window \\
holdpot' = tail\ holdpot \\
\hline
\end{array}$$

Key changes from the original *SameButNew* schema are the two variables, *interval* and *window*, the substitution of the line $newcol?.time > old.time$ with the new line $newcol?.time - old.time > window$), and the inclusion of the line $currentTime - newcol?.time \geq interval$. The new column can not be fused with the melting pot until the appropriate interval of time has passed. If the new column shares the same slotname, but not the same value, as an existing column, then providing that the new column was created at the same time as or later than the existing column, it is taken that it forms part of a new query. This is represented by the following schema:

$$\begin{array}{|l}
\multicolumn{1}{l}{NewQuery_{alt}} \\
NewQuery \\
currentTime : Number \\
interval : Number \\
\hline
currentTime - newcol?.time \geq interval \\
\hline
\end{array}$$

Here, the change is again the inclusion of the delay before fusing the column, allowing for the possibility of a delayed column to move ahead in the holding pot.

By introducing a time delay in the form of the variable *interval*, the system is given more time to process information, and therefore there is less chance of incorrect fusion taking place. All columns are checked to see that the stated interval has passed before they are allowed to be fused with the appropriate melting pot. Thus, columns that may have been delayed whilst being processed are given more time to reach the *HoldingPot* and be positioned in the correct order. This means that there is less chance of needing to use the *UndoFusion* schema, since the possibility of incorrect fusion has been decreased. This marks a change in strategy away from that implemented in the MATIS system as "eager" fusion, into one that allows a delay, a "lazy" fusion strategy.

The use of a delay may adversely affect the Presentation aspect of the system, since there may be a time lag between information being entered into the system, and that information being assigned to melting pots and thus appearing on the screen in the form of a query field. However, this is where the use of a specification to assess this usability element becomes most worthwhile. Since the interval value is not explicitly stated, it means that it can be defined for each system separately according to the characteristics of each system, rather than being defined as a constant value that cannot be changed and might be inappropriate. Also, the trade-off between immediate feedback and correct feedback can be examined without having to have a system already implemented. The potential problem has already been identified, and the designers of the final system can refer to the end users as to what solution may be most appropriate in resolving the criteria of observation and predictability.

By introducing another time-based variable, that of *window*, into the schemas, the temporal window of redundancy has been extended, and this allows information that previously would have been considered to be part of a new query to be considered as emphasising information already entered into the system. Whereas before only parallel redundancy was supported by the system, in that potentially parallel information had to be entered at the same time as the original data, now the system will disregard identical information that is received up to a specified limit after the original and will effectively support sequential redundancy. The designer of the system now has the option, which did not exist before, of modifying the variable window to suit a particular implementation and set of user characteristics. Again, the use of a specification has allowed a potential usability problem to become apparent, and to be examined and reasoned about in an earlier stage of the design cycle than if a designer was already committed to a particular implementation and was relying on more traditional usability assessment making use of heuristic analyses.

6 Conclusions

This paper has demonstrated that it is possible to reason about and modify aspects of a system through a specification written in a formal notation. Traditionally, usability has been determined by using experimental prototypes. This has drawbacks in that a system is far along in development, with all which that entails in terms of cost and commitment, before it can be assessed, and modifications as a result of assessment are consequently more difficult (and expensive) to effect. In this paper, we have shown that an implementation of a system is not immediately necessary in order to reason about its usability characteristics. Further, we have also shown that no special notations beyond those already used in a software engineering context need to be learned. A current, widely accepted notation, such as Z, can be used for this purpose.

Fusion of data is a complex procedure that is difficult to get right from a usability perspective, given that there are different, and sometime conflicting, criteria to take into account. A central compromise identified in this analysis is between accuracy and speed of fusion. Different systems will peg their trade-off at different levels, according to their user needs and domain. We have been able to modify the specification and reason about how changes might affect usability without having to actually implement a system. It must however be left to a later stage in design to determine the exact compromise - for example a suitable length of time for the temporal window identified in the modification. Cognitive engineering techniques and user evaluation could be used for this purpose.

The CARE properties were found to be a useful basis for establishing some of the usability criteria appropriate to multi-modal interfaces. However, we found it important to acknowledge that CARE, like all modelling approaches, has its own particular scope, and should be augmented by other usability criteria (for example, the use of Abowd's properties). There has been substantial research on multi-media system issues, but relatively little on the problems posed by multi-modal systems. The way in which users interact with multi-modal systems is still not fully understood, and the usability criteria of multi-modal systems is as yet incomplete. Humans make use of multi-modal communication when interacting in other situations, so there is a possibility of utilising research in other areas and applying it to human-computer interaction in order to gain a thorough understanding of the usability issues posed by multi-modal systems. Barnard and May examined the role of cinematography in this light [3] and state: "Interface designers are increasingly relying on craft based approaches to compensate for a perceived lack of relevant theory." In [4] certain usability aspects of multi-sensory-modality systems have been examined from a user's perspective, and aspects of this work have been integrated with formal system models of interfaces to create a 'syndetic' representation [8] that provides an integrated view of human-system interaction. However, much remains to be done on developing a broad theory of multi-modal blending and reasoning about its consequences for user interface design.

We conclude with a brief note about the use of formalism. Although the analysis was based on a formal specification, we have made no use of formal proof, although all of the properties used can be represented formally in a suitable model. Firstly, the problem of taking a formal property from an abstract model of interaction and instantiating it within a detailed specification of a particular system is non-trivial and was beyond the scope of the project reported here. Secondly, the exercise would have added little; it is now accepted that the use of proof is not fundamental, and often not appropriate, in the use of formalism [10]. The advantage of a mathematically-based model over other techniques, such as diagrammatic approaches, is that the model itself can accommodate rather more details of the behaviour of the artefact. The longer-term problem is to understand the trade-off between formality in the model against the type and extent of insight and design improvement that can be obtained in this way.

Acknowledgments

We would like to thank the anonymous referees for their helpful comments on the paper, and also thank Prof J. Coutaz and Dr L. Nigay of CLIPS, Université Joseph Fourier, Grenoble for their comments and assistance during the development of the paper.

References

1. M. Abadi and L. Lamport. An old-fashioned recipe for real time. Technical Report 91, DEC Systems Research Center, October 1992.
2. G. Abowd. Formal aspects of human-computer interaction. D.Phil Thesis, Oxford University Computing Laboratory: Programming Research Group, 1991. Available as Technical Monograph PRG-97.
3. P.J. Barnard and J. May. Cinematography and user interface design. In *Human-Computer Interaction: INTERACT'95*, pages 26–31. Chapman and Hall, 1995.
4. P.J. Barnard and J. May. Interactions with advanced graphical interfaces and the deployment of latent human knowledge. In *Eurographics Workshop on Design, Specification and Verification of Interactive Systems*, pages 15–49. Springer, June 1995.
5. J. Coutaz, D.J. Duke, G. Faconti, M.D. Harrison, and F. Paterno'. Formal methods and multimodal interactive systems. Technical Report SM/WP61, ESPRIT BRA 7040 Amodeus-2, 1995.
6. J. Coutaz, L. Nigay, and D. Salber. The MSM framework: A design space for multi-sensory-motor systems. In *Proc. EWHCI'93*, volume 753 of *Lecture Notes in Computer Science*, pages 231–241. Springer-Verlag, 1993.
7. J. Coutaz, L. Nigay, D. Salber, A.E. Blandford, J. May, and R.M. Young. Four easy pieces for assessing the usability of multimodal interaction: the CARE properties. In *Human-Computer Interaction: INTERACT'95*, pages 115–120. Chapman and Hall, 1995.
8. D.J. Duke, P.J. Barnard, D.A. Duce, and J. May. Systematic development of the human interface. In *APSEC'95: Second Asia-Pacific Software Engineering Conference*, pages 313–321. IEEE Computer Society Press, 1995.
9. D.J. Duke and M.D. Harrison. Interaction and task requirements. In P. Palanque and R. Bastide, editors, *DSV-IS'95: Eurographics Workshop on Design, Specification and Verification of Interactive Systems*, pages 54–75. Springer-Verlag, 1995.
10. A. Hall. Seven myths of formal methods. *Software*, pages 11–19, September 1990.
11. J.K. Hyde. Fusion engines and melting pots. Msc Project Dissertation, Department of Computer Science, University of York, 1995.
12. L. Nigay. Conception et modélisation logicielles des systèmes interactifs. Ph.D. Thèse de l'Université Joseph Fourier, Grenoble, 1994.
13. L. Nigay and J. Coutaz. A design space for multimodal systems: Concurrent processing and data fusion. In S. Ashlund, K. Mullet, A. Henderson, E. Hollnagel, and T. White, editors, *Proc. INTERCHI'93*, pages 172–178. Addison-Wesley, 1993.
14. L. Nigay and J. Coutaz. A generic platform for addressing the multimodal challenge. In *Proc. of CHI'95*. Addison-Wesley, 1995.
15. J.M. Spivey. *The Z Notation: A Reference Manual*. Prentice Hall International, second edition, 1992.

This article was processed using the LaTeX macro package with LLNCS style

Monolingual, Articulated Modeling of Users, Devices, and Interfaces

Thomas Moher[1], Victor Dirda[1], Rémi Bastide[2], and Philippe Palanque[2]

[1]EECS Department (M/C 154), University of Illinois at Chicago, Chicago, IL 60607, USA
[2]LIS/University Toulouse I, Place Anatole France, 31042 Toulouse Cedex, FRANCE
{moher | vdirda}@eecs.uic.edu, {bastide | palanque}@cict.fr

Abstract. This paper presents a framework for combining the discrete models of users and devices into a global system model suitable for analysis and simulation. It views a system as a composite of interacting subsystems, and describes how those subsystems must be structured to permit compositions in which responsibility for global behavior can be appropriately ascribed. The paper presents a human-device example (wrist watch) and develops a range of task and device models. The devices and tasks are modeled by colored Petri nets partitioned to cleanly distinguish submodel component visibility and interface affordances. The formality of Petri nets allows for axiomatic validation of isolated and interacting subsystems.

1 Introduction

Research in human-computer interaction abounds in various notations, formalisms and models that aim at capturing one or more aspects of a domain. One underlying problem is that various disciplines relating to HCI are complex in their own right, with no easy mechanism for transfer of results and expertise across diverse theories and models [3, 19]. As HCI encompasses concerns ranging from human factors to software engineering and even device mechanics, it is necessary, for each of those approaches, to define a more or less crisp boundary, stating what belongs to the modeling domain, and what is exterior to it.

In part the mixture of concerns may stem from a recognition that no one issue in HCI can be adequately treated *in vacuo*. It is generally clear that, with respect to the HCI paradigm, certain approaches lend themselves towards the "human" side, while others are mainly devoted to the "system" side. For example (and without any pretense at exhaustivity) some approaches may be ranked among the "interface modeling" formalisms [7,14,30]. Several proposals attempt to include in their investigation domain not only the interface, but also a model of the internals of the application which constitute its functional kernel [8,24]. On the other hand, several formalisms focus on the description and the analysis of the users, either attempting to capture the essence of their goals and tasks [13, 22, 25, 28,29] or undertaking the modeling of their cognitive behavior [1,2,4,5,6].

The fact that a formalism is dedicated to a given domain is generally considered as beneficial, since it allows for a straightforward mapping of concepts and notations, and allows for the building of concise and descriptive models. However, discrepancies between the approaches also make it difficult to achieve a coherent model encompassing the system, its interface and its potential users.

Bridging of approaches has been achieved by the manual combination of user and device models [10,18]. For the most part, this has been restricted to skilled and relatively simple tasks for which a static production rule or GOMS analysis is feasible [10], although more recent extensions of GOMS to more interactive [26] and dynamic tasks [16] demonstrate the potential for more modular modeling approaches. The limited availability of such approaches may stem from the laborious nature of user model construction and the absence of a standardizing framework. Our belief in the feasibility of generic user models is encouraged by work on general user modeling [17].

This paper is an attempt to provide an unified approach in which the same framework is used to describe the inner behavior of a system (whether software or hardware), the interface it offers to users, and the behavior of the system's users. The system and (one or more) user models may be built and analyzed in isolation, while keeping within the same formal notation. System and user models may thus be merged, and useful results may be obtained by formal analysis of the *global* model resulting from the merger.

2 Language requirements

The choice of a particular *lingua franca* for modeling is less important than the decision to use one which should meet the following four criteria:

Formality. The semantics of the language should be precise, so that the models be free of ambiguity and subjective interpretation. When a given approach has mathematically defined semantics, models may be verified by static analysis, proving the correctness before testing or implementation.

Executability. The ability to execute the design provides insights into system operation and depends on a solid formal base.

Graphical representation. While the relative value of graphical languages remains an open research area [20], they do offer certain benefits [12], particularly in animation and simulation. Colored Petri nets in particular support a range of representational *styles* in that they allow a relatively smooth tradeoff between graphical and textual modeling elements.

Parallelism and non-determinism. The ability to properly describe parallelism, concurrency and synchronization is necessary for modern interactive systems in which users and systems work on several tasks in parallel. While highly precise device models will usually be deterministic, tasks representing the "free will" of users will often be non-deterministic.

While several formalisms satisfy some of these criteria, few satisfy them all [23]. This led us to select *colored Petri nets* [15] as our formalism of choice.

3 Colored Petri nets

When modeling with Petri nets, a system is described in terms of state variables (called *places*, depicted as ellipses) and by state-changing operators (called *transitions*, depicted as rectangles), connected by annotated *arcs*. The state of the system is given by the *marking* of the net, which is a distribution of *tokens* in the net's places. In colored Petri nets, the tokens assume values from predefined types, or *colors*.

State changes result from the firing of transitions, yielding a new distribution of tokens. Transition firing involves two steps: (1) tokens are removed from *input* places and their values *bound* to variables specified on the input arcs, and (2) new tokens are deposited in the *output* places with values determined by *emission rules* attached to output arcs. A transition is *enabled* to fire when (1) all of its input places contain tokens, and (2) the value of those tokens satisfy the (optional) Boolean constraints attached to the input arcs. (In some Petri net variations, constraints and emission rules are bound to transitions; here the annotations are always associated with arcs.) Figure 1 depicts a small Petri net used to illustrate our notational conventions. The different shadings used for the tokens in Figure 1 reflect the fact that they have different domains of value (color).

The transition in Figure 1 is enabled to fire if the values of the tokens in place P (bound to variables x and y) and place Q (bound the variable z) satisfy the Boolean condition B1(x,y,z). Upon firing, the three input tokens are removed from places P and Q and new tokens deposited in places R and S. The values of the two new tokens in place R are defined by the emission rules E(x,y,z) and F(x,y,z). The value of the new token in place S is determined by a conditional rule: if B2(x,y,z) is true, the value is given by G(x,y,z), otherwise, the value is given by H(x,y,z). (Notation adapted from the C language for brevity.) Note that although in this case the number of tokens remained constant, in general token cardinality is not conserved, nor are there constraints on the fan-in or fan-out of transitions.

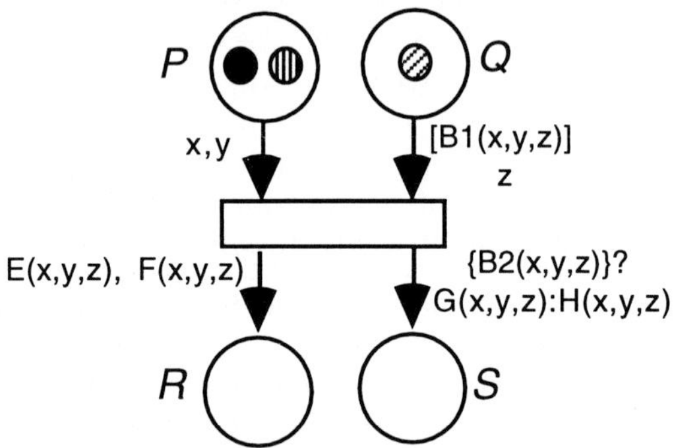

Fig. 1 Petri net notational conventions

While multiple transitions may be simultaneously enabled, simultaneous transition firing is disallowed within a single net. The selection among enabled transitions of the one to be fired reflects either a choice on the part of the user to execute a mental or physical operation, or a selection rule (deterministic or non-deterministic) embodied in the device model.

4 Device Model

In our examples we shall use a simple watch (Figure 2) with a conventional stem control for setting time and two (unlabeled) buttons: MODE and PHASE. The MODE button toggles between two states: RUN and SET. The PHASE button toggles between AM and PM (displayed on the watch face), but only when the MODE is SET; pressing the PHASE button in RUN mode has no effect.

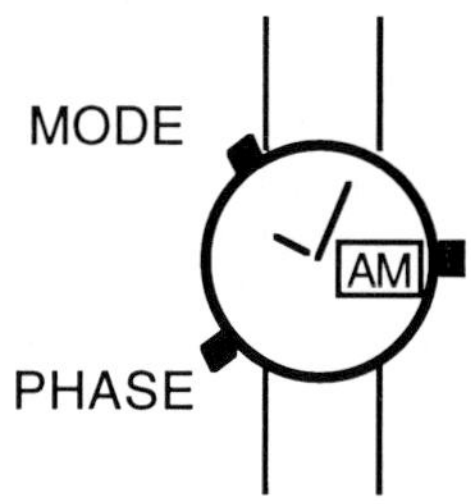

Fig. 2 Simple wrist watch

To model a concrete device such as the watch in Figure 2, the transitions of the net are used to model the physical actuators offered by the device, and the firing of a transition is considered initiated by a user action. Figure 3 depicts a Petri net model of the watch (considering only the phase setting), and includes two transitions corresponding to the PHASE and MODE buttons. The net of Figure 3 has two places, corresponding to the watch's mode and phase states.

Pressing the MODE button alters the state of the device by removing a token from the *MODE STATE* place, binding it to the variable mode, and depositing a new token into *MODE STATE*. We use the generic function other to "toggle" between the two values in a bivalued place domain, in this case RUN and SET.

The *PHASE BUTTON* transition has two input arcs and two output arcs. For economy of notation, the arcs between *MODE STATE* and *PHASE BUTTON* have been collapsed into a single bidirectional edge; the annotation, mode, represents both a binding (on the input arc to the transition) and an emission rule (on the output arc). However, this consolidation also has conceptual significance, as we can view the bidirectional edge as limiting operations to a non-destructive *inspection* of *MODE STATE*. The value of the new token in *PHASE STATE* is determined by the annotation on the output arc from *MODE BUTTON*, in this case the simple conditional function other which toggles the phase only if the mode is SET.

The *MODE STATE* and *PHASE STATE* are each marked with an initial token; the precise values of the tokens are unspecified as the watch may be in one of four states at the outset of the task. The presence of the tokens in the two places is sufficient to enable the button transitions, since no Boolean constraints are attached to their input arcs.

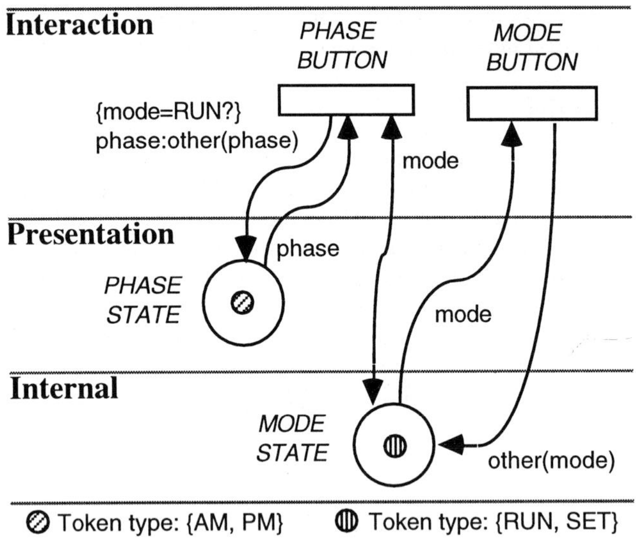

Fig. 3 Phase setting in the simple wrist watch

We have partitioned the graph nodes into three *bands* to reflect the role and visibility of model places and transitions. (The positions of labels and arcs within bands is not significant here.) The **Interaction** band includes the input affordances which the device provides to the user, the **Presentation** band represents device state information which is made visible to the user, and the **Internal** band contains places and transitions to which the user does not have direct access. This permits us to distinguish between, for example, the *PHASE STATE*, which is visible on the watch face, and the *MODE STATE*, which is not.

5 Modeling the impatient user

The next task is to develop a model of a watch user who wishes to set the phase to PM. Different users, of course, might approach the task in different fashions; we model a target user who is familiar with watches and anticipates modal operation, but who is unfamiliar with this particular watch. For such a user, a common strategy might be to simply push the buttons at random until the display registered PM. A Petri net model of this strategy is shown in Figure 4.

In the model, the place *DESIRED DISPLAY* is initially marked with a token whose value is the desired phase (AM or PM), while the *PHASE DISPLAY* place is marked according to the actual phase displayed on the watch. If the present phase (bound to the variable phase) matches the desired phase (bound to the variable d), the *QUIT* transition is enabled. Firing the *QUIT* transition (a mental operation) causes the *PHASE DISPLAY* to be re-seeded with the current phase, but does not replace the token in the *DESIRED DISPLAY* place, thus precluding any further transition firings. Thus, the "dead net" condition is synonymous with successful termination.

The *TOP BUTTON* and *BOTTOM BUTTON* transitions are always enabled, independent of whether the current phase is the one desired by the user. Firing either of these transitions (a physical action) results in the consumption and regeneration of a token in the *DESIRED DISPLAY* place. The value of that token remains constant (reflecting the assumption that the user's goal remains the same throughout the task).

Unknown to the user, who does not know how the watch works, is the effect that firing either the *TOP BUTTON* or *BOTTOM BUTTON* transitions may have on the internal state of the watch. In particular, the user must assume that firing these transitions has the potential to affect the value of the token in the *PHASE DISPLAY* place, otherwise pushing the buttons would be futile. Articulation of the user model reflects hypotheses concerning the input and output affordances provided by the device, as well as "theories" concerning the external behavior of the device. However, it is important to note that a knowledge of the internal operation of the device (the device model) is *not* required to formulate a user model.

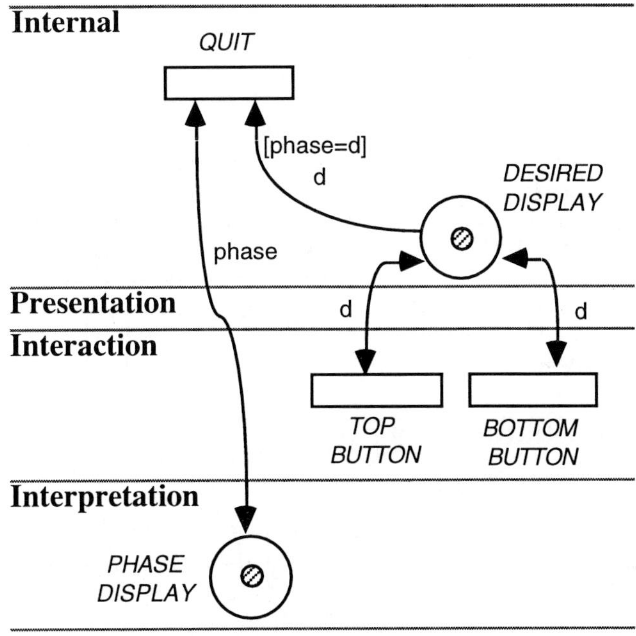

Fig. 4 The Impatient User Model

Two features of this model are worth emphasizing. First, at the outset, the user does not even know which role the two buttons assume (hence the use of the transition labels *TOP* and *BOTTOM* rather than *MODE* and *PHASE*). Second, pressing a button may or may not enable the *QUIT* transition, but even if it is enabled, the user may not choose to fire it.

As with the watch device model, we have added bands to reflect affordances and state visibility. The **Presentation** band of Figure 4 is empty, reflecting the fact that the user presents no sensible state information upon which the device can act. (In general, however, such places may well exist, as in a gesture-based input system.) The **Interpretation** band represents the user's expectation that the phase (but not the mode) will be visible in the user interface.

6 Merging the models

Both of the models presented above are incomplete; the device cannot operate without the user, and the user cannot complete the task without the device. While certain questions about each model may be answered in isolation, questions about the global system cannot be addressed unless we are able to effect a merger between the two components models.

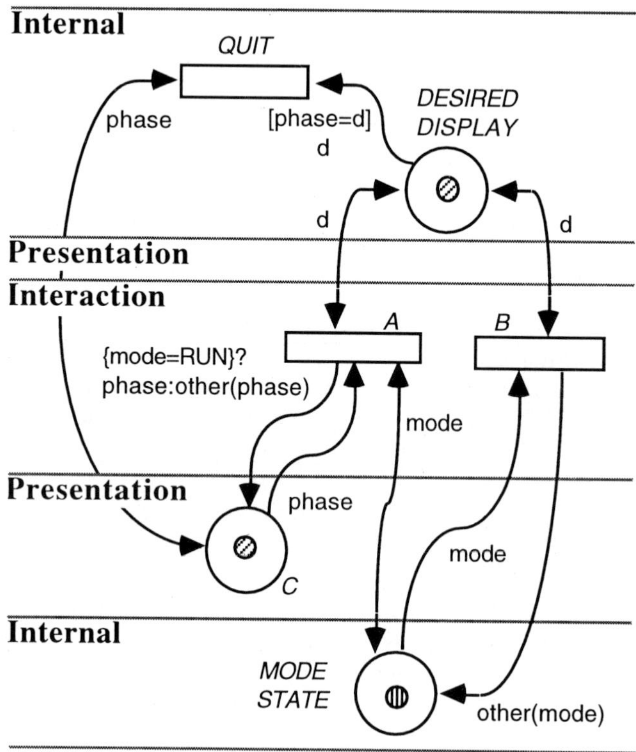

Fig. 5 Global model: simple watch and impatient user

Figure 5 depicts the global model resulting from the merger of the device and user models for the simple watch. As with the single agent models, the bands reflect interaction affordances and state visibility. The net place in the **Interpretation** band of the user was merged with the net place in the **Presentation** band of the device to form a device **Presentation** place in the global model.

The bands are not merely cosmetic; they impose constraints on the topology of the global net. Arcs between internal components of two agents are prohibited, and only inspection arcs are allowed between **Internal** or **Presentation** transitions in one agent and **Presentation** places in the other agent.

Merging the models requires *matching* operation between device and user interaction transitions, and between device presentation and user interpretation places. Table 1 shows the mapping used to construct Figure 5; later we shall discuss how such matching takes place.

Merging the models affords us opportunities both to analyze and to simulate the global system. By assigning initial markings to the watch state places, we can take advantage of a large body of existing Petri net theory to prove certain properties of the global system.

Global	Device Model	User Model
A	*PHASE BUTTON*	*TOP BUTTON*
B	*MODE BUTTON*	*BOTTOM BUTTON*
C	*PHASE STATE*	*PHASE DISPLAY*

Table 1 Device/user model component matching

Suppose that the watch state is initially in RUN mode and AM phase—the most difficult condition which the user might confront. While a formal proof will not be presented here, it should be apparent that (1) there exists a sequence of transition firings which will allow the user to complete the desired task, and (2) there is no guarantee that this will ever happen. (For example each sequence of action such as ((MODE MODE)*PHASE*)* will never change the phase on the watch).

7 Modeling the careful user

The strategy used by the impatient user is only one of many which might be employed. A more careful user might adopt a policy of strict alternation between the two buttons; in contrast to the impatient user, this strategy is guaranteed to reach the goal state in a maximum of four button pushes. (The worst case is when the watch is already in SET mode and the user pushes the *MODE BUTTON* first.) Figure 6 depicts a global model incorporating this strategy; note that the device model remains unchanged.

Turn-taking in Figure 6 is modeled by the addition of a black token in *DESIRED STATE,* which now contains tokens representing the user's desire (strayed) and turn (black). The turn token can hold two different values: T if the *TOP BUTTON* is the last one that has been pushed, and B if the *BOTTOM BUTTON* was pushed last. In order to describe the toggling of buttons in the model, the arc annotations between *DESIRED STATE* place and the buttons have been changed. A precondition has been added to each arc, the condition [turn=B] indicating that button *A* can be pushed only if the value of the turn token is B. Firing either button transition toggles the state of the turn token in *DESIRED STATE,* while leaving the desire value unchanged.

Analysis of the net in Figure 6 reveals that, regardless of the initial markings in the net places, the user will set the phase to PM in at most four transition firings. Replacing the user model, without modifying the device model, has produced a global model which analysis reveals to have distinct (provable) properties from the model of Figure 5.

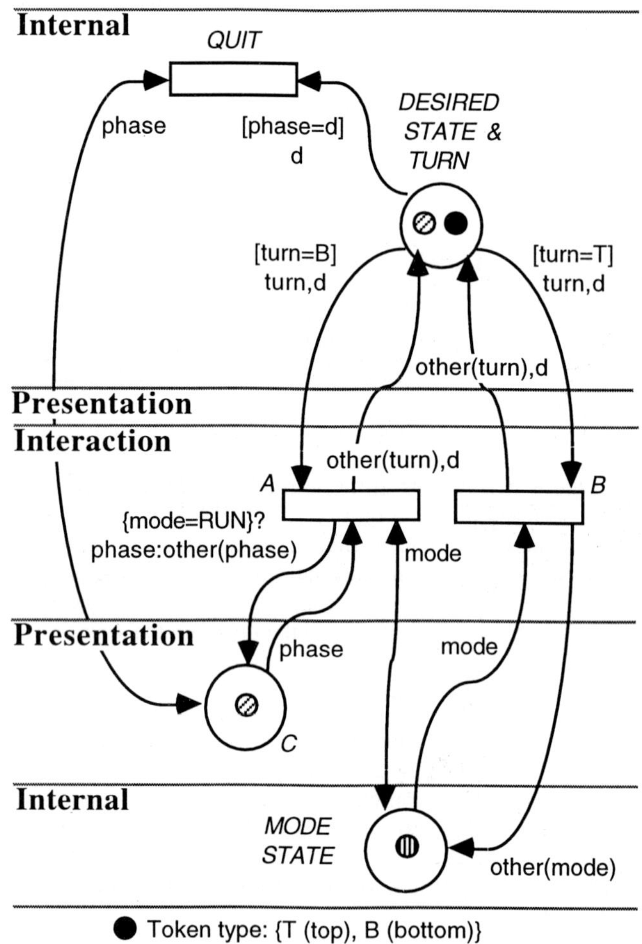

Fig. 6 Global model: simple watch and careful user

8 Modeling the learning user

The models presented in the previous section are adequate to describe a strategy for accomplishing the task of setting the watch phase to the PM phase, but do not provide for a description of how the user learns about the watch, in particular, which button is which. Effective models of human-computer interaction need to support the description of learning beyond the specifics of isolated task completion [6,31].

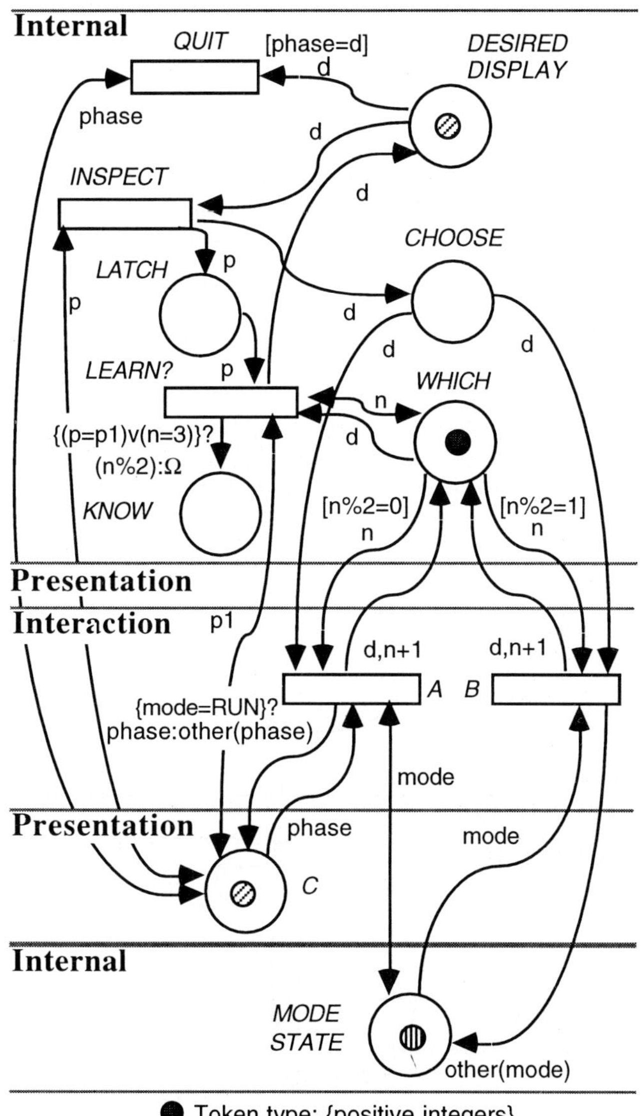

Fig. 7 Global model: simple watch and learning user

The device model of the simple watch dictates that no user strategy can guarantee that the watch phase will be set to PM in less than four button presses. However, a bit of reflection reveals that three button presses are sufficient to determine the identity of the two buttons.

Figure 7 depicts a model which reflects this user strategy. In the model, the number of button pushes is explicitly maintained in the place labeled *WHICH* (by a grayed token whose domain of value is Integer), which is assigned an initial marking of 0. Turn-taking is enforced based on whether the counter in *WHICH* (n) is even or odd. (The constraints on the arcs from the buttons to *WHICH* use the % symbol to signify the *modulo* operator.)

The strategy represented by the net in Figure 7 calls for the user to inspect the phase display, latch the result, press a button, and inspect the phase display again. If the display changed, or if three presses did not result in a change, a token with numeric value (0 or 1) is deposited in the *KNOW* place; otherwise, a dummy token (Ω) is deposited. The presence of the numeric token represents the acquisition of button identity knowledge on the part of the user.

9 Expectation failure

The examples thus far have illustrated perfect formal correspondence between the user's interpretations of the device and the device's presentation. Figure 8, in contrast, depicts a user model reflecting an expectation that phase changes depend strictly on a non-modal button press.

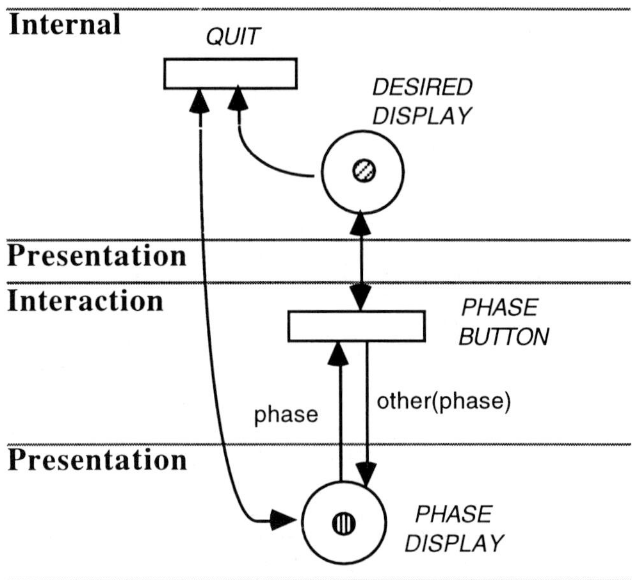

Fig. 8 User model: expecting a simpler watch

Merging the model of Figure 8 with the device model of Figure 3 introduces two classes of expectation failures (Figure 9). First, there is an *unmatched interaction:* the *MODE BUTTON* in the device model has no correspondent in the user model. Since the user has no knowledge of the existence of the *MODE BUTTON*, it can never be fired, even though it remains enabled within the device model. In the global model, we represent this discrepancy by filling the rectangle representing the transition, indicating that the transition is *dead* [9].

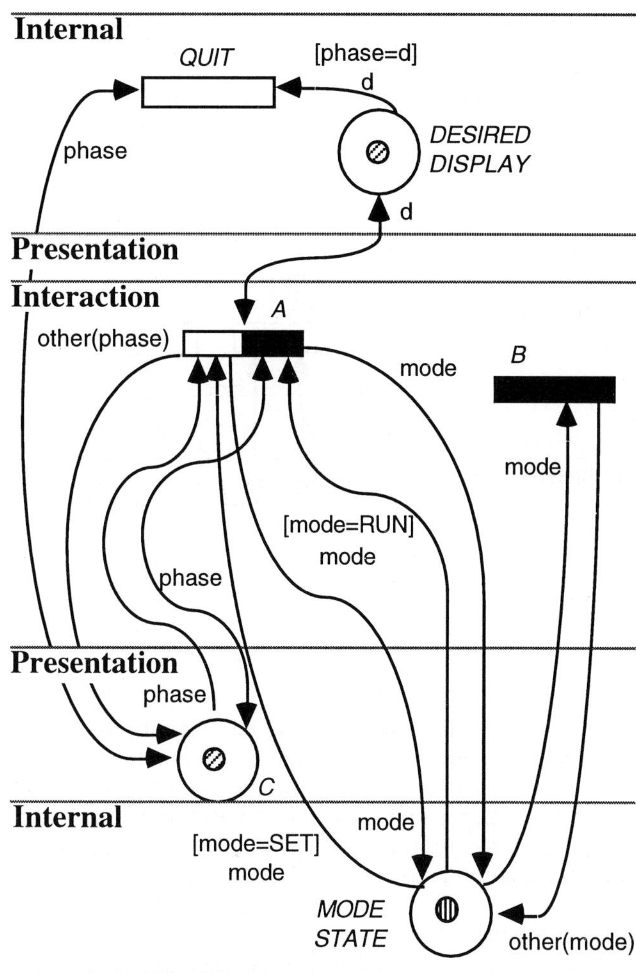

Fig. 9 Modeling expectation failures

The second complication involves a case of *inconsistent modeling* relating to the *PHASE BUTTON*. If the initial marking of the *MODE STATE* was SET, the *PHASE BUTTON* transition would be enabled, and firing it would change the contents of the *PHASE DISPLAY*. However, if the *MODE STATE* was RUN, then pushing *PHASE BUTTON* would have no effect; since the *MODE BUTTON* could never be pushed, the *PHASE BUTTON* transition would be dead.

This dual "personality" of the *PHASE BUTTON* is reflected by an *articulated* transition. Part of the transition (unfilled) acts as a normal transition dependent on its global inputs; the other part (filled) is depicted as dead. Device place inputs to the transition are attached to one or the other parts of the transition.

10 A better watch

All of the examples to this point have employed the simple watch model, in which the phase is made explicit in the interface, but in which the mode is hidden. In this section, we consider a watch in which the mode is made explicit by blinking the phase display when the watch is in SET mode, and in which the buttons are explicitly labeled. The device model for this watch, shown in Figure 10, is nearly identical to that presented in Figure 3; the only change is that the *MODE STATE* has been moved from the **Internal** band to the **Presentation** band to reflect its visibility to the user.

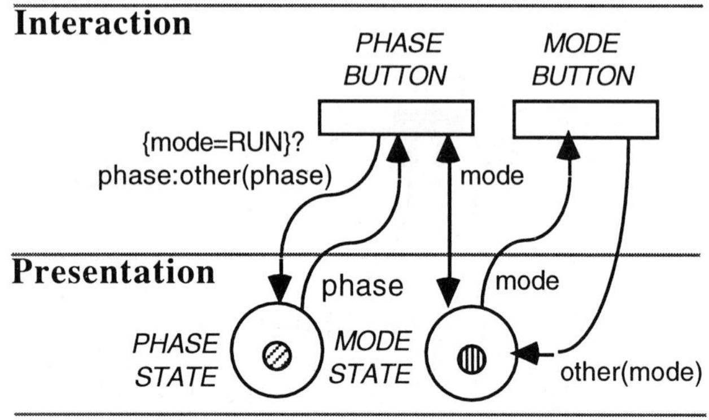

Fig. 10 Petri net model of a better wrist watch

The accessibility of the mode state in the user interface affords a strategy requiring a maximum of two button presses. In the global model employing this device model with an optimal user strategy (Figure 11), the user explicitly tests (senses) the mode. If the watch is in RUN mode, the user presses the *MODE BUTTON*, confirms the mode state change, and presses the *PHASE BUTTON* to complete the task. If the watch is already in SET mode, the user directly presses the *PHASE BUTTON*.

One limitation of the model presented in Figure 11 is that the distinction between the presence and absence of "hard" labels on the watch buttons is not represented within the net formalism. One way to remedy this might be to explicitly model the labels as token values within "label" places contained within the device model, and to include the label places as inputs to the button transitions. The positioning of the label places, either in the **Presentation** or **Internal** band of the device, would denote the presence or absence of hard labels, respectively. Other solutions, however, are also possible.

The user model implicit in Figure 11 reflects an expectation of a *MODE STATE* place in its **Interpretation** band. Attempting to merge that user model with the simpler watch model of Figure 3 would have resulted in an *unmatched interpretation,* since a corresponding place would not have been available in the device's **Presentation** band. This would have resulted in the formation of a *dead place* analogous to the dead transitions of the previous section. It is important to note that the identity of the buttons need not have resulted from explicit labels in the user interface; that knowledge may have come from prior experience with the device, as in Figure 7.

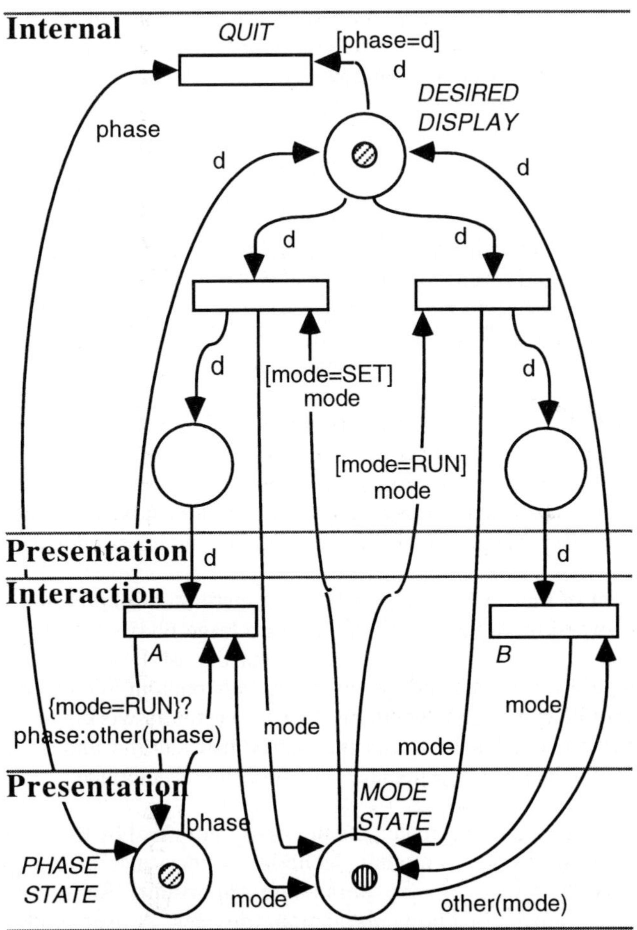

Fig. 11 Global model: optimal user and better watch

11 Issues and hesitations

While the emphasis in this paper is on the notion of a common language for representing device and user models and not on Petri nets *per se*, it would be impossible to ignore the complexity of the representation presented. The natural question arises as to whether Petri nets (or, for that matter, any graphical representation) is suitable for scaling to larger problems. Indeed, recent research in the comprehensibility of graphical vs. textual representations has found little support for the "intuitive" advantage of graphical representations [20].

While it would be tempting to write off Petri nets as overly complex and unscalable, the situation is actually more muddied. In the first place, the examples shown in this paper, for all the apparent simplicity of the watch interface, are not so simple as they appear. The "simple watch, impatient user" model of Figure 4 is easily understood, the more so when the component user and device models are considered in isolation. The "careful user" model of Figure 6 extends that model by simply making explicit the "remembering" of the most recent button press; while the notation may be unfamiliar, it is difficult to see how it could be expressed much more succinctly. The "learning user" case, for all its detail, reflects a fairly complex underlying mechanism requiring the remembering of several previous steps; while the device remains simple, the strategy is not. Certainly the rules represented here by Petri nets would permit textual representation as well, but at the cost of the two-dimensional canvas (and its opportunities for clarifying—or obscuring—associations among rules) afforded by a graphical representation. The impact of alternative graphical layouts (or *secondary representations* [27]) remain an active research area.

One mechanism not introduced in this paper is the notion of *hierarchical* Petri nets. Subnets afford opportunities for notational reduction analogous to the use of macros and subprograms in textual languages. Some of the criticism of graphical representations may be traced to an unbalanced comparison between monolithic (non-hierarchical) graphical structures on the one hand, and modularized text on the other.

Independent of their comprehensibility as a medium of expression, Petri nets are attractive for formal modeling because there is a large body of analytic methods for proving the properties of such models. The utility of Petri nets in this context ultimately depends on our ability to devise systems in which models can be transformed into Petri net representations. We are presently working on system which supports this approach, but in which the user constructs the underlying (Petri net) model through an quasi-natural language "conversational" interface.

The user models we have created are obviously informed by the device and task at hand; in no way are our user models "generic" in the sense of being reusable for arbitrary situations. However, one immediate opportunity for reuse arises in the modeling of the transfer of knowledge among devices. A user model representing skilled behavior on one device might be combined with the model of a similar (but not identical) device as a means of predicting performance errors (as in Section 8. above). In an earlier paper, we discussed the problem of modifying the user model to accommodate such expectation failures [21].

12 Conclusion

This paper has attempted to illustrate the desirability of a common framework for combining theories, models and knowledge from various disciplines, each of which may have something relevant to say about a global model, but may say it in different terms or formalisms. The purpose of the framework is to define a combinable unit of modeling. In particular, the framework we have articulated indicates what a specific model must define to be combinable (its form), and how concerns can be separated so that a its boundaries are unambiguously identified (its semantics).

The task models presented here represent instances of early interactions rather than skilled behavior, and embody no theories of how sequences of task models are generated by users, or of how task models and device models are composed. In general, users would use existing task and device knowledge and experience to evolve better models, and develop long term strategies. Our framework provides a formal representation for the fruits of such activities.

Acknowledgments

This research was funded in part by grants from the National Science Foundation (CDA-9303433), the GTECH Corporation, and the Centre d'Etudes de la Navigation Aerienne of Toulouse. The authors wish to thank the anonymous referees for their helpful suggestions on an earlier draft of this paper.

References

1. Anderson, J.R. The Architecture of Cognition. Harvard Press, Cambridge, 1983.

2. Barnard, P., Wilson, M., and Maclean, A. Approximate modeling of cognitive activity with an expert system: a theory-based strategy for developing an interactive design tool. The Computer Journal, 31(5), 1988, pp. 445-456.

3. Barnard, P. Bridging between Basic Theories and the Artifacts of Human-Computer Interaction, in J.M. Carroll (Ed.), Designing Interaction: Psychology at the Human-Computer Interface. Cambridge Univ. Press, Cambridge (1991), pp. 103-127.

4. Bibby, P.A. and Payne, S.J.. Internalization and Use Specificity of Device Knowledge. Human-Computer Interaction 8(1), 1993, pp. 25-36.

5. Card, S., Moran, T., and Newell, A. The Psychology of Human-Computer Interaction. Lawrence Erlbaum Associates, 1983.

6. Carroll, J.M. and Olson, J.R.. Mental Models in Human Computer Interaction, in M. Helander (Ed.), Handbook of Human-Computer Interaction, . Elsevier Science Publishers B. V., North-Holland (1988), pp. 45-65.

7. Coutaz, J. and Bass, L. Developing Software for the User Interface. Addison Wesley Publishing, 1991.

8. Foley, J.D. and Sukavirya, P. A Second generation User Interface Design Environment: The Model and The Runtime Architecture. Proc. INTERCHI'93 (Amsterdam, April 1993), pp. 375-382.

9. Genrich, H.J. Predicate/Transition Nets, in K. Jensen and G. Rozenberg (Eds.), High-Level Petri Nets: Theory and Application. Springer-Verlag, Berlin (1991), pp. 3-43.

10. Gray, W.D., John, B.E., and Atwood, M. Project Ernestine: Validating a GOMS Analysis for Predicting and Explaining Real-World Task Performance. Human-Computer Interaction 8(3), 1993, pp. 237-309.

11. Gray, W.D.. Why You Can't Program Your VCR. Poster presented at CHI'94 Conf. on Human Factors in Computing Systems. (Boston, MA, 1994).

12. Harel, D. Statecharts: A visual formalism for complex systems. Science of Computer Programming 8, 231-274 (1987).

13. Hix, D. and Hartson, H.R. Developing User Interfaces: Ensuring Usability Through Product & Process. Wiley Professional Computing, 1993.

14. Jacob, R.J.K. A Specification Language for Direct-Manipulation User Interfaces. ACM Transactions on Graphics 5(4), 1986, pp. 283-317.

15. Jensen K. Coloured Petri nets and the invariant method. Theoretical Computer Science 14, 1981, North-Holland, 317-336.

16. John, B.E. and Vera, A.H. A GOMS Analysis of a Graphic, Machine-Paced, Highly Interactive Task. CHI'92 Conf. on Human Factors in Computing Systems (1992), pp. 251-258.

17. Kass, R. and Finin, T. General User Modeling: A Facility to Support Intelligent Interaction, in J.W. Sullivan and S.W. Tyler (Eds.), Intelligent User Interfaces. Publisher (1991), pp. 111-128.

18. Kieras, D.E. and Bovair, S. An approach to the formal analysis of user complexity. International Journal of Man-Machine Studies, 22 (1985) pp. 365-394.

19. Kuutti, K. and Bannone, L. Searching for Unity Among Diversity: Exploring the "Interface" Concept. Proc. INTERCHI'93, ACM Press (1993), pp. 263-267.

20. Moher, T., Mak, D., Blumenthal, B., and Leventhal, L. Comparing the comprehensibility of textual and graphical programs: The case of Petri nets. Empirical Studies of Programmers: Fifth Workshop. Ablex Publishing Co., 1993, pp. 137-161.

21. Moher, T. and Dirda, V. Revising Mental Models to Accommodate Expectation Failures in Human-Computer Dialogues, in P. Palanque and R. Bastide (eds.), Design, Specification and Verification of Interactive Systems '95, Springer, 1995, pp. 76-92.

22. Moran, T.P. The Command Language Grammar: a representation for the user interface of interactive computer systems. Int. J; Man-Machine Studies, 15 (1981), pp. 3-50.

23. Murata, T. Personal communication (1994).

24. Palanque, P., Bastide, R., Sibertin, C., and Dourte, L. Design of User-Driven Interfaces using Petri nets and Objects, in Proceedings of 5th Conference on Advanced Information Systems Engineering (CAISE'93). Lecture Notes in Computer Science N° 685, Springer-Verlag.

25. Payne, S.J. and Green, T.R.G. Task Action Grammars: a model of mental representation of task languages. Human Computer Interaction, 2 (2) , 1986, pp. 93-133.

26. Peck, V. and John, B.E. Browser-Soar: A Computational Model of a Highly Interactive Task. CHI'92 Conf. on Human Factors in Computing Systems (1992), pp. 165-172.

27. Petre, M. and Price, B. Why Computer Interfaces Are Not Like Paintings: the user as a deliberate reader, in J. Gornostaev (ed.), Proceedings of East-West HCI92: The St. Petersburg International Conference on Human-Computer Interaction, 1 (pp. 217-224), ICSTI: Moscow 125353, Russia.

28. Singley, M.K. and Anderson, J.R. A Keystroke Analysis of Learning and Transfer in Text Editing. Human-Computer Interaction 3(3), 1987, pp. 223-274.

29. Tauber M. ETAG: Extended Task Action Grammar - A Language for the Description of the User's Task Language. 1990.

30. Van Biljon, W.R. Extending Petri nets for specifying man-machine dialogues. International Journal of Man-Machine Studies 28 (1988), 437-455

31. Woods, D.D. and Roth, E.M. Cognitive Systems Engineering, in M. Helanders (Ed.), Handbook of Human-Computer Interaction, Elsevier Science Publishers B. V., North-Holland (1988), pp. 3-43.

Evaluating the Interfaces of Three Theorem Proving Assistants

Nicholas A. Merriam* and Michael D. Harrison

Department of Computer Science
University of York
Heslington, York YO1 5DD, U.K.

Abstract. A first step in systematically engineering better interfaces for theorem proving assistants (TPAs) is to assess what has already been achieved in the domain. We examine three TPAs employing quite different styles of interaction. We consider the support provided by the interfaces for each of four mechanisms for efficient interactive proof: planning, reuse, reflection and articulation. Common themes are observed, as are strengths and weaknesses of the interfaces and we discuss the general issues, attempting to abstract away from the particular artifacts studied.

1 Introduction

As formal methods unburden themselves of notions of uniquitous proof, recognition is growing for the value of formal proof in certain areas, for example verifying algorithms, see [6] and exploring properties of complex interactions, see [17]. Furthermore the tools necessary to check and to manage large and complicated proofs are emerging allowing significant proofs to be performed interactively. Such theorem provins assistants (TPAs) are also equipped with decision procedures and other automated deduction which can make proof much quicker and easier.

However, there is universal agreement that formal proof is very hard and that TPAs neither make this activity easy, nor are themselves easy to use. That is, even when the user knows the correct proof, it is a complex task to communicate this proof to the TPA. Whilst theoreticians attempt to reduce the difficulty of formal proof itself, the builders of TPAs must bring expertise in interactive system design to bear on the difficulty of using TPAs.

To the HCI practitioner, interactive theorem proving assistants (TPAs) provide a useful forcing function: designing of better interfaces for them can be viewed as a kind of challenge problem in the domain of cooperative decision-making software.

The interfaces of TPAs have provided rich and interesting examples in our studies of human-computer interaction (HCI). In machine assisted theorem proving there is a complex and difficult cooperative decision-making process going on. The close coupling this requires between user and software places particular

* Employed on EPSRC grant GR/K09205

demands on the user interface, both to make information available effectively and to make interaction natural. In addition, we have a great deal of information flowing between the machine and the user. Thus the interface designer is challenged to make the best possible use of the available bandwidth.

A vital first step in the study of interfaces for TPAs is to consider what has already been achieved with current artifacts. Through this we can understand the important factors in TPA usability; we can identify gaps in support for user activities and locate "hot spots" where users are prone to time consuming errors. The present paper makes a contribution to this end, discussing three TPAs with very different interfaces. Additionally, we look for clues in the literature which can direct us to better interface designs in the future.

In Section 2, we define rigorous formal proof and briefly define what we have identified as key activity categories which can help us to analyse existing TPAs. In Sections 3.1, 3.2 and 3.3, we examine three TPAs, CADiℤ, IMPS and PVS, considering some aspects of the support that they provide for each of the kinds of activity. In Section 4, we attempt to draw out the important issues arising from the interface examinations and look at how our analysis can suggest future improvements to TPAs. Finally, Section 5 summarises this work and assesses future directions.

2 What is a TPA?

A proof of an argument might involve an "advocate" attempting to convince a "sceptic" by means of a dialogue. Logic allows proof to be made less subjective. A logic is published by the advocate of an argument so that, if the argument can be demonstrated using the rules of that logic, it will be universally accepted *with respect to this logic*. A logic is defined by a collection of axioms and inference rules. The set of theorems provable in that logic is considered to be the axioms plus all statements which can be constructed from those axioms using the rules of inference. A proof of a theorem is an unbroken chain of inferences which connect that theorem to axioms.

For example, if we have a logic with one axiom, A, and one inference rule, **&I** ("& introduction"), which allows us to construct $X \,\&\, Y$ given X and Y then the theorems of this logic are $A, A \,\&\, A, (A \,\&\, A) \,\&\, A, \ldots$ The proof of $A \,\&\, A$ uses **&I** exactly once with both X and Y as the axiom A.

The assembly of a formal proof is incrementally achieved in one of three styles:

- The inference rules can be used to build "forward" from axioms towards the goal statement. This is called *forward proof*.
- The inference rules can be used to link "backward" from the goal towards axioms of the logic. Nearly all theorem proving is done by such *backward proof*, since this is usually the most natural approach.
- The inference rules can be used to extend a guessed intermediate goal either forward or backward.

A goal to be proved is often expressed as a "sequent", a collection of antecedent (assumption) formulae followed by a collection of succedent (conclusion) formulae, usually separated by "$\vdash$". The informal meaning is that the antecedents imply at least one of the succedents, see [4].

Now the process of proof is no longer a dialogue but, at least initially, a game of solitaire. The advocate must solve a puzzle, discovering how to show that the argument can be constructed according to a given set of rules. The advocate's task has been made more onerous, since statements which another human might accept without challenge must be precisely formulated and proved from first principles. The large number of steps involved in proving interesting results makes the construction of such logical proofs relatively slow and laborious. Partly as a consequence, this level of rigour is rarely encountered. However, computers provide a means to make such proof more reliable and less arduous. The goal of the TPA is to fulfil both of these rôles, checking such formal proofs to make sure that the rules are followed correctly and helping the advocate to build large and complicated proofs. Then we can hope to build proofs which are both interesting and rigorous. Furthermore there is a need ultimately to extract higher level human-readable proofs so that they may be used to convince a sceptical community of the argument's validity.

2.1 System mechanisms

The system consists of the user and the TPA together. System mechanisms provide the means to achieve the goal of constructing a proof. Four such mechanisms have been identified as being important and requiring support from the interface: planning, reuse, reflection and articulation. This proposal of key mechanisms is an important contribution of this paper; it is our claim that they all must be working effectively to facilitate efficient theorem proving.

Planning In [18], it is noted that the essential activity of theorem proving is not the mechanics of applying proof steps but the discovery of the correct (high level) plan of the proof. We are concerned with how systems support or hinder the user's discovery of this plan.

We make a distinction between a plan, which is a simple sequence of actions, and the activity of planning, which involves a complex hierarchy of objective identification and attainment. A plan is *formed* by selecting a sequence of actions. Plans are *executed* by carrying out the appropriate actions in succession. Part or all of a plan may be *evaluated* either in advance, using thought experiments, or in retrospect, after execution. Planning consists of the formation, execution and evaluation of plans. In practice these three phases are interwoven, as part of a proof may be executed before other parts are even formed. Evaluation will occur continuously in an attempt to avoid pursuing unfruitful plans.

These ideas can be illustrated by an example narrative. Suppose the objective is to prove a goal. The user forms the initial part of the plan by choosing to apply induction. This is executed and evaluated before plans to prove the base and step case goals are formed. The user forms a plan for the base case involving expanding definitions and simplifying. This is then executed and, if successful, some strategy such as case analysis is selected to tackle the step case. The high level plan so far could be expressed as "apply induction, expand and simplify base case, apply case analysis to step case". The user continues until the goal is proven.

Reuse Theorem proving is a labour intensive activity and one way to make it more cost effective is to reuse existing proofs, either directly as theorems or by abstracting some more generally applicable heuristic and distilling it into an executable procedure, or "strategy". Reuse via theorems is very different in mechanised proof from informal proof. Whereas a mathematician must recall or look up the appropriate theorem, the user of a TPA must additionally then identify the theorem is such a way that the TPA can find an associated proof.

The creation and replaying of strategies is effectively a programming activity and the considerable literature on the usability of programming styles is pertinent here, see for example [7].

Reuse extends the options open to the user at any single point in a proof: in addition to applying single inference rules, a theorem may be introduced or a strategy invoked. Whilst this makes the plan shorter, plan formation is not necessarily simplified because the range of possible actions is now greater.

Reflection Proof reflection, is the reading and understanding of the current goals and the overall pattern of the proof. Reflection is important for plan formation, since a correct plan can only be formulated if the goal is understood. The ability to evaluate a plan is also determined by the support for reflection provided by the interface.

Articulation A single command may require cross-referencing between the goal being tackled and the available assumptions, theorems and other resources. The command may apply to only a small part of the goal, or to several parts of the goal, or to the entire goal. The need for efficient interaction demands that the articulation of commands with such complex arguments be as natural and error-resistant/tolerant as possible.

3 Existing TPAs

The subsequent descriptions of existing TPAs are each divided into five parts. The first is a brief overview, including the intended use of the TPA and its helper applications. The cost of building TPA interfaces is such that existing "off the shelf" software is used wherever possible. The choice of what to use is a key design decision.

We classify the different interfaces according to *interaction style*. Individual interaction styles can be grouped into *key-modal*, *direct manipulation* and *linguistic* style categories, after [13]. Here we will refer to the *menu* and *question and answer* key-modal styles, the *graphic direct manipulation* and *form fill-in* direct manipulation styles and the *command line* linguistic style. We have chosen to look at three different TPAs which employ graphic direct manipulation, menu and command line primary styles of interaction.

The remaining four parts look at the support for each of the key activities of machine assisted theorem proving: reflection, planning, reuse and articulation. We note the distinctive features of each system which support or hinder these activities.

3.1 CADiℤ

CADiℤ (Computer Aided Design In Z), is a tool for browsing documents written using the Z specification language, see [14, 23]. It incorporates a type checker and backward theorem proving capabilities, using a logic originally based on the sequent calculus "$\mathcal{W}$" logic, see [25, 26].

CADiℤ has a primarily graphical direct manipulation style user interface and uses a document preparation tool and a suitable viewer, modified to allow input. Interaction takes place in three windows. One shows the Z document containing the theorem being tackled, another shows a graph of the proof steps so far, and the third shows a list of the goals to which these steps have been applied. In Figure 1, these are at the right, lower-right and left, respectively.

Planning There is no explicit support for the formation of plans but the proof tree display may help the user to extract patterns for future plans and to monitor plan execution. However, the execution of a plan may be disrupted by the unanticipated length of apparently simple subproofs. This makes planning more difficult but has more to do with the way that Z and $\mathcal{W}$ are constructed than the software or its interface. One way to overcome these problems is by using decision procedures to automate such minor proofs. CADiℤ does provide one such, for Presburger arithmetic, see [22].

Reflection CADiℤ originated as a document viewing tool and uses appropriate powerful helper applications; all windows are high-resolution displays and have good Z formatting. In the goal window (left in Figure 1) goals are shown using the correct Z symbols rather than ASCII alternatives. Using the mouse, the user positions the pointer within a goal. Left-clicking highlights the smallest well-formed subterm containing the pointer. A further click causes the next smallest well-formed subterm to be selected instead and so on, until the entire sequent is highlighted, as in Figure 1. In this way the exact structure of the sequent can be uncovered. To see the need for this, consider the subterm $S \cup T \cap U$. If the user cannot remember the relative precedences of the $\cup$ and $\cap$ operators, the fact that the boxed $S \cup \boxed{T \cap U}$ can be selected shows that the expression is equal to $S \cup (T \cap U)$ and not $(S \cup T) \cap U$.

The tree display allows the user to see more or less at a glance how many goals are outstanding, since nodes where further steps are required are denoted with

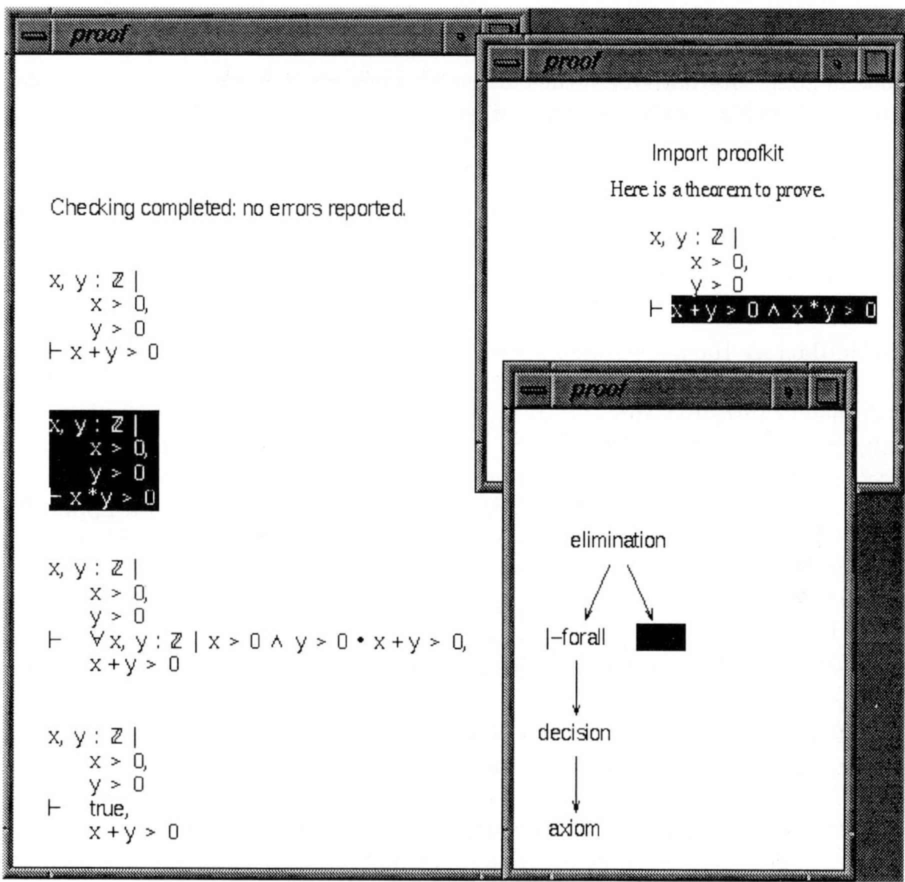

Fig. 1. Part of the screen from the CADiℤ interface

rectangles. The tree and the goal displays are linked. After a sequent has been selected in the goal window, the corresponding tree node can be highlighted using a proof command in the sequent's pop-up menu. A similar link allows the user to find the context sequent for a tree node. In practice the tree must be used to navigate around the proof; new goals are appended to the goal list as they arise, without being labelled, so this goal dump does not allow the structure of the proof to be directly inferred. The most recently selected tree node is highlighted as in the lower-right window of Figure 1, rather than empty. In addition, the proof tree provides confirmatory feedback when a step is applied, since it grows immediately.

Where more than one attempt has been made on a goal, just one of the further developments of the proof is shown, with the first step ringed. The different attempts can be brought up using the pop-up menu for the tree node.

Reuse CADi$\mathbb{Z}$ allows previously proved theorems to be used in constructing new proofs. This can be achieved by using a theorem as a rewrite rule on the current goal. Alternatively, if the current goal is in the same form as the theorem, the proof of that theorem can be replayed to prove the current goal.

Strategies can be encoded using a simple, functional language. Such strategies are written as text files using an external editor, there is no support for strategy construction from within CADi$\mathbb{Z}$. In order to apply a strategy, the user must supply the file name. Since there is no file store browsing capability, the user must again resort to external tools.

Articulation Having selected a well-formed subexpression as described above, the user can view the applicable rules in a pop-up menu. Selecting an item from this menu causes the appropriate rule to be applied. Unless this completes the subproof, new goals will be added to the goal dump and the proof tree is updated accordingly. Then the cycle of interaction repeats.

By demanding selection first and only showing rules which are applicable with that selection the menu is kept to a manageable length. The large number of rules which are applicable anywhere in a goal would otherwise make the menus unreadably long. When a rule requires more than one, the additional arguments must be selected from the displayed goal prior to rule application. A marking mechanism allows such selections to be retained when the user goes on to make further selections.

Of course, such selections alter the applicable and inapplicable sets, and this can lead to the mysterious appearance or non-appearance of rules in the pop-up menu. It can be extremely difficult to understand why a rule is not offered in a menu since the applicability of some rules depends on the order of selection, which is not apparent. Also it can be difficult to infer what the correct selection should be for an intended command, leading to a time consuming, trial and error process of selection.

3.2 IMPS

The Interactive Mathematical Proof System (IMPS) is intended to support rigorous mathematical reasoning, see [3]. Rather than being based on a single monolithic theory, theorems are proved within particular customised theories called "little theories", see [2]. For example, if a result about monoids is being proved, this will be done in a theory of monoids where the monoid laws are axioms. In a monolithic theory, the monoid laws would have to be included as assumptions in the goal.

Backward reasoning is supported by IMPS, with a sequent logic. In order to protect the user from very low level proof, interaction is performed at the the level of "proof commands" or using "macetes", small procedures which apply theorems as rewrites. Interaction in IMPS is primarily menu style.

The emacs editor, the TeX document preparation system and a dvi file (TeX output) previewer are all used by IMPS as helper applications. Figure 2 shows

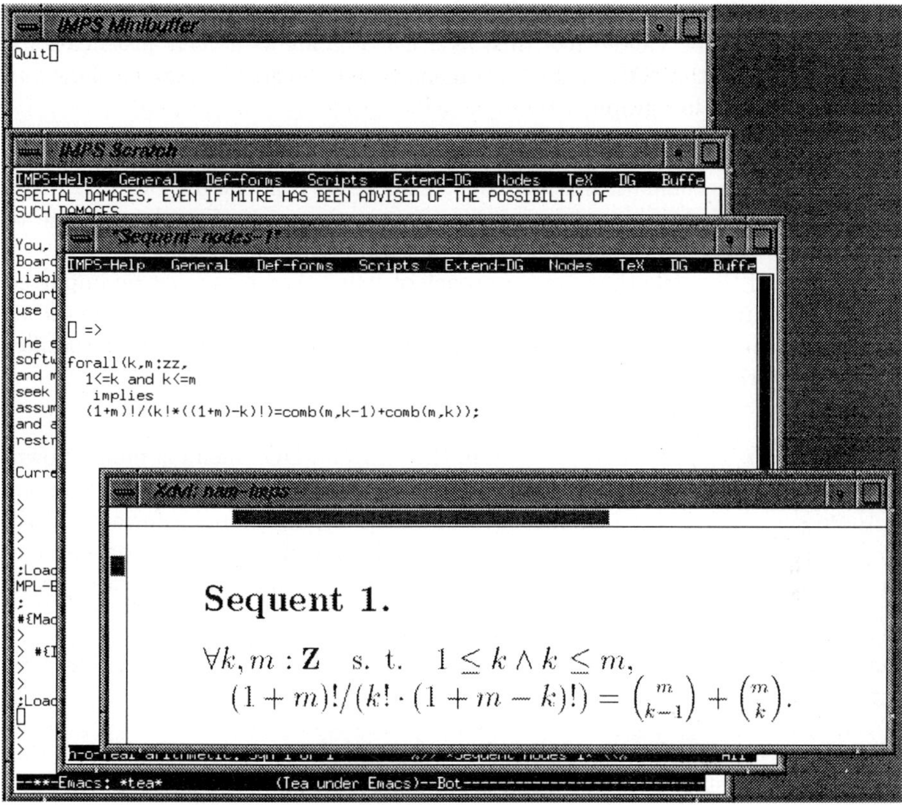

Fig. 2. Part of the screen from the IMPS interface

a TEX formatted goal sequent at the fore and its ASCII representation in the emacs window behind. The back window is the start-up window, from which proofs are launched, and the small emacs window at the top is used for entering menu item numbers plus some question and answer style input.

Planning There is no explicit support for plan formation in IMPS. Various displays of the proof tree or "deduction graph" can be called upon but none show the overall shape of the proof very clearly. The most useful is probably a textual display (not shown), which uses indentation to show the tree structure of the proof. Planning with IMPS seems to benefit from the choice of commands and macetes, in that they fit the mathematical style with which users will presumably be familiar.

Reflection An emacs window is devoted to displaying the current goal sequent in an ASCII representation of mathematical notation. A goal may also be viewed using more standard notation as a TEX formatted dvi file. Indeed a linear history of the proof may be viewed like this, as a list of goals with their justifications. The user can use the "DG" pull-down menu to set the goal display to show any goal in the deduction graph.

Reuse In the mathematical style, IMPS relies heavily on the use of proven theorems to develop the current proof. Macetes do the work of applying these to the current goal, performing the necessary variable instantiations.

Because of the little theorems approach of IMPS, theorems must be imported using theory interpretations. This is designed to be as easy and automatic as possible and it is claimed that the advantage of the simplified theorems justifies any extra work.

Articulation To perform a proof step, the "Extend-DG" menu is pulled down. Each item causes a menu for one kind of proof step to pop-up and selecting one of these steps causes it to be executed. If necessary, additional information will be prompted for in the minibuffer. The emacs display of the current goal changes to show one of the new goals generated by this step and the interaction cycle continues afresh.

3.3 PVS

PVS (Prototype Verification System), see [16, 21], is an industrial strength theorem proving system with an emphasis on debugging and proving properties of a specification in order to contribute to engineering good practice, see [20]. There is a PVS language based on higher-order logic, in which functions and axioms are declared and in which putative theorems are stated. Goals are represented as sequents and commands are entered at a command line from the keyboard. Proofs are developed backward, starting with the intended conclusion.

PVS has an emacs front end, using many different buffers for the proof interaction, the theories, help texts, *etc.*, plus an optional graphical proof tree window, implemented in Tcl/Tk. In Figure 3, the top emacs window shows the current theory, the lower emacs window shows the proof interaction window and the window to the right shows the proof tree.

Planning The usability of PVS for larger proofs relies on the existence of powerful decision procedures and strategies. Planning can thus be performed at a more intuitive level, with automation taking care of the low level proof. Forming and executing plans in PVS is therefore much easier than with some TPAs. However the complexity of the built-in procedures means that the outcome of commands cannot be predicted and some experimentation is nearly always required. Also there is no way, other than by thinking about it, to determine the *applicability* of a rule until it is attempted.

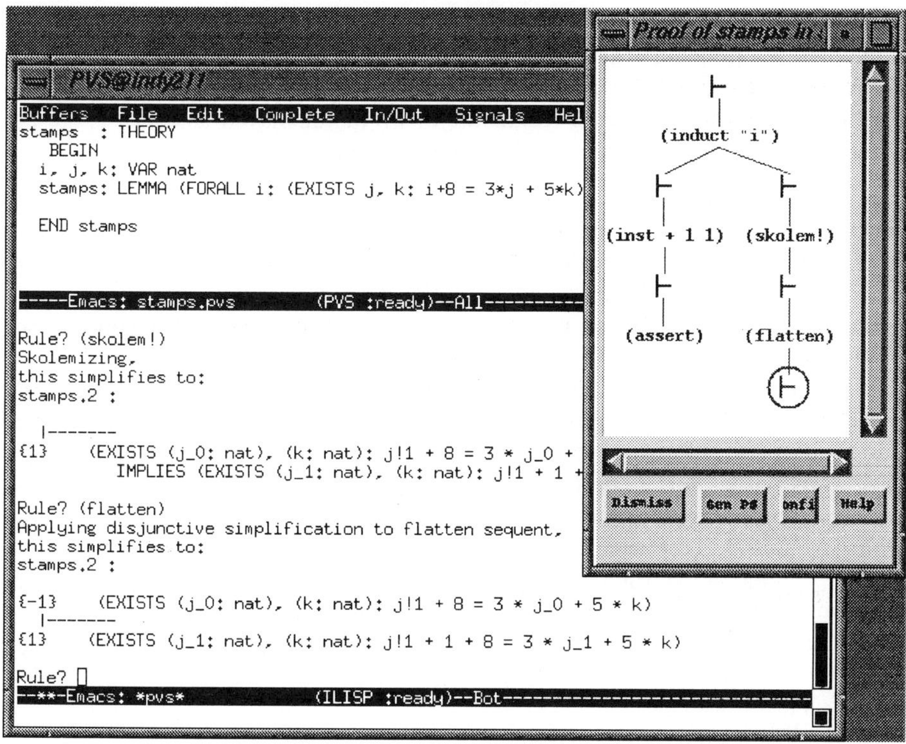

Fig. 3. Part of the screen from the PVS interface

Reflection Since PVS uses a low-resolution text editor as a front end, it cannot display standard mathematical symbols and so ASCII alternatives are used. At each step the current sequent is displayed and, if there is more than one open branch of the proof tree, the user can cycle between them. Numbering the sequents shows which goals are descended from which: in Figure 3 we can see that the current goal is labelled "stamps.2", showing that it is in the second branch.

It is possible to view the current proof steps as formatted text. Branching is indicated by indentation and this makes the tree structure quite clear. It is also possible to see the proof tree as in Figure 3. Only the steps are displayed in full, with all goals being denoted simply by "⊢". This display is for output only and a goal marker cannot be traced back to the full goal sequent in the proof window. The current goal can be identified by its colour and the circle ringing it, and complete branches of the tree are distinguished from incomplete ones using more colours.

Reuse Reuse of theorems is achieved either by introducing them as assumptions or using them as rewrite rules. The buffers containing relevant theorems can be browsed using emacs commands but there is no special support for identifying immediately applicable theorems.

A brief report of the activity being performed by the strategy is given. Strategies may be built from fragments of proofs displayed as text. There are several meta level commands available for writing strategies.

Articulation After displaying the current goal in the proof window, PVS supplies a command line prompt, "Rule?". The user enters a command together with any arguments required and then the display scrolls up to allow a brief commentary and the new current goal to be shown, see Figure 3, beginning the interaction cycle again.

Command arguments must be supplied in the form of various labels and switches. Mistakes are easy to make in calculating and typing such parameters and so alternative forms of some commands are provided which attempt to infer their own parameters. The formation of commands also benefits from the X-windows copy and paste facilities provided by the emacs editor, allowing text from any source to be used as input.

Help buffers allow the user to view either one or the whole range of possible commands and their correct usages. Unfortunately the descriptions are somewhat cryptic unless the reader is familiar with the LISP programming language.

4 Issues

By looking at existing TPAs, we have been able to examine design decisions and their consequences within the context of fully developed software, rather than purely with reference to theory. We wish to learn which factors are important and which are not so important. We want to know which problems have been successfully solved and where gaps in support remain to be filled.

4.1 Planning

In planning, the user is attempting to discover a sequence of instructions which will achieve their goals. In order to be able to plan efficiently, the user of a TPA needs to be able to determine which actions are available and predict their effects. For experienced users, PVS seems to lead the other two interfaces in this respect. Most proving is done with a small number of commands which are easily understood and remembered. The other interfaces make it explicit which steps are possible but this advantage is counterbalanced by the greater difficulty in putting things right when the desired step is not applicable. By profferring the possible steps in menus, CADiℤ and IMPS might seem to be easier for beginners to use. However, it seems to be that case that by the time the user has enough proficiency to understand what the commands mean, this advantage is negligible. That is, because the underlying task is so hard, the user will learn the text commands in less time than it takes to understand the theorem prover.

This does not mean that a command line style is the best way to interact with a theorem prover, merely that some advantages of graphical user interfaces are diminished.

Some PVS command names emphasise their effects on the proof structure rather than on the syntax of the goal. For example, (backward) and-introduction is performed by the PVS `(SPLIT)` command. The effect of this is to fork the current branch of the proof tree into two. Other TPAs, such as CADi$\mathbb{Z}$, provide a (backward) introduction command, which may or may not split the proof branch, depending on the connective being targetted. A good choice of commands and their names can have a large impact on the usability of a system.

A TPA must have a set of commands which maps well to the corresponding actions and is easy to understand and recall in order to facilitate plan formation. It must be possible to get an overview of the progress made so far, which reflects the structure of the plan at a suitable level, if plan evaluation is to be supported well. Plan execution must dovetail with articulation, *i.e.* the articulation of a single command should not disrupt the process of executing a sequence of actions.

4.2 Reflection

The decision to use existing tools to build an interface strongly affects the presentation of a proof which a TPA can provide. For example the emacs front end to PVS precludes enriching the text with non-ASCII mathematical symbols. The question of how best to present mathematics and logic is complicated by many factors. We can refer to the standard texts on typesetting mathematics, such as [24], for clues but typesetters would normally demand an expert consultant for expressions of the size that we encounter with TPAs. The engineering of pretty-printing, the insertion carriage-returns and indentation to make the structure of a document more readily apparent, is well understood, see [15]. However the question of what is "pretty", what is most effective for most readers, remains the subject of debate. Only certain issues have been tackled by psychologists, such as making tree structure apparent, see [5], and the value of mathematical notation over ASCII substitutions, see [9]. In a more general setting the use of layout and diagrammatic representation to enhance reasoning is investigated in [11] but would require much work to apply to the particular domain of TPAs. Lamport's suggestions for formatting large formulae, see [10], have not been tested in conjunction with TPAs but could be effective. Furthermore not all pretty-printers are state of the art and, in any case, under extreme conditions any pretty-printer will perform poorly. The pretty-printing exhibited by CADi$\mathbb{Z}$, IMPS and PVS all help considerably when they are working well but can all be found wanting in some circumstances.

The importance of spacing to show structure, see [12], conflicts with the computational expense of formatting the high-resolution displays necessary to fully exploit this and means that it is not widely used. For example, IMPS will only give a high-resolution presentation of the current goal on request, and even then the formatting is not particularly good. The TEX formatting system that it employs is only capable of automatically typesetting mathematical formulae which do not exceed one line.

A TPA which supports reflection effectively must make the key proof objects available to the user, allowing them to be displayed in ways which genuinely reflect their structure.

4.3 Reuse

Finding previously proven theorems for use in the current proof is supported at the interface level by making the current goal and the store of available theorems easy to browse. However, only IMPS searches for appropriate theorems. With the other interfaces, the user must search through theorems which are not applicable as well as those which are relevant. This may be very important. A study of users of the HOL theorem proving assistant were observed to reprove a theorem which already existed in the theory store, rather than look for it, see [1].

However, reuse is being overtaken by automation as theorem proving technology advances. With powerful decision procedures and strategies, the need to find and use modest theorems is removed; they can be re-proved at the press of a button. Now only more complex theorems need be invoked, and managing such a smaller store of theorems may be much easier and require less support. When such theorems must be located, we would ideally like to allow the user to search in terms meaningful to them, searching for the theorem itself rather than searching for its (perhaps somewhat arbitrary) name. The technologies of case-based reasoning and fuzzy matching could be employed to this end.

The creation and customisation of strategies to meet an individual users requirements merits proper support from an interface. However supporting such programming activity is beyond the scope of the present paper.

4.4 Articulation

In our analysis of articulation we will deal first which commands which develop the proof and second with those commands which allow us to "move around" in the proof, switching from one branch of the proof to another.

Articulating proof steps One of the major difficulties in using TPAs is executing a proof step. Clearly the choice of interaction style is one of the key design decisions in terms of command articulation. Each of the three main interaction styles used by CADi$\mathbb{Z}$, IMPS and PVS have their own advantages and disadvantages for articulating proof commands, summarised in Table 1.

The graphical direct manipulation style of CADi$\mathbb{Z}$ is sometimes difficult to use because the user does not know which well-formed subterm or subterms must be selected in order to make the desired step applicable. This problem is all the more acute if the user is not sure exactly which is the right step. In contrast, the menu style IMPS interface only requires the user to select a goal before a step becomes available, not the specific subterm. The price to be

	Advantages	**Disadvantages**
Graphical direct manipulation	efficient, immediate and less prone to errors, appealing to novices	hard to identify correct objects, hard to combine objects
Menu	only relevant commands are shown	slip can cause unintended command to be executed
Command line	vast range of commands available, redundancy limits slip error impact	must recall name correctly, no cue as to range of relevant commands

Table 1. Interaction styles for proof step articulation

paid is that the IMPS menus offer more steps which the user has no interest in. To limit the length of such menus, hierarchical menus are used, although the TEX formatted macete menu may contain many items and extend over several pages. This design decision means that the user does not have to explore for the correct selection and, very importantly, means that there is more constancy and predictability in the interaction. On the other hand, when information such as a specific occurrence *is* required by IMPS, it must be supplied as a number. Only with a graphical direct manipulation style does it become easy to specify a subterm of an expression.

Some IMPS menus are pull-down menus, whereas others are displayed in the dvi previewer. Selections from such menus are made by entering the number of the desired menu item in a different window, allowing errors which cannot occur with a more immediate style.

A verb-object interaction style, where the action to be performed is selected before the object(s) allows the user to be prompted for appropriate parameter objects. The object-verb style common in graphic direct manipulation interfaces is only effective when the user is clearly aware of the objects and their properties. Because a TPA command may well require several subtle objects to be supplied as parameters, the graphic direct manipulation style can become awkward.

Both CADiℤ and IMPS enjoy the advantage that a valid command is formulated for the user without the need for recall of command names and accurate typing. The disadvantage of the linguistic style employed by PVS is that the user is responsible for forming commands which may be long and complex: a rule-based behaviour. However, the redundancy involved in typing so many characters means that a typing slip will almost always result in the command being rejected. The user is then alerted to the problem and can correct it. With the other interfaces, a selection slip can result in an entirely valid but unintended command. This may lead to some confusion, if the user does not immediately recognise what has happened.

Because it is hard to specify a particular subterm without a graphical direct manipulation interface, PVS and IMPS attempt to infer parameters for rules. It is frequently the case, given a goal and an inference rule, that there is only one subterm at which to apply the rule. And where there is more than one, either the first one can be used (PVS) or the user can be prompted for more information (IMPS).

There is no simple recommendation which can be given to TPA interface designers on command articulation but the comments here highlight some of the issues and trade-offs to be borne in mind.

Navigation Since proofs have a tree or graph structure, the obvious way to present them is with a pictorial representation of the graph. If the user can interact with the graph display, as with CADiℤ, then they can navigate around the proof quickly and naturally. However, a graph display can become too large to fit into a convenient window or, if scaled down, become unreadably small. It is not clear how one can naturally interact with some a complex entity but the eliding of some completed or not yet examined branches and the compaction of now irrelevant sequences of steps may provide the key.

5 Conclusions

We have investigated the design of three TPA interfaces exhibiting three different interaction styles. By identifying planning, reuse, reflection and articulation as mechanisms vital to the usability of TPAs we have been able to analyse the support provided by the interfaces. We have been better able to understand the rôle of helpful features and to locate usability problems in these specific categories of activity.

We have paved the way for the modelling process which is required to properly understand interaction in theorem proving. Perhaps more than in any other domain, a formal model is needed in order to understand exactly what interactions with a TPA mean. The complexity and size of some interactions is such that a less rigorous approach risks confusion, whereas a formal model should allow us to structure information precisely, hiding and revealing it as necessary. With such a semantic understanding we aspire to significant improvements in TPA interface design. The next stage in our work will be to develop such a model and to use it to formalise the notions in this paper.

In our attempt to understand TPA interfaces, it may be useful to adopt Rasmussen's categorisation of cognitive mechanisms into skill, rule and knowledge based behaviours, see [19]. Briefly, skill-based behaviour includes sensory motor skill, rule-based behaviour is the application of straightforward "if ...then ..." rules, and knowledge-based behaviour requires deep, model-based knowledge and understanding of the task. In the domain of TPAs we suppose that skill-based behaviours include typing and using a mouse to point and click; rule-based behaviours include spelling command names and applying inference rules; and knowledge-based behaviours include planning and introducing new information.

Note that we have focussed on the initial discovery stage of proof development, more than on the evolution and maintenance of proofs. These latter stages are also important in mechanical verification and should be addressed in future work. In addition, we have not looked at the preparation process, where datatypes are defined, induction theorems are verified and goals are formulated. This may account for a great deal of the overall effort required for a proof and would ideally be supported by tools within the TPA. Although CADi$\mathbb{Z}$, HOL and PVS all provide immensely valuable type checkers for the specification, this is more or less the full extent of such support.

It is clear that the design issues arising from our examination of TPAs are applicable in other domains. The planning which is supported by the interface to a TPA is similar to the planning which must occur in other problem solving arenas, for example fault diagnosis and recovery, or multitask scheduling. The ability to present mathematics well, taking advantage of a non-static screen presentation, has applications in all mathematical and algebraic software. We feel that the exacting tests induced by TPAs provide a very suitable way to understand the key factors in these domains.

References

1. J. S. Aitken, P. Gray, T. Melham, and M. Thomas. A Study of User Activity in Interactive Theorem Proving. Submitted to J. Symbolic Computing.
2. W. M. Farmer, J. D. Guttman, and F. J. Thayer. Little theories. In Kapur [8], pages 567–581.
3. W. M. Farmer, J. D. Guttman, and F. J. Thayer. *The IMPS User's Maunual.* The MITRE Corporation, 1994.
4. G. Gentzen. Investigations into logical deduction. In M. E. Szabo, editor, *The Collected Papers of Gerhard Gentzen.* North-Holland, 1969.
5. D. J. Gilmore. Structural Visibility and Program Comprehension. In M. D. Harrison and A. F. Monk, editors, *People and Computers: Designing for Usability,* BCS Workshop series, pages 527–545. Cambridge University Press, 1986.
6. Li Gong, Patrick Lincoln, and John Rushby. Byzantine agreement with authentication: Observations and applications in tolerating hybrid and link faults. In *Dependable Computing for Critical Applications—5,* pages 79–90, Champaign, IL, September 1995. IFIP WG 10.4, preliminary proceedings.
7. J.-M. Hoc, T. R. G. Green, R. Samurçay, and D. J. Gilmore, editors. *Psychology of Programming.* Computers and People. Academic Press, 1990.
8. D. Kapur, editor. *Automated Deduction — CADE 11,* volume 607 of *Lecture Notes in Computer Science.* Springer-Verlag, 1992.
9. D. Kirshner. The Visual Syntax of Algebra. *Journal for Research in Mathematical Education,* 20(3):274–287, 1989.
10. Leslie Lamport. How to Write a Long Formula.
Online at http://www.research.digital.com/SRC/proofs/proofs.html, 1993.
11. J. H. Larkin and H. A. Simon. Why a Diagram is (Sometimes) Worth Ten Thousand Words. *Cognitive Science,* 11:65–99, 1987.

12. W. N.Dember and J. S. Warm. *Psychology of Perception*, chapter 7: Perception of Form. Holt, Rinehart and Winston, 1979.

13. W. M. Newman and M. G. Lamming. *Interactive System Design.* Addison-Wesley, 1995.

14. J. Nicholls. Z Notation: version 1.2. Technical report, Z Standards Panel, University of Oxford, Sept. 1995.

15. D. C. Oppen. Pretty-printing. *ACM Transactions on Programming Languages*, 2(4), Oct. 1980.

16. S. Owre, J. Rushby, and N. Shankar. PVS: A Prototype Verification System. In Kapur [8], pages 748–752.

17. Philippe Palanque. Towards an integrated proposal for Interactive Systems design based on TLIM and ICO. In F. Bodart and J. Vanderdonckt, editors, *Eurographics Workshop on Design, Specification and Verification of Interactive Systems: Informal Proceedings*, pages 69–85, Belgium, 1996. Computer Science Dept. U. Namur.

18. G. Polya. *How To Solve It.* Princeton University Press, second edition, 1957.

19. J. Rasmussen. Skills, Rules and Knowledge; signals, signs and symbols, and other distinctions in human performance models. *IEEE Transactions on Systems, Man and Cybernetics*, 13(3):257–266, 1983.

20. John Rushby. Design choices in specification languages and verification systems. In Phillip Windley, editor, *International Workshop on the HOL Theorem Proving System and its Applications*, pages 195–204. IEEE Computer Society, 1991.

21. N. Shankar, S. Owre, and J. Rushby. *The PVS Proof Checker: A Reference Manual.* SRI International, 1995.

22. R. Shostak. On the SUP-INF method for proving Presburger formulas. *Journal of the ACM*, 24(4):529–543, Oct. 1977.

23. J. M. Spivey. *The Z Notation.* Prentice Hall International, 2nd edition, 1992.

24. E. Swanson. *Mathematics into Type.* American Mathematical Society, 1971.

25. I. Toyn and J. Hall. *Proving Conjectures using Cadiz.* University of York, 1995.

26. J. C. P. Woodcock and S. M. Brien. $\mathcal{W}$: A logic for Z. In J. E. Nicholls, editor, *Z User Workshop, York 1991*, Workshops in Computing, pages 77–96. Springer-Verlag, 1992.

This article was processed using the LaTeX macro package with LLNCS style

Validating Properties of Component-based Graphical User Interfaces

Peter Bumbulis[1], P. S. C. Alencar[1], D.D. Cowan[1], C.J.P. Lucena[2]

[1] Computer Science Department, University of Waterloo, Waterloo, Ontario, Canada. Emails: {peter, alencar, dcowan}@csg.uwaterloo.ca

[2] Departamento de Informática, Pontifícia Universidade Católica do Rio de Janeiro, Rio de Janeiro, Brazil. Email: lucena@inf.puc-rio.br

Abstract. In this paper we describe a validation process for graphical user interfaces based on existing toolkits and higher-order logic as mechanized in the HOL system. The underlying approach uses a single specification for constructing both implementations (prototypes) for experimentation and models for formal reasoning. The formal models allow the designer to verify mechanically specific requirements imposed on the user interface such as those found in safety- or security-critical applications. We illustrate our approach with an example that shows how the proof process works for behavioral properties that have been expressed in a rule-based fashion.

1 Introduction

Various toolkits are available which allow user interface designers to build graphical user interfaces (GUIs) rapidly. Unfortunately verification that these GUIs behave as intended (in the sense that they possess certain formally expressed properties) is difficult, and yet verification of behaviour is of significant concern for safety- and security-critical applications. In this paper we show how to construct a single formal model for GUI-based applications from which we can both validate behavioral properties and construct implementations. The approach that we propose can be adapted for use with most existing toolkits.

Our approach specifies GUIs as a hierarchy of interconnected component instances [16]. From such descriptions we automatically derive both prototypes (implementations) for experimentation and a variety of corresponding formal models suitable for mechanical reasoning.

Deriving formal models directly from most GUI implementations which can then be used for mechanical reasoning, is a difficult task. We introduce an alternative approach which uses the following three steps: (i) devise a formalism which depends on the GUI toolkit, and which describes GUIs as based on a set of primitive components and an interconnection language (IL); (ii) construct implementations (and corresponding models) for each of the primitive components introduced in the first step; (iii) specify the GUIs using IL and the components; and from these descriptions generate implementations. The advantage of this approach is that we also can automatically generate models suitable for formal reasoning from the IL description.

In this paper we present one approach to ensuring that GUIs behave as intended by applying techniques of program verification rather than the more common approach based on concurrency analysis [10]. First, we present a description of the language used to model interface behavior, and show how it can be used to model both components and complete GUIs. We use higher order logic as mechanized by the HOL system to model behavior, and express properties to be validated where the HOL deductive apparatus is used to support the validation process. We then follow this with an illustration of how we might reason about these HOL descriptions. The approach that we propose is to construct a special purpose logic for reasoning about a particular type of behavioral property and then use this to prove specific properties. We illustrate how this might be used to verify that a simple user interface satisfies a state invariant.

2 Structuring GUI-based Applications

In this section we sketch our proposed approach to structuring GUI-based applications. Our approach involves devising a component-oriented formalism for describing GUIs and fashioning a set of presentation primitives for it from the presentation primitives provided by some existing (callback-based) toolkit. We use the name IL to denote one possible formalism that we have devised for describing GUIs.

2.1 IL

IL is a interconnection language, much like those proposed for structuring distributed systems (in particular Darwin [12].) However, IL components will represent widgets, not processes, and IL connections will represent the binding of procedures to call sites, not FIFO communications channels.

We give a taste of IL by example. As our aim is primarily to illustrate the concepts involved, we will use simple examples that are easily described and manipulated; they are not meant to be considered as real user interface descriptions. Our first example is the user interface described in Figure 2. It consists of a dial and a slider connected so that they track each other. To simplify matters, Sliders and Dials provide only layout parameters; more realistic components would provide additional parameters for specifying properties such as their initial, minimum and maximum value. This example makes use only of some of IL's features; however, the features presented will be sufficient fo illustrate our approach to modeling GUI implementations.

An IL description of a GUI consists of one or more component definitions. A component definition minimally consists of a name. For example `component Button` defines a component named `Button`. Components act as templates from which instances are created. A unique name must be provided for each instance as it is declared. For example`b:Button` declares a `Button` named `b`. Each component instance makes a number of named *ports* available for binding. Each port has a *polarity*: *input* (<) or *output* (>); only ports of opposite polarity may be bound together. The ports provided by a component are specified when the component is defined. For example

```
component Button clicked>
component TextField clear< setValue< entered> changed>
```

defines two components `Button` and `TextField`. `Buttons` provide an output port named `clicked` and `TextFields` provide input ports named `clear` and `setValue` as well as output ports named `entered` and `changed`. Instances `b` and `t` of these components can be declared with the `clicked` port of `b` and the `clear` port of `t` bound as follows:

```
b:Button t:TextField
b.clicked --> t.clear
```

IL component definitions only determine how instances can be interconnected; they do not provide a description of instance behavior. We will implement component instances using widgets, with output ports corresponding to callbacks and input ports corresponding to methods. Given this interpretation, IL can also be viewed as a dataflow language with bindings representing the flow of events between components. Events are introduced as a result of an action by the user (or environment); these events then flow from one component to another, being transformed as they go. IL's binding rules allow a number output ports to be connected to a single input port but disallow the converse: at most one input port can be connected to an output port. Events can be modified as they flow through a binding with the use of functions called *filters*. For example,

```
b:Button t:TextField
b.clicked -[x => "default"]-> t.setValue
```

applies a constant function (with value `"default"`) to the values that flow from `b.clicked` to `t.setValue`: whenever a value `x` is produced at `b.clicked` the value `"default"` is presented to `t.setValue`.

The body of a filter is not restricted to constants; simple expressions can also appear. These expressions are built from constants, variables, a tuple building operator, and function applications. All IL functions are unary (nullary functions can be expressed as functions on the unit type, functions of arity greater than one can be expressed as functions on tuples.) IL provides no predefined functions: as with primitive components, functions must be defined using other means. Besides ports, components can make available a number of functions called *observers*. Observers only provide a view of an instance's state: they are not allowed to modify it. Further, observers can only be referenced in filters. The observers and ports provided by an instance form its interface. Observers are specified when a component is defined and they are denoted by a name followed by the symbol '`@`'. For example

```
component TextField clear< setValue< entered> changed> value@
```

provides another definition of `TextField`; one which provides an observer named `value`. When queried, the `value` observer provided by a text entry field returns the field's value.

Composite components consist of an interconnected collection of simpler components; they provide a means of structuring IL descriptions. The definition of a composite component consists of an interface definition, just as for primitive components, along with a body consisting of a number of component instantiations and port bindings. The definition of a simple composite component is shown below:

```
component Main {
    b:Button t:TextField
    b.clicked --> t.clear
}
```

The ports and observers provided by a composite component are defined in terms of the ports and observers provided by its constituent components through bindings. Figure 1

```
component D o1> o2> i1< i2< v@ {
    c1:C c2:C
    i1   --> c1.i
    c1.o --> c2.i
    c2.o --> o1
    i2   --> o2
    v x => c1.v(x)
}
```

Fig. 1. Possible binding styles.

illustrates the possible binding styles. Adapting Darwin terminology, bindings are either *inbound* (`i1 --> c1.i`), *internal* (`c1.o --> c2.i`), *outbound*(`c2.o --> o1`), or *forwarded* (`i2 --> o1`.)

An IL description of a user interface simply consists of a collection of component definitions. By convention, the user interface described is an instance of the component named `Main`.

We address the concern of whether IL can be used to describe GUIs in practice by noting that it is similar to formalisms implicitly provided by a growing number of commercially available GUI construction tools such as PARTS Workbench, Visual Age, and Visual AppBuilder. IL differs from the existing formalisms primarily in the choice of relationships explicitly represented with connections and containment; these differences primarily were motivated by the desire to simplify modeling and reasoning. Patterning IL on these commercially available formalisms has a number of advantages including: (1) there is empirical evidence that GUI designers find such formalisms easy to understand and use, and (2) manuals for (tools that use) these formalisms provide a wealth of documentation on how to construct practical GUIs using them.

There are three different roles associated with the development of IL-based user interfaces: the *user interface designer*, or just *designer*, the *developer*, and the *verifier*. The description of these roles give us an indication about how our approach scales to larger examples. The tasks of the designer and developer can be characterized as using and constructing primitive components, respectively. The designer typically requires greater problem domain understanding but less programming skill than the developer. The designer constructs IL descriptions of user interfaces using primitive components supplied by the developer. User interface construction essentially consists of selecting and connecting components. If a required component is not available the designer provides the developer with its specification; the developer then uses traditional software development techniques to construct the actual implementation. Designing the primitive components with reuse in mind reduces the amortized cost of development.

The verifier works in concert with the designer and the developer and is responsible for ensuring that prototypes meet formally expressed requirements. The verifier generates formal models from the IL descriptions and uses these as the basis for validation. To have confidence in the results obtained, the verifier must ensure that models are accurate, and that reasoning is sound. The task of ensuring that models accurately reflect implementations can be reduced to ensuring that the primitive components are accurately modeled. If the software development technique used to construct the primitives does not provide the necessary assurance, testing can be used. To ensure that reasoning is sound, mechanical validation is used. Although proofs are not automatically generated, they can be reused.

3 Modeling User Interface Behavior

Models as well as implementations must be supplied for each of the primitive components, and some notation must be provided for expressing these models. We express user interface behavior using a simple non-deterministic language, modeling components essentially as a collection of code fragments. We define the semantics of this language by defining HOL predicates corresponding to the various statements.

We model behavior as an alternating sequence of states and events (actions) called *runs*. We identify programs (or program statements) with sets of runs, each run representing a possible computation. A variety of sequence-based approaches to modeling programs have appeared in the literature [11, 13, 14, 2, 17]. Our approach to constructing sequences is most closely related to those presented in [13, 14, 17].

In the following text $\langle s_0, e_0, s_1, \ldots, e_n, s_{n+1}\rangle$ and $\langle s_0, e_0, s_1, \ldots\rangle$ will denote finite and infinite runs, respectively. We define a number of functions and operations on runs: $\mathsf{start}\, e$ returns the first state of the run e, $\mathsf{idle}\, s$ returns the run $\langle s\rangle$ consisting of a single state s, $\mathsf{step}\, s\, e\, s'$ returns the run $\langle s, e, s'\rangle$, and $\mathsf{diverge}\, s$ returns the run $\langle s, \tau, s, \tau, \ldots\rangle$. The distinguished event τ is used as a placeholder for internal (hidden) events. We say that two runs e_1 and e_2 are $\mathsf{composable}$ if the first run is finite and the initial state of the second run is identical to the final state of the first. The operator '$\bullet$' is used to splice two runs together.

3.1 A Command Language and its Mechanization

The Command Language: We express behavior using a notation based on Nelson's extension [15] of Dijkstra's guarded command language [6], which we use as a simple, non-deterministic, programming language. Each statement is associated with a predicate or *guard* that describes the possible states in which the statement can be activated. Three primitive statements and some fundamental operators are defined namely: skip (do nothing); abort (loop forever); $\mathsf{assign}\, E$ (when activated in state s, results in a state $E\, s$); $c_1 \,[]\, c_2$ (activate either c_1 or c_2); $c_1 \fatsemi c_2$ (activate c_1, then activate c_2); $P \longrightarrow c$ (activate c if in a state where P is true or fail otherwise); $\mathsf{do\_od}\, c$ (activate c until it fails). More conventional programming language constructs can be built from these primitives.

For example, "if P then c_1 else c_2" can be expressed as $(P \longrightarrow c_1) \,[]\, (\neg P \longrightarrow c_2)$ and "while P do c" can be expressed as $\mathsf{do\_od}(P \longrightarrow c)$.

We define the semantics of this notation in terms of a program executed by a machine. Executing a statement in the program causes a change to the machine's internal state. The meaning assigned to a program is the set of all state sequences (finite or infinite) that may result from the program's execution.

To handle events, we augment this notation with an operator for flagging their occurrence:

atomic $x\ c$ flag the occurrence of event x (the command c expresses the effect of x)

Our execution sequences are now state sequences interspersed with events. To simplify manipulation, we represent these execution sequences with an alternating sequence of states and events, using a distinguished event τ for padding. We refer to these sequences as *runs*. A run consists of a start state followed by a (finite or infinite) sequence of actions, each action consisting of an event-state pair. We represent runs in the obvious fashion: a pair whose first element holds the start state and whose second element holds the subsequent actions.

Rather than providing a concrete syntax for our notation we identify statements with the HOL predicates that model them. The predicates that we use to model statements are viewed as defining sets of runs. We explicitly construct these sets for each of the primitive statements; the various operators for constructing more complex statements are defined in terms of set operations.

Commands: We can now define the semantics of commands. We model commands as sets of runs. If we model commands as sets of runs, each run describing a possible behavior, then the guard of a command is just the set of all states that start some run in the command. The function grd maps a command to its guard:

Definition 1 $\vdash \forall c\, s.\, \mathsf{grd}\, c\, s = (\exists e.\, c\, e \wedge (\mathsf{start}\, e = s))$

The semantics of the primitive statements are defined by explicitly describing the sets of runs that model them.

Theorem 1 $\vdash \mathsf{skip} = (\lambda e.\, \exists s.\, e = \mathsf{idle}\, s)$

Definition 2 $\vdash \forall e.\, \mathsf{abort} = (\lambda e'.\, \exists s.\, e' = \mathsf{diverge}\, s)$

Definition 3 $\vdash \forall E.\, \mathsf{assign}\, E = (\lambda e.\, \exists s.\, e = \mathsf{step}\, s\, \tau\, (E\, s))$

We define the semantics of the various operators by modeling them as operations on sets.

Definition 4 $\vdash \forall c_1\, c_2.\, c_1 \,[]\, c_2 = c_1 \vee c_2$

Definition 5 $\vdash \forall P\, c.\, P \longrightarrow c = (\lambda e.\, P\, (\mathsf{start}\, e) \wedge c\, e)$

Definition 6

$\vdash \forall c_1\, c_2\, e.\ \$\,;\, c_1\, c_2\, e = \quad c_1\, e \wedge \neg \mathsf{finite}\, e$
$\vee(\exists e_1\, e_2.\, c_1\, e_1 \wedge c_2\, e_2 \wedge \mathsf{composable}\, e_1\, e_2 \wedge (e = e_1 \bullet e_2))$

The command $\mathsf{do\_od}\, c$ activates c until no longer possible. We define the semantics of $\mathsf{do\_od}$ in parts:

Definition 7

$\vdash (\forall c.\, \mathsf{iter}\, c\, 0 = \mathsf{skip}) \wedge (\forall c\, n.\, \mathsf{iter}\, c\, (\mathsf{SUC}\, n) = \mathsf{iter}\, c\, n\, ;\, c)$
$\vdash \forall c.\, \mathsf{finrep}\, c = (\boldsymbol{\exists} n.\, \mathsf{iter}\, c\, n\, ;\, (\boldsymbol{\neg}\, (\mathsf{grd}\, c) \longrightarrow \mathsf{skip}))$
$\vdash \forall c\, e.\, \mathsf{iterw}\, c\, e = (\exists f.\, (\forall i.\, c\, (f\, i) \wedge \mathsf{composable}\, (f\, i)\, (f\, (\mathsf{SUC}\, i))) \wedge$
$(e = \mathsf{inf\_fjoin}\, f))$
$\vdash \forall c.\, \mathsf{infrep}\, c = \mathsf{iterw}\, c\, ;\, \mathsf{abort}$
$\vdash \forall c.\, \mathsf{do\_od}\, c = \mathsf{finrep}\, c \,[]\, \mathsf{infrep}\, c$

In the above definition '$\boldsymbol{\exists}$' is just '$\exists$' "lifted" to work on predicates: for all b, $(\boldsymbol{\exists} i.\, R\, i)\, b = (\exists i.\, R\, i\, b)$. In general, a bolded logical connective denotes the corresponding boolean connective lifted to work on predicates.
The statement $\mathsf{finrep}\, c$ contains all of the runs that result from a finite number of activations of c and $\mathsf{infrep}\, c$ contains all of the runs that result from an infinite number of activations. The use of abort in the definition of finrep is needed to handle cases such as $\mathsf{do\_od\ skip}$. The atomic operator introduces events and elides intermediate states.

Definition 8

$\vdash \forall x\, c.\, \mathsf{atomic}\, x\, c = (\lambda e.\, \exists e'.\, c\, e' \wedge (e = (\mathsf{size}\, e' \prec \omega \Rightarrow \mathsf{step}\, (\mathsf{start}\, e')\, x\, (\mathsf{final}\, e')$
$\mid \mathsf{diverge}\, (\mathsf{start}\, e'))))$

The command $\mathsf{atomic}\, x\, c$ is modeled as the set of all runs of the form $\langle s, x, s' \rangle$ such that $\langle s, \ldots, s' \rangle$ is a run of the original command c together with runs of the form $\langle s, \{\tau, s\}^\omega \rangle$, one for each state s that starts an infinite run in c.

3.2 Generating Models from IL Descriptions:

We first generate models for each of the composite components and then instantiate the model of a distinguished component, `Main`, to compose a model for the user interface. We use simple examples that are are easily described and manipulated. Our example is the user interface described in Figure 2 that consists of a dial and a slider connected so they are synchronized.
Modeling Components: We model each component `c` with a predicate C. The predicate C has parameters, including one for each port and observer provided by the component. In particular, if `c` has interface `component C(`$p_1, \ldots, p_m$`)` $x_1 \ldots x_n$ (each x_i being of the form 'x_i^*`<`', 'x_i^*`>`' or 'x_i^*`@`') then the predicate modeling `c` will be of the form $\mathsf{C}\, i\, c\, s\, q\, e\, p_1 \ldots p_m\, x_1^* \ldots x_n^*$. Ignoring all but the first two parameters for the moment, C is defined so that $\mathsf{C}\, i\, c$ will be true iff i and c specify the initial state and

```
component Window(width,height)
component Dial(parent,x,y,width,height) set< changed>
component Slider(parent,x,y,width,height) set< changed>

component Main {
    f:Window(170,220)
    d:Dial(f,5,5,160,160)
    s:Slider(f,5,165,160,60)
    s.changed --> d.set
    d.changed --> s.set
}
```

Fig. 2. A simple IL description.

subsequent behavior of instances of that component, respectively. Given such a predicate for the component `Main`, we can easily construct a command that expresses user interface behavior. For example, many of the theorems of interest will be of the form

$$\vdash \forall i\, c.\ \mathsf{Main}\, i\, c \Rightarrow \mathsf{P}\,((\lambda s.\, s = i) \longrightarrow \mathsf{do\_od}\, c)$$

for some predicate P. Such a theorem states that if i and c describe the initial state and subsequent behavior of the user interface described by `Main`, then $(\lambda s.\, s = i) \longrightarrow \mathsf{do\_od}\, c$ possesses the property P, i.e., P is a property of the user interface described by `Main`.

We express the predicates representing components essentially as a set of code fragments: one for each input port and observer (in the case of `Sliders`, just `set`) and one for expressing the behavior of an instance. For example, Figure 3 contains the definition

$$\begin{aligned} &\mathsf{Slider}\, i\, c\, s\, q\, e\, \mathit{parent}\, x\, y\, \mathit{width}\, \mathit{height}\, \mathit{set}\, \mathit{changed} = \\ &\quad (\mathit{set} = (\lambda v.\, \mathsf{if}\, (q\, (\lambda n.\, \neg(n = v)))\, (\mathsf{assign}\, (s\, (\lambda n.\, v))\, ⨾\, \mathit{changed}\, v))) \\ &\quad \wedge (i = 0) \\ &\quad \wedge (c = (\exists v.\, \mathsf{atomic}\, (\mathsf{numEv}\, e\, v)\, (\mathit{set}\, v))) \end{aligned}$$

Fig. 3. A model for `Sliders`.

of a predicate suitable for modeling `Sliders`[3].

Generating Models for Composite Components: In our approach the definitions for the predicates modeling composite components must be generated from their IL descriptions. We illustrate this by example. The HOL code generated from the IL description of Figure 2 defines the predicate shown in Figure 4. While the term defining

[3] The function numEv takes a string (identifying the type of event) and a number and produces an event.

$$\begin{array}{l}\mathsf{Main}\, i\, c\, s\, q\, e = \exists i_1\, c_1\, i_2\, c_2\, set_2\, i_3\, c_3\, set_3 . \\ \quad \mathsf{Window}\, i_1\, c_1\, (\lambda f.\, s\, (\lambda(v_1, v_2, v_3).\, (f\, v_1, v_2, v_3)))\, (\lambda P.\, q\, (\lambda(v_1, v_2, v_3).\, P\, v_1)) \\ \qquad (\mathsf{CONS}\, 1\, e)\, 170\, 220 \\ \quad \wedge\, \mathsf{Slider}\, i_2\, c_2\, (\lambda f.\, s\, (\lambda(v_1, v_2, v_3).\, (v_1, f\, v_2, v_3)))\, (\lambda P.\, q\, (\lambda(v_1, v_2, v_3).\, P\, v_2)) \\ \qquad (\mathsf{CONS}\, 2\, e)\, \mathsf{VOID}\, 5\, 5\, 160\, 160\, set_2\, set_3 \\ \quad \wedge\, \mathsf{Dial}\, i_3\, c_3\, (\lambda f.\, s\, (\lambda(v_1, v_2, v_3).\, (v_1, v_2, f\, v_3)))\, (\lambda P.\, q\, (\lambda(v_1, v_2, v_3).\, P\, v_3)) \\ \qquad (\mathsf{CONS}\, 3\, e)\, \mathsf{VOID}\, 45\, 165\, 160\, 60\, set_3\, set_2 \\ \quad \wedge\, (i = (i_1, i_2, i_3)) \\ \quad \wedge\, (c = c_1 \,[]\, c_2 \,[]\, c_3)\end{array}$$

Fig. 4. A model for `Main` components.

Main looks unwieldy, it is simple to generate and can be easily manipulated using HOL. The terms that we use to define predicates such as Main are expressed as a number of conjuncts: one for each of the constituent instances, one for each of the ports and observers provided by the composite component (in this case there are none), and one each for describing the initial state and subsequent behavior of instances of the composite component. The state of a composite instance is represented with a tuple, each component holding the state of a different constituent instance. In the current example, the state of a `Main` instance is represented with a triple (v_1, v_2, v_3), with the components v_1, v_2, and v_3 holding the states of the constituent `Window`, `Slider` and `Dial` instances, respectively. The third and fourth parameters passed to the predicates Window, Slider and Dial enforce this representation. The functions passed as the third parameters are used to extend state assignments.[4] In general, we will use a function of the form $\lambda f.\, s\, (\lambda(v_1, \ldots, v_n).\, (v_1, \ldots, f\, v_k, \ldots, v_n))$ for this purpose. If f is a state assignment function then $\lambda(v_1, \ldots, v_n).\, (v_1, \ldots, f\, v_k, \ldots, v_n)$ extends f to act on the kth component of an n-tuple. The function s (provided as the third parameter of the predicate modeling the composite component) extends this function to act on the global (user interface) state. The functions passed as the fourth parameters are used to extend predicates similarly. In general, these functions will be of the form $\lambda P.\, q\, (\lambda(v_1, \ldots, v_n).\, P\, v_k)$.

4 The HOL Validation of User Interface Properties

We express the properties to be validated as predicates on sets of runs. Validating that a user interface model possesses a property P entails proving a theorem of the form

$$\forall i\, c.\ \mathsf{Main}\, i\, c\, (\lambda f.\, f)\, (\lambda P.\, P)\, [\,] \Rightarrow \mathsf{P}\, ((\lambda s.\, s = i) \longrightarrow \mathsf{do\_od}\, c)$$

[4] State assignment functions appear, for example, as parameters of assign operators.

where Main is a predicate modeling the user interface. The approach to constructing proofs for such theorems involves mechanizing a logic for each property family of interest. This consists of devising a suitable representation for those properties and deriving a set of inference rules from the run-based semantics given to commands.

The proofs that we construct use a forward reasoning predicate transformer sp that has the following operational semantics: if c is a command, then for any set of states (i.e. state predicate) P, $\mathsf{sp}\, c\, P$ is the set of states in which execution of c can terminate if started from a state in P. This predicate transformer was first introduced by Francez in [8] and is referred to as the strongest postcondition predicate transformer by Dijkstra and Scholten [7]. Given our representation of commands, sp can be expressed as follows:

Definition 9 $\vdash \forall c\, P.\, \mathsf{sp}\, c\, P = (\lambda s.\, \exists e.\, P\, (\mathsf{start}\, e) \wedge c\, e \wedge \mathsf{finite}\, e \wedge (\mathsf{final}\, e = s))$

A state s satisfies $\mathsf{sp}\, c\, P$ iff we can find a finite run e in c that starts in P and terminates in s. We should note that equations comprising inductive definitions of sp are used for simplifying terms involving sp. Unfortunately the equation for assignment statements is complicated: $\mathsf{sp}\,(\mathsf{assign}\, E)\, P = (\lambda s.\, \exists s'.\, P\, s' \wedge (s = E\, s'))$. However, we can simplify this expression by carefully choosing our representation of state predicates. We introduce a function constant xs

Definition 10 $\vdash \forall s\, P.\, \mathsf{xs}\,(s, P) = (\lambda t.\, (t = s) \wedge P)$

for representing state predicates. Note that a predicate of the form $\mathsf{xs}(s, P)$ is true of at most one state, s, depending on whether P is true or not. In such predicates we refer to the terms s and P as path expressions and path conditions, respectively. For states represented in this fashion, the equation for assignment statements simplifies to $\mathsf{sp}\,(\mathsf{assign}\, E)\,(\mathsf{xs}\,(s, P)) = \mathsf{xs}\,(E\, s, P)$. Note that all state predicates can be expressed in terms of xs.

Theorem 2 $\vdash \forall P.\, P = (\exists s.\, \mathsf{xs}\,(s, P\, s))$

4.1 State Invariants

We now illustrate one approach to proving that our construction of a user interface model maintains a state invariant. We first define what it means for a predicate to be held invariant by a command. Proving that commands maintain invariants directly from this definition is impractical, and so we introduce a number of theorems that support a syntax-directed approach to constructing such proofs. We then provide an example that illustrates our approach.

We say that a state predicate Q **holds** on a run e if it is true of every state in e.

Definition 11 $\vdash \forall Q\, e.\, \mathsf{holds}\, Q\, e = (\forall i.\, \neg \mathsf{size}\, e \prec i \supset (i \prec \omega \supset Q\, (\mathsf{st}\, e\, i)))$

The definition of holds is complicated because we use lnums, not nums (natural numbers) to index the states in a run. A state predicate Q is held invariant by a command c if it holds on all runs in the command. This notion of invariance can be expressed using the predicate sinv:

Definition 12 $\vdash \forall Q\,c\,P.\, \mathsf{sinv}\,Q\,c\,P = (\forall e.\, c\,e \supset (P\,(\mathsf{start}\,e) \supset \mathsf{holds}\,Q\,e))$

The formula $\mathsf{sinv}\,Q\,c\,P$ is true iff Q holds for all runs in c that start in P. We will refer to the predicates P and Q as the pre-condition and invariant of such a formula. A state predicate Q is held invariant by a command c iff $\mathsf{sinv}\,Q\,c\,\mathsf{T}$ is valid; in particular, proving that a user interface maintains an invariant Q reduces to proving a goal of of the form:

$$\forall i\,c.\, \mathsf{Main}\,i\,c\,(\lambda f.\, f)\,(\lambda P.\, P)\,[\,] \supset \mathsf{sinv}\,Q\,(\mathsf{xs}\,(i, \mathsf{T}) \longrightarrow \mathsf{do\_od}\,c)\,\mathsf{T}$$

where Main is a predicate modeling the user interface.

Rather than trying to prove goals of the form $\mathsf{sinv}\,Q\,c\,P$ directly from the definition of sinv, we use a collection of inference rules. These rules will be used in a backwards fashion to decompose each complex goal into a number of simpler goals. We write these rules as follows:

$$\frac{G_1 \cdots G_n}{G}$$

Such a rule can be interpreted as *"To prove that G holds, it is sufficient to prove that G_1 through G_n hold."* We express these rules as HOL theorems, with the HOL logical (boolean) connectives serving as meta-logical operators for the object logic. For example, we can express the previous rule with HOL theorem

$$\vdash G_1 \supset \ldots \supset G_n \supset G\ .$$

Rather than postulating such theorems as axioms, we prove their validity using the definition of sinv; this eliminates the chance of introducing an inconsistency.
The Rules: We introduce rules for strengthening the invariant and weakening the precondition of goals involving sinv.

Theorem 3 $\dfrac{[\,] \vdash (Q \supset R) \qquad \mathsf{sinv}\,Q\,c\,P}{\mathsf{sinv}\,R\,c\,P}$

$\vdash \forall c\,P\,Q\,R.\, [\,] \vdash (Q \supset R) \supset (\mathsf{sinv}\,Q\,c\,P \supset \mathsf{sinv}\,R\,c\,P)$

Theorem 4 $\dfrac{[\,] \vdash (P \supset R) \qquad \mathsf{sinv}\,Q\,c\,R}{\mathsf{sinv}\,Q\,c\,P}$

$\vdash \forall c\,P\,Q\,R.\, [\,] \vdash (P \supset R) \supset (\mathsf{sinv}\,Q\,c\,R \supset \mathsf{sinv}\,Q\,c\,P)$

Another way to simplify goals involving sinv is to simplify the command that appears in the goal. The following rule eliminates a top-level guard by strengthening the precondition of the goal.

Theorem 5 $\dfrac{\mathsf{sinv}\,Q\,c\,(P \wedge R)}{\mathsf{sinv}\,Q\,(R \longrightarrow c)\,P}$

$\vdash \forall c\,P\,Q\,R.\, \mathsf{sinv}\,Q\,(R \longrightarrow c)\,P = \mathsf{sinv}\,Q\,c\,(P \wedge R)$

However, the majority of our rules are based not on sinv, but on a closely related predicate inv.

Definition 13 $\vdash \forall Q\, c\, P.\, \mathsf{inv}\, Q\, c\, P = (\forall e.\, c\, e \supset P\,(\mathsf{start}\, e) \supset Q\,(\mathsf{start}\, e) \supset \mathsf{holds}\, Q\, e)$

The formula $\mathsf{inv}\, Q\, c\, P$ is true iff Q holds for all runs in c that start in both P and Q. If we can show that the pre-condition of a goal involving sinv is stronger than the invariant, then we can reduce it to one involving inv.

Theorem 6 $$\frac{[\,] \vdash (P \supset Q) \qquad \mathsf{inv}\, Q\, c\, P}{\mathsf{sinv}\, Q\, c\, P}$$
$\vdash \forall c\, Q\, P.\, [\,] \vdash (P \supset Q) \supset (\mathsf{inv}\, Q\, c\, P \supset \mathsf{sinv}\, Q\, c\, P)$

The advantage of using inv instead of sinv is that it allows us to derive simpler rules. In particular, we do not have to verify that the invariant holds for start states; this is assumed to be the case. Note that reducing a goal involving sinv to one involving inv with the previous rule requires ensuring that the invariant holds for all start states satisfying the pre-condition. As with sinv, we can weaken the pre-condition of goals involving inv.

Theorem 7 $$\frac{[\,] \vdash (P \supset R) \qquad \mathsf{inv}\, Q\, c\, R}{\mathsf{inv}\, Q\, c\, P}$$
$\vdash \forall R\, Q\, P\, c.\, [\,] \vdash (P \supset R) \supset (\mathsf{inv}\, Q\, c\, R \supset \mathsf{inv}\, Q\, c\, P)$

We now introduce rules for simplifying the commands that appear in goals involving the invariant inv.

Theorem 8 $$\frac{}{\mathsf{inv}\, Q\, \mathsf{skip}\, P}$$
$\vdash \forall Q\, P.\, \mathsf{inv}\, Q\, \mathsf{skip}\, P = \mathsf{T}$

Theorem 9 $$\frac{}{\mathsf{inv}\, Q\, \mathsf{abort}\, P}$$
$\vdash \forall Q\, P.\, \mathsf{inv}\, Q\, \mathsf{abort}\, P = \mathsf{T}$

Theorem 10 $$\frac{[\,] \vdash (\mathsf{sp}\,(\mathsf{assign}\, E)\, P \supset Q)}{\mathsf{inv}\, Q\,(\mathsf{assign}\, E)\, P}$$
$\vdash \forall E\, Q\, P.\, [\,] \vdash (\mathsf{sp}\,(\mathsf{assign}\, E)\, P \supset Q) \supset \mathsf{inv}\, Q\,(\mathsf{assign}\, E)\, P$

Theorem 11 $$\frac{\mathsf{inv}\, Q\, c_1\, P \qquad \mathsf{inv}\, Q\, c_2\, P}{\mathsf{inv}\, Q\,(c_1 \,[\!]\, c_2)\, P}$$
$\vdash \forall c_1\, c_2\, Q\, P.\, \mathsf{inv}\, Q\,(c_1 \,[\!]\, c_2)\, P = \mathsf{inv}\, Q\, c_1\, P \wedge \mathsf{inv}\, Q\, c_2\, P$

Theorem 12 $$\frac{\forall v.\, \mathsf{inv}\, Q\,(c\, v)\, P}{\mathsf{inv}\, Q\,(\$\exists\, c)\, P}$$
$\vdash \forall c\, Q\, P.\, (\forall v.\, \mathsf{inv}\, Q\,(c\, v)\, P) \supset \mathsf{inv}\, Q\,(\$\exists\, c)\, P$

Theorem 13 $\dfrac{\mathsf{inv}\, Q\, c\, (P \wedge R)}{\mathsf{inv}\, Q\, (R \longrightarrow c)\, P}$

$\vdash \forall c\, P\, Q\, R.\, \mathsf{inv}\, Q\, (R \longrightarrow c)\, P = \mathsf{inv}\, Q\, c\, (P \wedge R)$

Theorem 14 $\dfrac{\mathsf{inv}\, Q\, c_1\, P \qquad \mathsf{inv}\, Q\, c_2\, (\mathsf{sp}\, c_1\, P)}{\mathsf{inv}\, Q\, (c_1 \mathbin{;} c_2)\, P}$

$\vdash \forall c_1\, c_2\, Q\, P.\, \mathsf{inv}\, Q\, c_1\, P \wedge \mathsf{inv}\, Q\, c_2\, (\mathsf{sp}\, c_1\, P) \supset \mathsf{inv}\, Q\, (c_1 \mathbin{;} c_2)\, P$

Theorem 15 $\dfrac{[\,] \vdash (\mathsf{sp}\, c\, P \supset Q)}{\mathsf{inv}\, Q\, (\mathsf{atomic}\, x\, c)\, P}$

$\vdash \forall c\, x\, Q\, P.\, [\,] \vdash (\mathsf{sp}\, c\, P \supset Q) \supset \mathsf{inv}\, Q\, (\mathsf{atomic}\, x\, c)\, P$

Theorem 16 $\dfrac{[\,] \vdash (\mathsf{sp}\, c\, P \supset P) \qquad \mathsf{inv}\, Q\, c\, P}{\mathsf{inv}\, Q\, (\mathsf{do\_od}\, c)\, P}$

$\vdash \forall c\, Q\, P.\, [\,] \vdash (\mathsf{sp}\, c\, P \supset P) \supset (\mathsf{inv}\, Q\, c\, P \supset \mathsf{inv}\, Q\, (\mathsf{do\_od}\, c)\, P)$

Theorem 17 $\dfrac{\mathsf{inv}\, Q\, c\, Q}{\mathsf{inv}\, Q\, (\mathsf{do\_od}\, c)\, Q}$

$\vdash \forall c\, Q.\, \mathsf{inv}\, Q\, c\, Q \supset \mathsf{inv}\, Q\, (\mathsf{do\_od}\, c)\, Q$

An Example: We now give a simple example of how these rules can be applied by verifying that the slider and dial of the user interface of Figure 2 are synchronized; whenever one is changed the other changes accordingly. We first generate a model of the user interface from its IL description in Figure 2 and the models of its constituent components. The model for `Sliders` is in Figure 3; `Dials` are modeled similarly. `Windows` are modeled as essentially having no state and no associated command: $\mathsf{Window}\, i\, c\, s\, q\, e\, width\, height\ =\ (i\ =\ \mathsf{VOID}) \wedge (c\ =\ \mathsf{F})$. The resulting predicate is shown in Figure 4. Next, we express the property to be validated in terms of this model. The second and third state components of the model hold the value of the slider and dial, respectively. We need to show that in all states of all possible behaviors of the user interface these two values remain the same. That is, we have to show that the predicate $\lambda(t, u, v).\, u = v$ is held invariant by any command that models the behavior of the user interface. In particular, we have to prove the validity of the following formula:

$$\forall i\, c.\, \mathsf{Main}\, i\, c\, (\lambda f.\, f)\, (\lambda P.\, P)\, [\,] \supset \mathsf{sinv}\, (\lambda(t, u, v).\, u = v)\, (\mathsf{xs}\, (i, \mathsf{T}) \longrightarrow \mathsf{do\_od}\, c)\, \mathsf{T}$$

We construct the proof using the HOL subgoal package. We start by initializing the goal stack with the previous formula.

```
#g "Main i c (\f.f) (\P.P) [] ==>                                    1
      sinv (\(t,u,v).u=v) ((xs (i,T)) --> (do_od c)) TT";;
"Main i c(\f. f)(\P. P)[] ==>
 sinv(\(t,u,v). u = v)((xs(i,T)) --> (do_od c))TT"
```

Next we rewrite the goal with the definitions of Main, Window, Slider and Dial and beta reduce. Transforming (the conjuncts of) the antecedent of the resulting implication into assumptions, and eliminating unecessary variables leaves us with a goal whose assumptions are simply the definitions of the various input ports: set_2 and set_3 correspond to the `set` ports of the slider and dial, respectively.

```
#e(REWRITE_TAC[Main]);;                                          2
#e(CONV_TAC (REDEPTH_CONV GEN_BETA_CONV));;
#e(REPEAT STRIP_TAC);;
#e(SMART_ELIMINATE_TAC);;
#e(UNDISCH_ALL_TAC);;
#e(REWRITE_TAC[Window;Slider;Dial]);;
#e(CONV_TAC (REDEPTH_CONV GEN_BETA_CONV));;
#e(REPEAT STRIP_TAC);;
#e(SMART_ELIMINATE_TAC);;
OK..
"sinv
 (\(t,u,v). u = v)
 ((xs((VOID,0,0),T)) -->
  (do_od
   (FF ||
    ((?? v. atomic(numEv[2]v)(set2 v)) ||
     (?? v. atomic(numEv[3]v)(set3 v))))))
 TT"
  2  ["set2 =
       (\v.
         if
         (\(v1,v2,v3). ~(v2 = v))
         ((assign(\(v1,v2,v3). (v1,v,v3))) ;; (set3 v)))" ]
  1  ["set3 =
       (\v.
         if
         (\(v1,v2,v3). ~(v3 = v))
         ((assign(\(v1,v2,v3). (v1,v2,v))) ;; (set2 v)))" ]
```

Next we reduce the goal with the theorem `sinv_start`:

$$\vdash \forall c\, Q\, P.\, [\,] \vdash (P \supset Q) \supset (\mathsf{inv}\, Q\, c\, Q \supset \mathsf{sinv}\, Q\, (P \longrightarrow \mathsf{do\_od}\, c)\, \mathbf{T})$$

Splitting the resulting conjunct results in two subgoals.

```
#e(MATCH_IMP_MP_TAC sinv_start);;                                    3
Theorem sinv_start autoloading from theory `sinv` ...
sinv_start =
|- !c Q P. [] |= (P ==>> Q) ==> inv Q c Q ==> sinv Q(P -->
                 (do_od c))TT

#e(STRIP_TAC);;
OK..
2 subgoals
"inv
 (\(t,u,v). u = v)
 (FF ||
  ((?? v. atomic(numEv[2]v)(set2 v)) ||
   (?? v. atomic(numEv[3]v)(set3 v))))
 (\(t,u,v). u = v)"
   ... assumptions elided ...

"[] |= ((xs((VOID,0,0),T)) ==>> (\(t,u,v). u = v))"
   ... assumptions elided ...
```

The first is to show that the initial state of user interface satisfies the invariant; this is easily solved using the tactic `gc_solve_TAC`.

```
#e(gc_solve_TAC);;                                                   4
OK..
goal proved
.. |- [] |= ((xs((VOID,0,0),T)) ==>> (\(t,u,v). u = v))

Previous subproof:
"inv
 (\(t,u,v). u = v)
 (FF ||
  ((?? v. atomic(numEv[2]v)(set2 v)) ||
   (?? v. atomic(numEv[3]v)(set3 v))))
 (\(t,u,v). u = v)"
   ... assumptions elided ...
```

The tactic `gc_solve_TAC` is used to reduce lifted sequents of the form $[] \vdash \mathsf{xs}\,(s, P) \supset A$. It first rewrites the current goal with the following theorem

$$\vdash \forall s\, P\, A.\, ([\,] \vdash \mathsf{xs}\,(s, P) \supset A) = (P \supset A\, s)$$

and then beta reduces and applies some basic (HOL-supplied) rewrites to the result. This is enough to solve the current subgoal. The second subgoal is to show that the invariant still holds after a single interaction with the environment.

To make this subgoal more amenable to manipulation, we first express its precondition in terms of **xs** by rewriting with the theorem `prop_lemma` (proved ahead of time):

$$\vdash (\lambda(t,u,v).\, u = v) = (\exists t\, v.\, \mathsf{xs}\,((t,v,v),\mathsf{T}))$$

Applying the rule for the choice operator to the result and splitting the resulting conjuncts into separate goals leaves us with three subgoals.

5

```
#e(GEN_REWRITE_TAC (RAND_CONV) [] [prop_lemma]);;
#e(REWRITE_TAC[choose_inv]);;
Theorem choose_inv autoloading from theory 'sinv' ...
choose_inv = |- !c1 c2 Q P. inv Q(c1 || c2)P = inv Q c1 P /\
                            inv Q c2 P

#e(REPEAT STRIP_TAC);;
OK..
3 subgoals
"inv
  (\(t,u,v). u = v)
  (?? v. atomic(numEv[3]v)(set3 v))
  (?? t v. xs((t,v,v),T))"
     ... assumptions elided ...

"inv
  (\(t,u,v). u = v)
  (?? v. atomic(numEv[2]v)(set2 v))
  (?? t v. xs((t,v,v),T))"
     ... assumptions elided ...

"inv(\(t,u,v). u = v)FF(?? t v. xs((t,v,v),T))"
     ... assumptions elided ...
```

These correspond to the possible interactions the user might have with the window, slider, and dial, respectively. The first (bottom) subgoal is trivial to solve: our model for windows does not provide for any user interaction.

```
#e(REWRITE_TAC[inv_FF]);;                                          6
Theorem inv_FF autoloading from theory `sinv' ...
inv_FF = |- !Q P. inv Q FF P = T

OK..
goal proved
|- inv(\(t,u,v). u = v)FF(?? t v. xs((t,v,v),T))

Previous subproof:
2 subgoals
... subgoal elided ...

"inv
 (\(t,u,v). u = v)
 (?? v. atomic(numEv[2]v)(set2 v))
 (?? t v. xs((t,v,v),T))"
    ... assumptions elided ...
```

The second subgoal covers the possible interactions a user could have with the slider. Applying the rule for the **atomic** operator reduces this goal to showing that setting the slider to an arbitrary value does not result in invariant being violated.

```
#e(gc_exists_TAC);;                                                7
#e(MATCH_MP_TAC inv_atomic);;
Theorem inv_atomic autoloading from theory `sinv' ...
inv_atomic = |- !c x Q P. [] |= ((sp c P) ==>> Q) ==>
                inv Q(atomic x c)P

#e(CONV_TAC
     (REDEPTH_CONV (REWR_CONV exists_sp2 ORELSEC GEN_BETA_CONV)));;
Theorem exists_sp2 autoloading from theory `sp' ...
exists_sp2 = |- !c. sp c($?? P) = (?? v. sp c(P v))

#e(pred_TCL (REPEAT STRIP_TAC));;
OK..
"[sp(set2 v)(xs((v',v'',v''),T))] |= (\(t,u,v). u = v)"
    ... assumptions elided ...
```

Before we can proceed further we need to rewrite the goal with the definition of set_2 (one of the two assumptions.) The tactic `gc_unwind_TAC` simply rewrites the conclusion of the goal with its assumptions and then beta reduces the result.

```
#e(gc_unwind_TAC);;                                              8
OK..
"[sp
  (if
   (\(v1,v2,v3). ~(v2 = v))
   ((assign(\(v1,v2,v3). (v1,v,v3))) ;; (set3 v)))
  (xs((v',v'',v''),T))] |=
 (\(t,u,v). u = v)"
   ... assumptions elided ...
```

The tactic gc_if_TAC splits the current goal into two subgoals, based on whether the condition holds or not. It applies the following theorem

$$\vdash \forall c\, b\, P.\, \mathsf{sp}\,(\mathsf{if}\, b\, c)\, P = (P \wedge \neg b) \vee \mathsf{sp}\, c\, (P \wedge b)$$

(easily derived from the rules for skip and the guard and choice operators), splits the resulting goal, and then applies some simple rewrites to the subgoals. These rewrites are enough to solve the subgoal corresponding to the case that the new value of the slider is the same as the existing value. Thus, as there is no state change the invariant is maintained. As a result, we are left with only one subgoal. The remaining subgoal corresponds to the case that the new value of the slider differs from the existing value. Note how the path condition has been updated to take this into account.

5 Conclusions

We have described an approach to building GUI-based applications from a single IL-based formal description from which we automatically derive a corresponding formal model suitable for mechanical reasoning. This formal model is based on HOL and can be used to validate properties of the design. However, similar ideas can in principle be applied in the context of other formal approaches, e.g., temporal logic [1]. The overall description of our approach, including the mechanic derivation of implementations, is described in earlier work [3, 4]. Here, we concentrate on the language for describing behaviour, and illustrate how we would prove that a user interface obeys a certain behavioral formal property.

Mechanization of the validation process is important for two reasons: not only does it increase confidence in the validations performed [5, 9], but it also has the potential for reducing the amortized cost of the validation effort. GUIs are not static, they evolve over time, and usually a significant fraction of the total GUI development effort will be expended after the initial implementation. If models are automatically generated then producing new models as GUIs evolve will require little effort. If reasoning is also mechanized, then there is a chance that subsequent validations will be able to reuse at least portions of previous ones.

References

1. P.S.C. Alencar, D. Cowan, C.J.P. Lucena, and L.C.M. Nova. Formal Specification of Reusable Interface Objects. In *Proceedings of Symposium on Software Reusability*, Seattle, USA, April 1995.
2. C. Brink and I. Rewitzky. Modelling the algebra of weakest preconditions. *South African Computer Journal*, 6:11–20, July 1992.
3. P. Bumbulis. Combining Formal Techniques and Prototyping in User Interface Construction and Verification. Ph.d. thesis, Computer Science Department, University of Waterloo, 1996.
4. P. Bumbulis, P.S.C. Alencar, D.D. Cowan, and C.J.P. Lucena. Combining Formal Techniques and Prototyping in User Interface Construction and Verification. In *2nd Eurographics Workshop on Design, Specification, Verification of Interactive Systems (DSV-IS'95)*. Springer-Verlag Lecture Notes in Computer Science, 1995.
5. Avra Cohn. The notion of proof in hardware verification. *Journal of Automated Reasoning*, 5(2):127–140, June 1989.
6. Edsger W. Dijkstra. *A Discipline of Programming*. Prentice-Hall, Englewood Cliffs, New Jersey, 1976.
7. Edsger W. Dijkstra and Carel S. Scholten. *Predicate Calculus and Program Semantics*. Springer-Verlag, New York, 1990.
8. N. Francez. A case for a forward predicate transformer. *Inf. Proc. Lett.*, 6(6):196–198, December 1977.
9. Stephen J. Garland, John V. Guttag, and James J. Horning. Debugging larch shared language specifications. *IEEE Transactions on Software Engineering*, 16(9):1044–1057, September 1990.
10. M.D. Harrison and D.J. Duke. A review of formalisms for describing interactive behavior. In Richard N. Taylor and Joëlle Coutaz, editors, *Software Engineering and Human-Computer Interaction; ICSE'94 Workshop on SE-HCI: Joint Research Issues*, volume 896 of *Lecture Notes in Computer Science*, pages 49–75, Sorrento, Italy, 16–17 May 1994. Springer-Verlag.
11. C.A.R. Hoare. Some properties of predicate transformers. *Journal of the ACM*, 25(3):461–480, July 1978.
12. Imperial College of Science, Technology and Medicine. *Darwin Overview*, 1994.
13. Ruurd Kuiper. An operational semantics for bounded nondeterminism equivalent to a denotational one. In J.W. de Bakker and J.C. van Vliet, editors, *Algorithmic Languages*, pages 373–398. IFIP, North Holland, 1981.
14. Johan J. Lukkien. An operational semantics for the guarded command language. In *Mathematics of program construction : international conference*, volume 669 of *Lecture Notes in Computer Science*, pages 233–249. Springer-Verlag, 1992.
15. Greg Nelson. A generalization of Dijkstra's calculus. *ACM Transactions on Programming Languages and Systems*, 11(4):517–561, October 1989.
16. Oscar Nierstrasz, Simon Gibbs, and Dennis Tsichritzis. Component-oriented software development. *Communications of the ACM*, 35(9):160–165, September 1992.
17. G. Tredoux. Mechanizing execution sequence semantics in HOL. *South African Computer Journal*, 7:81–86, July 1992. Proceedings of the 7th Southern African Computer Research Symposium, Johannesburg, South Africa. Also available as part of the HOL distribution: ftp://lal.cs.byu.edu/pub/hol/holsys.tar.gz.

This article was processed using the LaTeX macro package with LLNCS style

Specifying and Reasoning About CSCW

Steve Reeves

Department of Computer Science, University of Waikato
Hamilton, New Zealand

Abstract In this paper we introduce a pair of logics which, taken together, can be seen as a first step towards a formal specification language for CSCW systems. We show the development of the logics and give some simple examples of their use. We also make a distinction between the computational part of the system and the people, i.e. we do not follow a simple action and agent analysis. Since people bring knowledge to a system we treat them differently. We also propose the use of situation theory as a way of capturing requirements.

Keywords

Interactive systems, CSCW, formal specification, modal logic, intensional logic, situation theory

1 Introduction

The ideas presented in this paper arose in the course of trying to answer the question: "What does it mean to be able to specify formally, at some level of generality, a CSCW system?". (By "CSCW system" we mean the software that supports several people in some collaborative task.)

A first step towards an answer was to look at several example systems and implementations to see what was common amongst them. This gave us some idea of what is likely to appear in any CSCW system and so be a general property of such systems. Having seen some such properties (which we might expect to form a rudimentary theory of CSCW systems) we then needed to step back and consider how such properties could be formalized. As we shall see this led to the requirement for two sorts of language—one to express what the functions of the software of the system are and one to specify what the users need to know about other users and their activities in order to collaborate. These two languages can be used to build theories of (i.e. collections of sentences which describe) the system required.

The paper presents a necessarily brief look at the development of a logic for specifying CSCW systems. This work, as will quickly become clear, is at its initial stages and very much consists of using simple examples to give confidence that what we are doing at least allows the possibility of arriving at a logic for CSCW. That is, the examples act as filters—if we could not even do these simple examples, then we could be sure we are thinking along the wrong lines. Hence, they may leave the reader feeling unsatisfied. In that case it needs to be remembered that the examples are meant as simple feasibility proofs, not full-blown examples to convince of the usefulness of the logic. They will come later.

From this it will be clear that we are taking the standard view (as taken in, for example, Diller (1994)) that a specification is a theory of, i.e. a set of sentences which talk about, the system being specified. If those sentences are in some suitable language then we can use them as a basis for reasoning about the system, just as a theory of mechanics, say, allows us to reason about how planets move without having to watch them. We can, for example, explore the implications of the specification by seeing what facts follow from it; we can also refer back to the specification as we develop an implementation and try to prove that the implementation meets the demands of the specification; we can also try to develop refinement rules which, in meaning preserving transformations, take us from the specification towards some target implementation.

In this paper we present the outcome of some of that work (which is continuing) and mention some further problems that have arisen.

The picture that has emerged came from reading about and using several existing systems for CSCW and also from reading about and reflecting on the (rather smaller number of pieces of) work done by others in the area of specification of CSCW systems.

In relation to this, the work of Devlin (1994) was particularly useful. It gave a very general but completely formal way of thinking about CSCW at a level prior even to specification. It also gave some reassurance about the rôle of formalism and theorizing in what is still a young and very experimental area. Of course, what we are aiming for in this paper is not a theory of CSCW as a whole but arguing for building theories which fit within the designation "CSCW". In this sense Devlin's words are relevant. He says

Designers are well known to carry out their craft with little or no "concern" for theory. But what this means is that design is a holistic, creative process, involving both experience and considerable trial and error, rather than something that can be codified by some theory... But just because designers often make little or no explicit use of a theory does not mean that theory is not important to the design process. Indeed, theory is crucial to design, since the entire approach to and way of thinking about a particular domain are conditioned by the pertinent theories of which designers are aware.

We return to Devlin's work later when presenting our conclusions.

We take an abstract view of a typical CSCW system. Nevertheless, we need to say what the functions of the system are, perhaps how the system should look and, crucially, how people can cooperate when using the system.

The most abstract view just asks what the static and dynamic aspects of the system are:

the static aspects include how the system fits together, what communication possibilities there are when working within it, what things look like. We formalize this by specifying what users must know about the system in order to use it;

the dynamic aspects include how people join and leave, messages, notes, changing representations, reporting change amongst co-workers. We formalize this by using an action-and-agent logic.

Of course, many people are trying to do all these things but in all the other treatments we have seen the level at which specification happens (if at all) appears to be either informal or via prototyping. Also, different parts of the problem are specified in different ways: perhaps standardly for functions of the system; via informal descriptions, diagrams or charts for structures and communication. The closest to what we have in mind (in that it is not simply a mixture of informal language and implementation) is given by CSDL (De Paoli and Tisato, 1994), which we will mention again later.

However, we want to explore the possibilities for doing all of this in a uniform and formal way. A lot of the informal ways will be seen as concrete manifestations of this uniform way.

Interestingly, many of the formal ideas that appear in this paper also appear in the context of agents in distributed AI and this idea of agents has been suggested many times as useful components of CSCW systems (Wooldridge and Jennings, 1995).

A final point: although there is not yet any general agreement of what a CSCW system is (beyond the sort of definition given in the first paragraph) it is still fruitful to stand back and try to take a more abstract view of what has generally be a purely experimental area of computer science. The fruitfulness comes from the fact that since, ultimately, software necessarily entails formalization, problems discovered when formalization is tried at an early stage (and when many complicating details are abstracted away from) may give guidance for future development. Trying to give a way of having a more abstract view also turns out to be useful since inconsistencies or infelicities in theory and design are likely to be seen more clearly if the detail that necessarily surrounds implementation is cleared away. Finally, the quote from Devlin above is relevant here, too.

2 An Action-and-Agent Logic

We extend the syntax of the standard modal logic (see, for example, Reeves and Clarke (1990) for an introduction) by indexing the modalities with *actions*, which will be left uninterpreted, though the name should suggest what we expect to find as an action. We assume that there is a set Act of actions and write [a]S to mean 'after action a has successfully been completed, S is true'.

In order to give a semantics to this we introduce a family of *frames*, indexed by elements of Act. A frame is $<P, \{R_i\}_{i \in Act}>$ where P is a set of *possible worlds* (which we can think of as a state) and $R_i \subseteq P \times P$ (an *accessibility relation*) for $i \in$ Act.

Pictures can be useful for giving examples of frames:

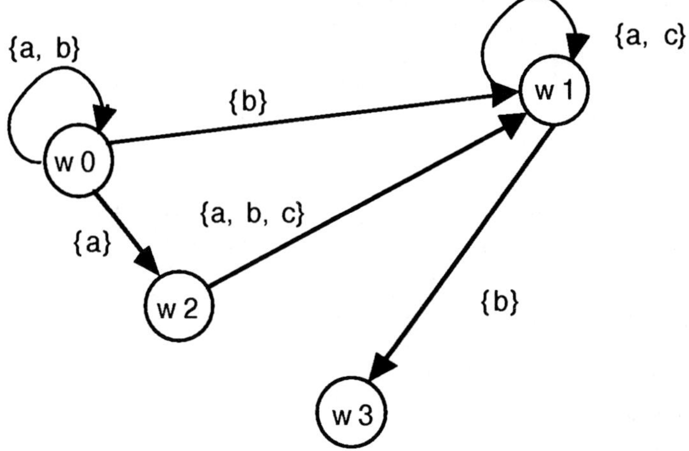

Here a, b, c $\in$ Act. An arrow between w and v (both elements of P, where P = {w0, w1, w2, w3}) is marked with i $\in$ Act iff $R_i(w,v)$.

For the indexed modality above we have

$\models_w$ [a]S iff for all v $\in$ P such that $R_a(w,v)$ we have $\models_v$ S

which is to say that [a]S is true in world w if and only if in all worlds v accessible due to the action a we have S true in v.

Next, we assume that there is a set of individuals over which we want to quantify.

First, the frame is extended to a triple by adding a function $I : P \rightarrow 2^D$, where D is the set of individuals (which in our case is going to contain things like numbers, characters, lists, buttons etc.). This function I simply tells us which individuals exist in which worlds, i.e. given some world w, I(w) is the set of all the individuals that exist there.

The language is extended to include the quantifiers and variables as usual.

Given this, we have the following

$\models_w \forall x S$ iff $\models_w S(x/b)$ for all b $\in$ I(w)

(where S(x/b) is S with all free occurrences of x replaced by b).

Clearly, the framework above is very general and could be used in many specific cases. To make it more particular, we now consider the set of possible worlds to be a set of states in which the program that is being shared can be during execution.

Consider the program when it displays a dialogue box:

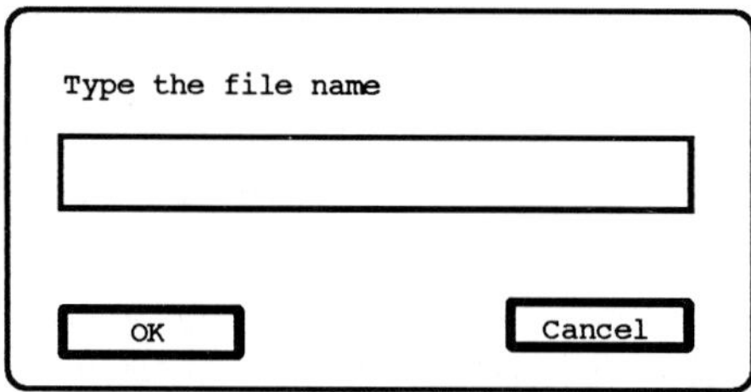

Here are two things we might want to have as part of the statement of the specification of the program that, on execution, displays this box:

Pressing "cancel" opens no more files

We could represent this by

$$\models \text{select}(\text{"Cancel"}) \rightarrow \forall l(\text{open}(l) \leftrightarrow [\text{push}]\text{open}(l))$$

assuming that we have other facts that say that the predicate 'select' is true just when its argument is the label on the area within which the mouse cursor is currently positioned, 'open' is a predicate which is true just when its argument is a list of the names of all the files currently open and 'push' is the action of pressing the mouse button;

Pressing "OK" keeps open all the files that were open and in addition opens the one whose name was typed in the text field if it is the name of some existing file

given by

$$\models \text{select}(\text{"OK"}) \rightarrow \forall x(\text{text}(x) \rightarrow \text{name}(x) \rightarrow \forall l(\text{open}(l) \rightarrow [\text{push}]\text{open}(x{:}l)))$$

assuming the definition as above and also that 'text' is a predicate which is true just when its argument is a piece of text (a string, perhaps), 'name' is a predicate which is true just when its argument is the name of a file which exists in the current environment and ':' adds an element to the front of a list.

So far, the logic that we have described is fairly standard and similar to that developed by many people (Goldblatt (1987) is a standard text on this). The application to which we are putting it is new, though has similarities, in that it talks about interactive systems, to the work of Duke and Harrison (1995).

3 Extensions to the language

The language as it stands does not allow us to talk explicitly about any agents (here we use the word "agents" to mean people, usually, rather than the DAI sense of things which are probably other programs behaving in some autonomous fashion. However, since we have left the notions of agent and action unconstrained, all that we talk about in this paper could be used to specify DAI-type agent systems) that might be around—so far we can give properties of parts of the system but we cannot yet denote the agents which cause, say, a button to be pressed or some text to be typed.

Given our aim of specifying and reasoning about systems which several agents co-operate in using, this is clearly important.

In order to do this we extend the language further by first extending the syntax to

$[a,\alpha]S$

where a is an action, just as before, and α is an agent, an element of the set Agt. The intended meaning is that when action a is successfully completed by agent α then S is true.

To give the semantics we extend the idea used in the previous extension and index the family of accessibility relations with action and agent, so the frame is now $<P, \{R_{i,\iota}\}_{i \in Act,\ \iota \in Agt}, I>$ where $R_{i,\iota} \subseteq P \times P$ for $i \in Act$, $\iota \in Agt$. Now, the accessibility relation tells us what worlds are accessible from what other worlds if a is done by α.

A picture of the frame now would be

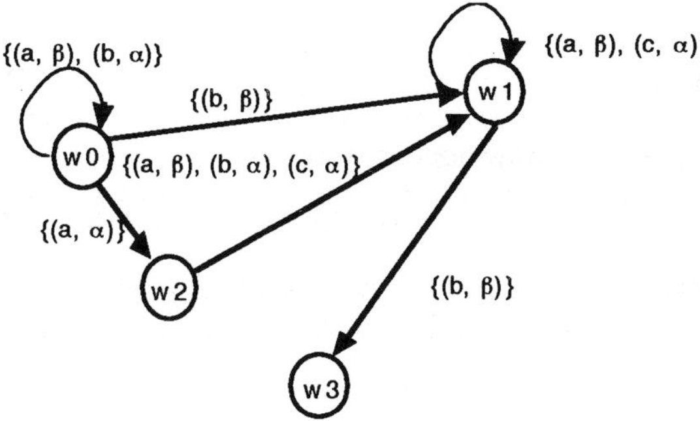

As before, all the non-modal parts of the language extend without change and we have

$\models_w [a,\alpha]S$ iff for all $v \in P$ such that $R_{a,\alpha}(w,v)$ we have $\models_v S$

We can also extend the language to allow both actions and agents to be quantified over. First, we have to extend the function I. Currently it maps worlds to sets of individuals. If we are to quantify over elements of Act and Agt we need I to be indexed on the sort of set it should deliver. So, we have

$I_{ind} : P \rightarrow 2^D$, as I was before

$I_{act} : P \rightarrow 2^{Act}$, takes a possible world to the actions which can be done there, so

$I_{act}(w) = \{ a \mid \exists \alpha \in Act.\ \exists v \in P.\ R_{a,\alpha}(w,v)\}$

$I_{agt} : P \rightarrow 2^{Agt}$, takes a possible world to the agents that can perform there, so

$I_{agt}(w) = \{ \alpha \mid \exists a \in Agt.\ \exists v \in P.\ R_{a,\alpha}(w,v)\}$

It is likely to be useful, also, to allow only some agents to only some actions. There are several ways to allow this, but the most straightforward would be to provide a further extension for I which tells us which agents can do which actions in each world, recorded as a binary relation, i.e. we have

$I_{allow}: P \rightarrow 2^{(Agt \times Act)}$

Then we can define the quantifiers over actions and agents in the obvious way:

$\models_w \forall x S$ iff for all $a \in I_{act}(w)$ we have $\models_w S(x/a)$

$\models_w \forall \delta S$ iff for all $\alpha \in I_{agt}(w)$ we have $\models_w S(\delta/\alpha)$

and a frame is now the triple

$\langle P, \{R_{i,\iota}\}_{i \in Act,\ \iota \in Agt}, \langle I_{ind}, I_{act}, I_{agt}, I_{allow}\rangle\rangle$.

As a first example, consider describing a system after initialization, perhaps after the 'init' button has been pressed. (Note that an undecorated turnstile like '$\models$' means that the sentence which follows it is true for all elements of P.):

$\models \forall \delta [init, \delta] Initialized$

$$\models Initialized \leftrightarrow (\forall v(var(v) \rightarrow v = 0) \wedge \forall x \forall \alpha (can(x, \alpha) \wedge \ldots\ldots)$$

Next, some facts which describe how buttons work:

$\models \forall x(button(x) \rightarrow \forall \alpha(\exists \beta(\alpha \neq \beta \wedge can(click(x),\beta)) \rightarrow \neg can(click(x),\alpha)))$

for any button x, for any agent α, if there is an agent β different from α and β is permitted to click x then no α can

$\models \forall x(button(x) \rightarrow \forall \alpha((can(click(x),\alpha) \wedge enabled(x)) \rightarrow [click(x),\alpha]\neg enabled(x)))$

for any button x, for any agent α, if α is permitted to click the enabled button x then if α clicks x successfully, x will then be disabled

Finally, consider the dialogue box from above:

$\models \forall \alpha \forall b(\exists \beta(\alpha \neq \beta \wedge button(b) \wedge select(\beta,b)) \rightarrow \neg select(\alpha,b))$

for any agent α and button b, if there is an agent β different from α and β has selected b then no other agent α can select b

We can also add some standard deontic predicates ('can' and 'must', say) to this formal framework.

4 A Larger Example

Imagine that we have a group editor. What properties might we want to specify for it? One technique that has been experimented with is that of multiple scrollbars. They can be used (by the owner) in the usual ways to move through a file. They can also be used (by everyone else) as a way of seeing where in a file other people are working.

Let scrollbar(s,α) be true when s is the scrollbar belonging to α. The basic fact about ownership of a scrollbar is given by

$\models \forall\alpha\forall s(\mathrm{scrollbar}(s,\alpha) \leftrightarrow (\mathrm{can}(\mathrm{up}(s),\alpha) \wedge \mathrm{can}(\mathrm{down}(s),\alpha))$

(SB1)

Let viewing(α,top,bottom) be true when α can see all lines between these numbered top to bottom.

We can give properties of the actions up and down by

$$\models \forall s\forall\alpha\forall t\forall b((\mathrm{can}(\mathrm{down}(s),\alpha) \wedge \mathrm{viewing}(\alpha,t,b)) \rightarrow [\mathrm{down}(s),\alpha]\mathrm{viewing}(\alpha,t+10,b+10))$$

(DN1)

$$\models \forall s\forall\alpha\forall t\forall b((\mathrm{can}(\mathrm{up}(s),\alpha) \wedge \mathrm{viewing}(\alpha,t,b)) \rightarrow [\mathrm{up}(s),\alpha]\mathrm{viewing}(\alpha,t\text{-}10,b\text{-}10))$$

(UP1)

Now let Steve launch the group editor on the file document X. We have the most basic fact that Steve is 'present' and working on the file, recorded by

$\models_{w0}$ present(Steve)

(we return to the meaning of 'present' later) for some world w0. Following our discussion above we also want, as a basic fact, that

$\models \forall\alpha(\mathrm{present}(\alpha) \rightarrow \exists s\,\mathrm{scrollbar}(s,\alpha))$

from which it follows that

$\models_{w0}$ scrollbar(s1,Steve)

for some scrollbar called s1.

We also expect the scrollbar to be positioned so that a standard amount (say the first 20 lines) can be seen:

$\models_{w0}$ viewing(Steve, 0, 19)

We can show that

$\models_{w0}$ can(down(s1), Steve)

SB1 gives us

$\models_{w0} \forall\alpha\forall s(\mathrm{scrollbar}(s,\alpha) \leftrightarrow (\mathrm{can}(\mathrm{up}(s),\alpha) \wedge \mathrm{can}(\mathrm{down}(s),\alpha)))$

so

$\models_{w0}$ scrollbar(s1,Steve) $\leftrightarrow$ (can(up(s1),Steve) $\wedge$ can(down(s1),Steve))

so since

$\models_{w0}$ scrollbar(s1,Steve)

we have

$\models_{w0}$ can(up(s1),Steve) $\wedge$ can(down(s1), Steve) and hence the required conclusion.

We would also expect to have other general facts, like

$\models \forall\alpha\forall s\forall a((\text{scrollbar}(s,\alpha) \wedge a \neq \text{close}) \rightarrow [a,\alpha]\text{scrollbar}(s,\alpha))$

—scrollbars persist

$\models \forall\alpha\forall s(\text{can}(\text{close},\alpha) \leftrightarrow \text{scrollbar}(s,\alpha))$

—a user can only close a file that they are editing (which fact is denoted by the existence of their scrollbar)

$\models \forall\alpha(\text{can}(\text{close},\alpha) \rightarrow [\text{close},\alpha]\neg\exists s\ \text{scrollbar}(s,\alpha))$

—after a user closes a file, they no longer have a scrollbar

Note that this specification is not as general as it might be. We have used the predicate 'scrollbar' which will be true (presumably) just when the window referred to in its first argument has a scrollbar on it and that scrollbar signifies the presence as a user of the system of the agent referred to in its second argument.

However, we might be happy (as a first most general specification) to leave the exact method of indication of presence unconstrained and so have a predicate 'indicator', say, to denote presence. This may then be refined at a later stage to a scrollbar which still indicates presence, but also scope of view, and so will still satisfy the original specification. This (standard) idea of refinement will, as usual, give rise to proof obligations. In this case we have to prove that the scrollbar has at least the properties of an indicator.

5 People or Processes?

We now come to the crux of our problem. The material above gives a way of describing interaction and the examples show that the logic we have is an expressive and, with practice, convenient way of describing how the program reacts to people using the actions that it supports. However, this is an almost entirely program-centred point-of-view and can be seen as nothing more than a specialization of the idea that people are just other programs and so the whole example reduces to a network of concurrent computations.

What, then, makes the above example (and all CSCW examples) more than 'just' a network of concurrent processes? The point is that people engage in this sort of work because they wish to cooperate over some task. When cooperating, people clearly need to form models of their colleagues' knowledge of the task. They do this partly through their own basic knowledge but they also use knowledge they derive from the program itself; for example, the existence of a scrollbar tells Steve that Chris is now working on this document and the position of Chris's view becomes known to Steve when he looks at the position of the shaded part of Chris's scrollbar.

So, we contend that knowledge is what people have that makes them different from the program they are sharing. (Of course, they also have intelligence, motivation, intention and other properties that the program will not have, but knowledge seems to be enough for now.)

Our plan, then, is to add a way of describing knowledge to the specification of the system. To do this we first need a language and logic for knowledge.

As a first attempt to illustrate the idea we can carry-on in the possible-worlds tradition and use the classic example of S4.

This is a logic which, in terms of its semantics, simply has a transitive and reflexive accessibility relation over a set of possible worlds that model states of knowledge. Necessity is interpreted as 'is known', so a sentence of the form $\Box$ S is read as 'S is known'.

It turns out, remarkably, that the various properties of the accessibility relation for modal logics can be exactly captured by a few simple axioms. In the case of S4 the following is a sound and complete axiomatization which exactly expresses the semantics which are captured by a reflexive and transitive accessibility relation:

0. Any axiom of classical propositional logic
1. $\vdash \Box S \rightarrow S$
2. $\vdash \Box S \rightarrow \Box\Box S$
3. $\vdash \Box(S \rightarrow T) \rightarrow (\Box S \rightarrow \Box T)$

   (where S and T are any sentences)

together with the rules modus ponens and

4. if $\vdash S$ then $\vdash \Box S$ (where S is any sentence)

The first axiom is there so that we have classical propositional logic to build on. The second characterizes an important property of knowledge: if S is known then S is true. Contrast this with belief, where something can be believed without it being true. We view knowledge as 'justified true belief'.

The third axiom says that if something is known that it is known that it is known. The last says that if you know S implies T and you know S then you know T. The rule 4 says that the axioms and theorems are known.

Since we want to specify situations in which there are several people working together, we clearly need to be able to attribute what is known to the knower, rather than having a big pool of knowledge with no differentiation as to who knows what.

To do this we can use the same device as we did for incorporating actions and agents, namely we index the modality, which means indexing the accessibility relation that moves us between states of knowledge. In order to make clear that we are dealing with a different set of possible worlds here (where they are states of knowledge rather than states of a computation) we use the notation K_α to denote the modality, where α is some agent, instead of $[\alpha]$.

Given this new language we can now go on to specify some properties of our editor which make it clear that we are talking about a system where people are cooperating.

A very basic requirement would be that

$$\forall\alpha(\text{present}(\alpha) \rightarrow \forall\beta(\text{present}(\beta) \rightarrow K_\beta\text{present}(\alpha)))$$

i.e. that for any people present, they know who else is present (and clearly, during subsequent refinement, this could be satisfied by using scrollbars, for example).

In order for this sentence to have meaning we need to give a meaning for the predicate 'present'. This is where an analysis of typical CSCW systems enters the picture.

It turns out that all CSCW systems have some notion of *coordinator*. The CSDL group, for example, see such a thing as having three parts:

specification—this defines the cooperation policy of the system;

body—this controls the communications channels;

context—this defines the coordinator's interaction within a modular system.

CSDL, though, describes systems in a far less abstract way that we are attempting. For our purposes the main rôle of a coordinator is to act as a place where presence is registered. This usually means that users have to record their name and all sorts of information about where they are to be found in a network (so that the messages relevant to their interactions find their way to the right computer, for example). Within our theory we formalize the coordinator (at this level of abstractness) in the definition of the predicate 'present'. Simply, 'present(α)' is true iff α has notified the coordinator of their presence within the system. It turns out that this very simple definition gives us a basis upon which to build the rest of the specification for the system.

The designer's job is to specify a system where these sorts of requirements are met, i.e. they must specify that certain facts hold so that the above requirements are provable.

So, we now have two logics: one to describe the functioning of the software and one to describe what people have to know about the system and the other people using it. This pair of logics and the associated language is what we view as the specification language and logic for specifying and reasoning about a CSCW system.

The modal logic both ties together the system as a whole and also allows us to see the special rôle that people have in the system - their knowledge supplies the glue that makes all the disparate interacting parts into a system.

6 Problems

While the language above allows us to specify our systems there is a fundamental problem with its semantics which is to do with the (appealingly simple) possible-world foundations. This problem has two aspects: that our agents know everything that is a logical consequence of what they know (i.e. they are logically omniscient) and that they cannot distinguish between logically true statements. To illustrate these we give two simple examples (and use the un-indexed modality without loss of generality).

First, consider the case where we have $\models_{w0} \Box P$ and $\models P \leftrightarrow Q$. Since $\models_{w0} \Box P$ we will also have $\models_{w0} P$ and so $\models_{w0} Q$. If w1 is any other possible world and w1 is accessible from w0 then we will also have $\models_{w1} P$. Since $\models P \leftrightarrow Q$ we will also have

$\models_{w1}$ Q so since $\models_{w0}$ Q and $\models_{w1}$ Q we have $\models_{w0} \Box Q$. This is illustrated in the following diagram:

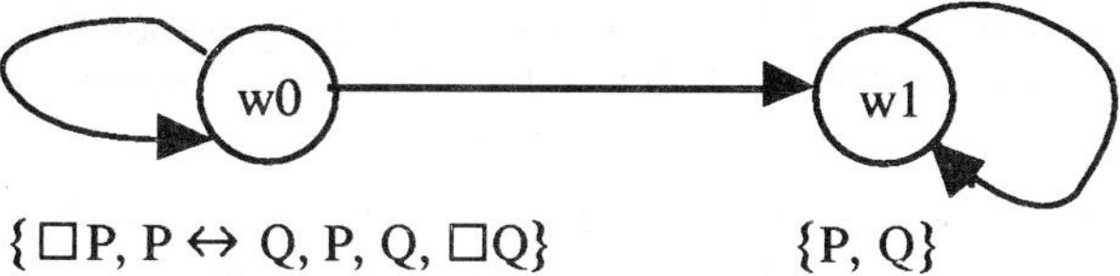

This is a general result which means that our agents will not be able to distinguish between logically equivalent statements.

Second, consider the case where we have $\models_{w0} \Box P$ and $\models_{w0} \Box(P \rightarrow Q)$. Then we will have $\models_{w0}$ P and $\models_{w0} P \rightarrow Q$ and so $\models_{w0}$ Q. Also, if w1 is a possible world accessible from w0 we will have $\models_{w1} P \rightarrow Q$ and $\models_{w1}$ P and so $\models_{w1}$ Q. So, since we have $\models_{w0}$ Q and $\models_{w1}$Q we will also have $\models_{w0} \Box Q$. This is illustrated in the following diagram:

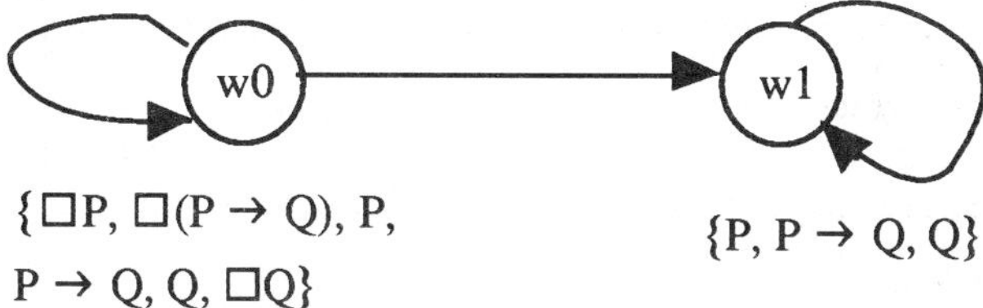

This is again a general result which means that our agents will know all the logical consequences of what they know.

Clearly we cannot expect our theories to be expressed in a language with such a semantics since it does not reflect reality—people are not like this. This means that we have to reject the possible-worlds semantics for our language. If the syntax we used in previous sections is really to form part of a language (rather than just a highly suggestive but nonetheless meaningless system of symbols) we have the problem of finding another basis for a semantics.

One possible solution to our current problem is to move to a language and logic like that proposed by Moore (1985) in the context of a logic for reasoning about knowledge and action in planning systems.

He views the sentence $\vdash_{w} K_{\alpha}P$ as meaning that P is true in all worlds that are compatible with α's knowledge in w. An action, then, is something which modifies knowledge (and so affects what possible-worlds are compatible). This approach has much in its favour. However, we choose to look at another way, which leaves possible worlds behind altogether—the theory of truth and propositions of Turner (1990).

Here, propositions are related directly to the sentences that we use to represent them—hence this is know as a *representationalist* view. The relation is, almost, that there is a one-to-one correspondence between them, i.e. that the structure of sentences exactly mirrors the structure of propositions. In fact (not surprisingly) things are more complicated than this because having all sentences representing propositions leads to inconsistencies. However, arranging matters so that some subset of sentences are in one-to-one correspondence with the propositions turns out to work and this is the view that Turner's theory takes.

First, the standard language of predicate calculus is extended by adding the clause "if t is a term then ^t is a wff" and we say that t is the representing term for the wff ^t. We also add all of the lambda-calculus as the term language.

Of course, such an expressive language immediately leads to inconsistencies (like the Russell paradox). However, if only certain sentences are allowed to denote propositions, we can avoid the inconsistencies, as Turner shows.

The theory is presented by giving axioms for the notion of being a proposition and then, further, giving axioms which define truth. Finally, on top of all this, we can give axioms for the modalities. It turns out, in the end, that the propositions are in a one-to-one relation with the sentences that can be either true or false. This, for example, rules out the sentence that expresses the Russell paradox from representing a proposition.

If we let P(t) mean "^t is a proposition" and T(t) mean "^t is true", for any term t, then the following axioms hold (Turner, 1990) where A and B are any terms:

Axioms of Propositions

$P(A) \wedge P(B) \rightarrow P(A \wedge B)$

$P(A) \wedge P(B) \rightarrow P(A \vee B)$

$P(A) \wedge (T(A) \rightarrow P(B)) \rightarrow P(A \rightarrow B)$

$P(A) \rightarrow P(\neg A)$

$\forall x P(A) \rightarrow P(\forall x A)$

$\forall x P(A) \rightarrow P(\exists x A)$

$P(s = t)$

$P(A) \rightarrow P(T(A))$

$\Box P(A) \rightarrow P(\Box A)$

Axioms of Truth

$P(A) \wedge P(B) \rightarrow [T(A \wedge B) \leftrightarrow T(A) \wedge T(B)]$

$P(A) \wedge P(B) \rightarrow [T(A \vee B) \leftrightarrow T(A) \vee T(B)]$

$P(A) \wedge (T(A) \rightarrow P(B)) \rightarrow [T(A \rightarrow B) \leftrightarrow (T(A) \rightarrow T(B))]$

$P(A) \rightarrow [T(\neg A) \leftrightarrow \neg T(A)]$

$\forall x P(A) \rightarrow [T(\forall x A) \leftrightarrow \forall x T(A)]$

$\forall x P(A) \rightarrow [T(\exists x A) \leftrightarrow \exists x T(A)]$

$T(s = t) \leftrightarrow s = t$

$P(A) \rightarrow [T(T(A)) \leftrightarrow T(A)]$

$\Box P(A) \rightarrow [T(\Box A) \leftrightarrow \Box T(A)]$

<u>Axioms of modality</u>

$\Box A \rightarrow A$

$\Box(A \rightarrow B) \rightarrow (\Box A \rightarrow \Box B)$

$\Box A \rightarrow \Box\Box A$

$\forall x \Box A \rightarrow \Box \forall x A$

if $\vdash A$ and $P(A)$ then $\vdash \Box A$

From the axioms for modality we can see that this is a form of S4 (which Turner calls S4') and hence it should be no surprise that it is this theory's version of what might be a modal logic of knowledge.

However, we still have problems—this time of a more philosophical nature. Are the axioms for the modality really ones which we would think of as characterizing knowledge? The first is no problem—we all agree that something which is known is actually true. The second axiom is a bit more worrying—it says that people know all consequences of the use of modus ponens. The final rule says that people know all the axioms and theorems of the logic. These axioms do not seem to reflect the world as it is.

However, what we can do is to keep all the advantages of the language which protect us from inconsistencies and give us the power to refer to propositions by terms but drop the wish to treat knowledge as a modality, i.e. we move to an intensional logic. This logic is called TP by Turner and is obtained by dropping all axioms of modality and the last two axioms from each of the axioms of propositions and the axioms of truth.

Now, since (via their representing terms) propositions are terms of our logic, we can talk about their properties and so reflect what we mean by 'know' directly via some axioms, rather than relying on the (flawed) possible-worlds interpretation. Also, we can quantify over propositions via quantification over their representing terms. So, we are using the power and security of TP as an implementation language for a logic of knowledge.

Of course, there is still much discussion over exactly what axioms should be used for knowledge, but at least the language we now have is both consistent and also expressive enough.

The example sentence from the previous section

$$\forall\alpha(\text{present}(\alpha) \rightarrow \forall\beta(\text{present}(\beta) \rightarrow K_{\beta}\text{present}(\alpha)))$$

would now be something like

$$\forall\alpha(T(\text{present}(\alpha)) \rightarrow \forall\beta(T(\text{present}(\beta)) \rightarrow T(\text{knows}(\beta,\text{present}(\alpha)))))$$

where 'present(x)' is a term rather than a sentence. Also, some of the modal axioms would be acceptable in straight translation:

for all a and t $\quad T(\text{knows}(a,t)) \rightarrow T(t)$

$T(\text{knows}(a,t)) \rightarrow T(\text{knows}(a,\text{knows}(a,t)))$

And, importantly, since we can talk about propositions as terms we can quantify over them safely, so a sentence like "Steve knows everything that Chris knows" is represented by

$$\forall p(T(knows(Chris,p)) \rightarrow T(knows(Steve,p))),$$

and in general this allows our agents to reflect on and reason about their own and other people's knowledge.

7 Future work

The main bulk of future work will be to use the language described above to specify some realistic systems—there are certain to be problems and this is a good way to find them. In particular, we need to concentrate on identifying some more basic properties of CSCW systems and see how well these can be expressed in our language.

Some ideas like trust between agents and how that affects interaction are beginning to be used in CSCW research—it would certainly be interesting, and may be useful, to consider how this sort of relationship might be formalized.

We also need to work on concrete syntax since not everyone is happy with the plain logical syntax we have been using in this paper. Simply following the usual convention of dropping the universal quantifiers at the beginning of sentences and reading all free variables as universally quantified would probably be a large improvement. Also, using some system of keywords (or perhaps Z's two-dimensional notation) to make clear the division between the declaration of actions and the sentences which describe the properties of those actions would be appropriate. It might also, then, be natural to add types to our language, so avoiding some of the relative universals that keep appearing in long strings within our sentences.

This point about acceptable syntax is very important: as Russell said in the Preface to Wittgenstein's *Tractatus Logico-Philosophicus*

A good notation has a subtlety and suggestiveness which at times almost seems like a live teacher.

8 Conclusion

We now have a pair of languages and their associated logics which allow us to specify typical CSCW systems. We can think of a specification as having two parts: an encompassing part that represents the static structure of the system by saying what the users should know as they use it and a part that represents the dynamic parts of the system by saying what the various parts of the software do in response to interaction with the users. We can picture this as

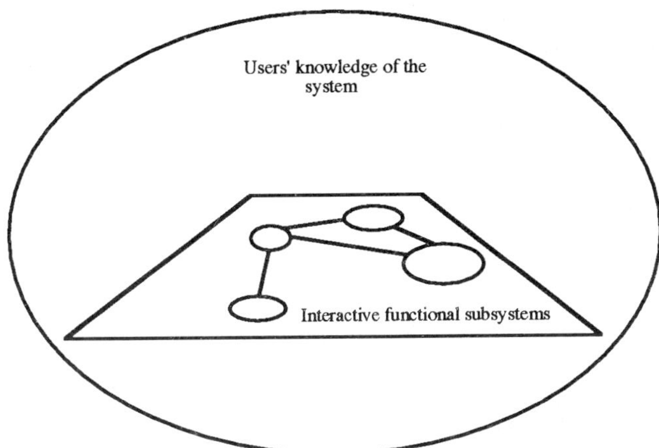

Finally, referring back to Devlin's work, it would be pleasing to be able use his situation theory in a rôle analogous to conventional systems analysis, which seeks to analyse a system into parts which can then be specified in some language like Z.

Devlin's theory is working towards a formal theory of information via situations—which are roughly the things that the word denotes in everyday English. In particular his theory does not assume that all reasoning is linguistically based. This makes it more general than formal logic, which starts with a language and so cannot help being linguistic. In our case it is a good place to start since it will allow us to capture non-linguistic information, e.g. the "look" of a screen (positioning of elements, colour, size), the "behaviour" of a menu, the knowledge of a user about a system or other users. In general, existing work seeks to capture these things by extending existing formal logics, and so are necessarily linguistic—everything has to be reduced to sentences in some language. This, it seems, might be a root cause of the undeniably cumbersome languages that now exist.

Clearly, since such non-linguistic features also occur in any interactive system (not just CSCW ones) the comments here about situation theory are also likely to be relevant to this wider field of systems. The approach suggested in Reeves (1996) can be seen as one which tries to combine the linguistic and non-linguistic, without the benefit of situation theory.

Of course, this paper also follows the linguistic line—it could not do otherwise given that it starts from existing logics. However, what we propose is that situation theory be used to capture (in a non-linguistic and so perhaps more natural form) the requirements of a system and so form a stepping-stone towards the later linguistic form of the system that gets us back to more traditional program development techniques.

Of course, it may turn out that all of our work will be subsumed by situation theory; at the moment we are taking the conservative view that formal logics are not the best vehicles for capturing the requirements for the non-linguistic parts of a system, which the requirements aspects of a problem are most likely to be, and then go on to treat the specification (which current methods assume is entirely linguistic) by conventional means, and on down to implementation.

The overall picture (by analogy with the development of non-CSCW systems) can be captured by the diagram

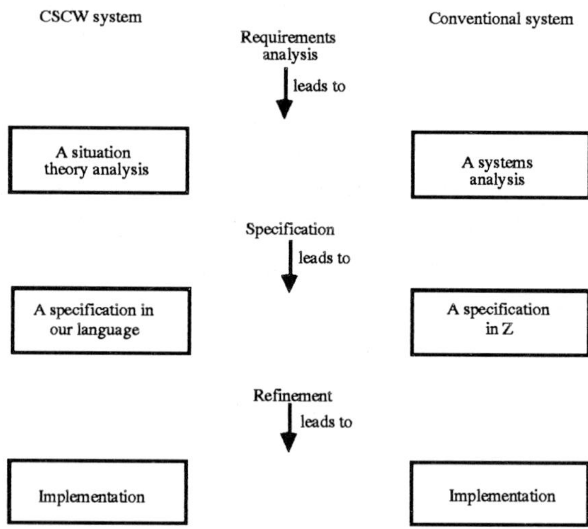

Acknowledgments

Thanks are due to: Mark Apperley, John Grundy and John Venable for providing the room within which (on a more general CSCW project) I was able to pursue this work; the Foundation for Research in Science and Technology for the grant which supports the CSCW project at Waikato; to the referees for their comments which have allowed me to improve this paper beyond its original form.

References

De Paoli F. and Tisato, F. (1994) CSDL: A Language for Cooperative Systems Design, IEEE Transactions on Software Engineering, volume 20, number 8, pp. 606-616.

Devlin, K.(1994) Situation Theory and the Design of Interactive Information Systems, in *Design Issues in CSCW*, D. Rosenberg and C. Hutchison (eds.), Springer-Verlag, 1994.

Diller, A. (1994) *Z: An introduction to formal methods*, John Wiley and Sons Ltd.

Duke, D.J. and Harrison, M.D. (1995) Interaction and Task Requirements, in *Design, Specification and Verification of Interactive Systems*,P. Palanque and R. Bastide (eds.), Springer-Verlag.

Goldblatt, R., (1987) *Logics of Time and Computation*, CSLI Lecture Notes, Number 7.

Moore, R. C. (1984) A Formal Theory of Knowledge and Action, in J.R. Hobbs and R.C. Moore (eds.), *Formal Theories of the Commonsense World*, Ablex, pp. 319-358.

Reeves, S. (1996) A Teaching and Support Tool for Building Formal Models of Graphical User-Interfaces, Proceedings of SE:E&P'96, IEEE Computer Society Press.

Reeves, S. and Clarke. M (1990) *Logic for Computer Science*, Addison-Wesley.

Ryan, M., Fiadeiro, J. and Maibaum, T. (1991) Sharing Actions and Attributes in Modal Action Logic, in Ito and Meyer (eds.), *Theoretical Aspects of Computer Software*, Lecture notes in Computer Science 526, Springer-Verlag.

Turner, R (1990) *Truth and Modality for Knowledge Representation*, Pitman.

Wooldridge, M and Jennings, N.R. (1995) Intelligent Agents: Theory and Practice, to appear in Knowledge Engineering Review.

SpringerEurographics

Bodo Urban (ed.)
Multimedia '96

Proceedings of the Eurographics Workshop in Rostock, Federal Republic of Germany,
May 28–30, 1996
1996. 71 figures. VII, 178 pages.
Soft cover DM 85,–, öS 595,–, US $ 67.00
ISBN 3-211-82876-1

Remco C. Veltkamp, Edwin H. Blake (eds.)
Programming Paradigms in Graphics '95

Proceedings of the Eurographics Workshop in Maastricht, The Netherlands, September 2–3, 1995
1995. 41 partly coloured figures. VIII, 172 pages.
Soft cover DM 85,–, öS 595,–, US $ 59.00
ISBN 3-211-82788-9

Philippe Palanque, Rémi Bastide (eds.)
Design, Specification and Verification of Interactive Systems '95

Proceedings of the Eurographics Workshop in Toulouse, France, June 7–9, 1995
1995. 153 figures. X, 370 pages.
Soft cover DM 118,–, öS 826,–, US $ 95.00
ISBN 3-211-82739-0

Martin Göbel (ed.)
Virtual Environments '95

Selected papers of the Eurographics Workshops in Barcelona, Spain, 1993,
and Monte Carlo, Monaco, 1995
1995. 134 partly coloured figures. VII, 307 pages.
Soft cover DM 108,–, öS 756,–, US $ 85.00
ISBN 3-211-82737-4

P.O.Box 89, A-1201 Wien • New York, NY 10010, 175 Fifth Avenue
Heidelberger Platz 3, D-14197 Berlin • Tokyo 113, 3-13, Hongo 3-chome, Bunkyo-ku

SpringerEurographics

Demetri Terzopoulos, Daniel Thalmann (eds.)
Computer Animation and Simulation '95

Proceedings of the Eurographics Workshop in Maastricht, The Netherlands, September 2–3, 1995
1995. 156 partly coloured figures. VIII, 235 pages.
Soft cover DM 89,–, öS 625,–, US $ 69.00
ISBN 3-211-82738-2

Riccardo Scateni, Jarke J. van Wijk, Pietro Zanarini (eds.)
Visualization in Scientific Computing '95

Proceedings of the Eurographics Workshop in Chia, Italy, May 3–5, 1995
1995. 110 partly coloured figures. VII, 161 pages.
Soft cover DM 85,–, öS 595,–, US $ 69.00
ISBN 3-211-82729-3

Patrick M. Hanrahan, Werner Purgathofer (eds.)
Rendering Techniques '95

Proceedings of the Eurographics Workshop in Dublin, Ireland, June 12–14, 1995
1995. 198 partly coloured figures. XI, 372 pages.
Soft cover DM 118,–, öS 826,–, US $ 98.00
ISBN 3-211-82733-1

Martin Göbel, Heinrich Müller, Bodo Urban (eds.)
Visualization in Scientific Computing

1995. 150 figures. VIII, 238 pages.
Soft cover DM 118,–, öS 826,–, US $ 85.00
ISBN 3-211-82633-5

Wolfgang Herzner, Frank Kappe (eds.)
Multimedia/Hypermedia in Open Distributed Environments

Proceedings of the Eurographics Symposium in Graz, Austria, June 6–9, 1994
1994. 105 figures. VIII, 330 pages.
Soft cover DM 118,–, öS 826,–, US $ 79.00
ISBN 3-211-82587-8

P.O.Box 89, A-1201 Wien • New York, NY 10010, 175 Fifth Avenue
Heidelberger Platz 3, D-14197 Berlin • Tokyo 113, 3-13, Hongo 3-chome, Bunkyo-ku

Springer-Verlag and the Environment